HANDBOOK FOR BLAST-RESISTANT DESIGN OF BUILDINGS

HANDBOOK FOR BLAST-RESISTANT DESIGN OF BUILDINGS

Edited by

Donald O. Dusenberry

WILEY

JOHN WILEY & SONS, INC.

To my wife, Alice

This book is printed on acid-free paper. ♾

Published by John Wiley & Sons, Inc., Hoboken, New Jersey.
Published simultaneously in Canada.

Library of Congress Cataloging-in-Publication Data:

Handbook of blast resistant design of buildings / edited by Donald O. Dusenberry.
p. cm.
Includes index.
ISBN 978-0-470-17054-0 (cloth)
1. Building, Bombproof. I. Dusenberry, Donald O.
TH1097.H36 2010
693.8′54–dc22

2009019203

ISBN: 978-0-470-17054-0

Printed in the United States of America

10 9 8 7 6 5 4 3 2 1

CONTENTS

PREFACE

The need for protection against the effects of explosions is not new. The use of explosive weaponry by the military necessitated resistive entrenchments ages ago. Industrialization of our societies well over a century ago meant that we intended to manufacture, store, handle, and use explosives in constructive ways. To support these military and industrial purposes, a relatively small group of designers have worked to devise ways to strengthen the blast resistance of our structures.

Early attempts at blast-resistance design necessarily relied on judgment, test, and trial-and-error construction to find the best solutions. As technology improved, designers became better able to predict the influences of explosions and the resistive responses that they strove to impart into their designs. More recently, in the past several decades chemists, physicists, blast consultants, and structural engineers have been empowered by technologies and computational tools that have enhanced the precision of their analyses and the efficiency of their designs.

At the same time, the need has increased. The small contingent of designers skilled in the art and science of creating structural designs that will resist explosive forces has been joined by a larger group of architects, engineers, blast consultants, and security consultants who are trying to respond to the increasing concern from a broader group of clients who fear an exposure that they did not anticipate before and frequently did not bring upon themselves. Consultants who have never before had to assess risks, devise risk-reduction programs, provide security systems, establish design-base threats, calculate the pressures and impulses from explosions, and create cost-effective structural designs are being thrust into the process. Many are ill-trained to respond.

There are several good references on some of the aspects of designing for blast resistance. Some of these references support military purposes or for other reasons have government-imposed restrictions against dissemination. As such, they are not widely available to consultants working in the private sector. Nearly all those references and the references that are public each treat an aspect of blast phenomenology, security systems, and structural design for blast resistance, but few, if any, bring together in one place discussions of the breadth of the issues that are important for competent designs. Consultants are forced to collect a library of references and extract from each the salient information that they then synthesize into a comprehensive design approach.

In addition, practitioners who do receive the limited-distribution references for the first time or who find references that are public usually discover immediately that designing for blast resistance is completely different from designing for any environmental load they encountered previously. Designers often realize quickly that they are embarking on design process for which they do not have the knowledge or experience for adequate competency. Those who do not have this realization might be operating at risk if they are not careful and thorough students.

The purpose for this handbook is to bring together into one publication discussions of the broad range of issues that designers need to understand if they are to provide competent, functional, and cost-efficient designs. The contributors to this book are among the most knowledgeable and experienced consultants and researchers in blast resistant design, and contribute their knowledge in a collaborative effort to create a comprehensive reference. Many of the contributors to this handbook are collaborating in the development of the first-ever public-sector standard for blast resistant design, being developed contemporaneously with this handbook by the Structural Engineering Institute (SEI) of the American Society of Engineers. While there undoubtedly will be some differences between the SEI standard and this handbook, many readers will consider these publications as companions.

This handbook is organized into four parts, each addressing a range of aspects of blast-resistance design.

Part 1: Design Considerations provides an overview of basic principles. It has five chapters dealing with general considerations and the design process; risk analyses, reduction, and avoidance; criteria that establish acceptable performance; the science of materials performance under the extraordinary blast environment; and performance verification for technologies and solution methodologies.

Part 2: Blast Phenomena and Loadings, in three chapters, describes the explosion environment, loading functions to be used for blast response analysis, and fragmentation and associated methods for effects analyses.

Part 3: System Analysis and Design has five chapters that cover analysis and design considerations for structures. This part instructs on structural, building envelope, component space, site perimeter, and building system designs.

Part 4: Blast-Resistant Detailing addresses detailing structural elements for resistance. Chapters on concrete, steel, and masonry present guidance that is generally applicable for new design. The fourth chapter addresses retrofits of existing structures.

I wish to thank all the contributors for their commitment to this work, their collaborative spirit, and, of course, their willingness to share the blast-related expertise that they have presented in their chapters. I wish to thank Steven Smith of CTLGroup in particular, for organizing and harmonizing the four chapters of Part 4. William Zehrt of the Department of Defense Explosives Safety Board improved the quality of this handbook by reviewing the chapters of Part 2.

I also wish to thank James Harper, Editor of John Wiley & Sons for supporting this effort; Daniel Magers, Senior Editorial Assistant, and Amy Odum for her able supervision of the copyediting and production; and the copyeditors, compositors, typesetters, and others of the publisher's staff who have professionally assembled this book and brought it to publication.

DONALD O. DUSENBERRY
Wakefield, Massachusetts

CONTRIBUTORS

John Abruzzo, P.E.
Principal
Thornton Tomasetti
555 12th Street, Suite 600
Oakland, CA 94607
Tel: (510) 433-9370
JAbruzzo@ThorntonTomasetti.com

Christopher Arnold, FAIA, RIBA
President, Building Systems Development Inc
Palo Alto CA
1248 Waverley Street
Palo Alto, CA 94301
Tel: 650-462-1812
chrisarno@aol.com
chr.arnold@gmail.com

Curt P. Betts, P.E.
Chief, Security Engineering Section
US Army Corps of Engineers
Protective Design Center
1616 Capitol Avenue
Omaha, Nebraska 68102-4901
Tel: (402) 995-2359
Curt.p.betts@usace.army.mil

Darrell D. Barker, P.E.
Vice President
Extreme Loads and Structural Risk
ABS Consulting
14607 San Pedro Ave., Suite 215
San Antonio, Texas 78232
Tel: (210) 495-5195
dbarker@absconsulting.com

William Bounds, P.E.
Fluor
PO Box 5014
Sugar Land, Texas 77487
william.bounds@fluor.com

Scott Campbell, Ph.D., P.E.
Structural Analysis Consulting Group
PO Box 91364
Louisville, KY 40291
Tel: (502) 762-9596
scott@str-analysis.com

Charles Carter
American Institute of Steel Construction
One East Wacker Drive
Suite 700
Chicago, Illinois 60601-1802
Tel: (312) 670-2400
carter@aisc.org

W. Gene Corley, Ph. D.
CTLGroup
5400 Old Orchard Road
Skokie, Illinois 60077-4321
Tel: (847) 972-3060
Fax: (847) 965-6541
gcorley@ctlgroup.com

John E. Crawford
Karagozian & Case
2550 N. Hollywood Way, Suite 500
Burbank, CA 91505-5026
Tel: (818) 240-1919
crawford@kcse.com

James D. Cunningham
Williamsburg, Virginia
Tel: (757) 645-4057
jdcham@cox.net

Robert Ducibella
Senior Principal
Ducibella Venter & Santore
Security Consulting Engineers
Sturbridge Commons – Franklin House
250 State Street
North Haven, CT 06473
Tel: (203) 288-6490
rducibella@dvssecurity.com

Donald O. Dusenberry, P.E.
Senior Principal
Simpson Gumpertz & Heger Inc.
41 Seyon Street, Building 1, Suite 500
Waltham, MA 02453
Tel: (781) 907-9000
dodusenberry@sgh.com

Chuck Ellison, P.E.
Senior Security Engineer
Applied Research Associates, Inc.
119 Monument Place
Vicksburg, MS 39180
Phone: (601)638-5401
cellison@ara.com

Andrew Hart, Ph.D., MSc, BEng (Hons), Aff.M.ASCE
Martinez, CA 94553
Tel: (925) 370-3866
Hartliliana@aol.com

Eve Hinman, Eng.Sc.D., P.E.
President
Hinman Consulting Engineers, Inc.
One Bush Street, Suite 510
San Francisco, CA 94104
Tel: (415) 621-4423
ehinman@hce.com

Kim W. King, P.E.
Director of Engineering
2195 Redwoods Crest
San Antonio, TX 78232
Tel: (210) 213-3737
kimwking@yahoo.com

L. Javier Malvar
Naval Facilities Engineering Service Center
1100 23rd Avenue
Port Hueneme, CA 93043-4370
luis.malvar@navy.mil

Shalva M. Marjanishvili, Ph.D., P.E., S.E.
Technical Director
Hinman Consulting Engineers, Inc.
One Bush Street, Suite 510
San Francisco, CA 94104
Tel: (415) 621-4423
shalva@hce.com

Paul F. Mlakar, Ph.D., P.E.
U.S. Army Engineer Research and Development Center
3909 Halls Ferry Road
Vicksburg, MS 39180
Tel: (601) 634-3251
Paul.F.Mlakar@usace.army.mil

MeeLing Moy, P.E.
President
The Link CE, PLLC
New York
Tel: (646) 385-5096
mmoy@thelinkce.com

Charles J. Oswald, Ph.D, P.E.
Protection Engineering Consultants
4203 Gardendale, Suite C112
San Antonio, TX 78229
Tel: (512) 380-1988
coswald@protection-consultants.com

James Angelo Ruggieri, P.E.
10710 Timberidge Road
Fairfax Station, VA 22039
Tel: (703) 250-3671
ForensicX@cox.net

Robert Smilowitz, Ph.D., P.E.
Weidlinger Associates, Inc.
375 Hudson Street
12th Floor
New York, New York 10014-3656
Tel: (212) 367-3090
smilowitz@wai.com

Joseph L. Smith, PSP
Senior Vice President
Applied Research Associates, Inc.
119 Monument Place
Vicksburg, MS 39180
Tel: (601) 638-5401
jsmith@ara.com

Steven Smith
CTLGroup
10946 Eight Bells Lane
Columbia, MD 21044
Tel: (410) 997-0400
sjsmith@ctlgroup.com

Darren Tennant
Weidlinger Associates, Inc.
6301 Indian School Road, NE, Suite #501
Albuquerque, NM 87110
Tel: (505) 349-2820
tenant@wai.com

Andrew Whittaker, Ph.D., S.E.
University at Buffalo
Buffalo, NY 14260
Tel: (716) 645-4364
awhittak@buffalo.edu

I Design Considerations

1 General Considerations for Blast-Resistant Design

Donald O. Dusenberry

1.1 INTRODUCTION

Until recently, relatively few engineers and architects have had to design structures and their systems to resist the effects of explosions. Military engineering personnel, consultants to the federal government, and consultants to industries that use explosive or volatile materials constituted the primary population of designers routinely analyzing blast effects.

Following the explosion that demolished the Alfred P. Murrah Federal Building in Oklahoma City in 1995, members of the structural design and construction industries have been increasingly quizzed by owners about blast-related hazards, risks, and methods of protection. The types and numbers of clients seeking blast resistance in their structures have expanded.

The terrorist events of the recent past and the fear that others may occur in the future have led many businesses, particularly those with an international presence, to consider their vulnerability. And, of course, as their neighbors work to enhance the performance of their buildings, owners and tenants who do not envision themselves as targets of malevolent acts nevertheless begin to wonder if their structures might be damaged as a consequence of their proximity to targets. Some have argued that adding blast resistance and enforcing standoff for one building on a block unfortunately increases the threat for others, because it encourages aggressors to attempt to assemble bigger bombs and detonate them closer to the target's neighbors.

There seems to be a sense of anxiety about the vulnerability of our buildings, bridges, tunnels, and utilities in the midst of numerous recognized international social and political instabilities, and given the potential for domestic groups and individuals to seek influence and create disruption by resorting to violent means. As a result, consultants designing rather pedestrian buildings now are expected to provide advice and sometimes specific enhancements in response to quantifiable threats, as well as perceived vulnerabilities.

In this environment, engineers need training and information so that they can provide designs that effectively enhance a building's response to explosions.

1.2 DESIGN APPROACHES

Most engineers and architects serving clients with growing interest in blast resistance are uninitiated in the relevant design practice. Blast loading is very different from loadings commonly analyzed by structural engineers. Peak pressures are orders of magnitude higher than those associated with environmental loads, but their durations generally are extremely short compared to natural periods of structures and structural components. In addition, given that the risk of an explosion at any one facility normally is very low and the costs to achieve elastic response often are prohibitive, designs usually engage the energy-dissipating capability of structural and enclosure elements as they are deformed far into their inelastic ranges. This forces engineers to account for geometric and material nonlinearities.

At first, designing for blast resistance might sound similar to designing for seismic resistance because neither is static and both rely on post-yield response. But even those similarities are limited. The dominant frequencies of seismic excitations are on the order of the lowest natural frequencies of building response, not much faster, as is generally the case for blast loadings. Blast loading usually is impulsive, not simply dynamic.

While we tolerate some damage in earthquakes, to dissipate energy, we usually allow more damage for blast events. We expect facades to sustain severe damage. In fact, blast-resistant design often tolerates breaching of the building enclosure (with attendant risk of fatalities) and even sometimes partial collapse of buildings.

Many blast-resistant designs require very sophisticated approaches for the analysis of building response to explosions (National Research Council 1995). There are techniques for accurate assessment of blast pressures and impulses in complicated environments, modeling the influence of those blast loadings on surfaces, and structural response to those loads. There are critical facilities and blast conditions that warrant the use of these techniques. However, much blast-resistant design is performed following simplified procedures (U.S. Department of Defense 2008) that approximate actual conditions, and therefore lack high fidelity. This often is appropriate because of, and at least in part follows from, inevitable uncertainties that mask the phenomenon and the structure's response. In addition, there are practical matters of prudence, economics, and risk acceptance that drive analyses of blast response.

Risk analyses are important components of the design for blast resistance (Federal Emergency Management Agency 2003). Among the products of such analyses are estimates of the threat for which a structure should be designed. The magnitude of intentional, nonmalevolent explosions and industrial explosions sometimes can be estimated with precision commensurate with that of other common loadings (Center for Chemical Process Safety 1996). The quantity of explosive materials can be estimated, the potential locations of the design-base explosion can be isolated, and often there are relatively few nearby objects that significantly affect the shock front advance.

This is not the case for many accidental explosions and most malevolent explosions. The assessment of the threat in these instances often does not have a probabilistic base. When sufficient data do not exist, consultants are forced to use judgment rather than hard science to establish the threat.

When data are not available, consultants often establish the magnitude of the threat of a malevolent explosion by assessing the probable size of the container (e.g., letter, satchel, package) in which a bomb is likely to be delivered (U.S. Department of Defense 2002a), and then selecting a design-base explosive mass based on a fairly arbitrary assignment of the quantity of explosive that could reasonably be accommodated in that container. In these cases, there is relatively high uncertainty about the intensity of the explosion that might actually occur. Obviously under these circumstances, there is a commensurate level of uncertainty about the outcome.

1.3 THE BLAST ENVIRONMENT

Engineers skilled in the design of buildings for occupancy-related and environmental loads (e.g., dead, live, wind, snow, and seismic loads), but faced with a new challenge to design for blast loads, often find themselves ill-equipped for the challenge. Designers are used to treating all other common loadings as either static or quasi-static, because the rise time and duration for the equivalent load are on the order of, or longer than, the longest natural periods of the structure. Designing for blast loading generally cannot follow this approach.

Conventional design for common time-varying loads, including wind and seismic, includes techniques that allow conversion of these dynamic phenomena into quasi-static events that recognize and simplify the dynamics. Wind loads defined in one of the most common references (American Society of Civil Engineers 2005) are based on an acknowledgment of the range of natural frequencies of common structural frames, and are calibrated to those values. When the frequency of a subject building falls outside of that default range, common design approaches provide for specified adjustments to the quasi-static design loads to account for dynamic response.

Common seismic design (American Society of Civil Engineers 2005) involves a very elaborate conversion of the dynamic loading environment into a quasi-static analysis problem. Building systems are characterized for stiffness and ductility, and site conditions are evaluated for seismic exposures and characteristics of shaking. On the basis of extensive research into building performance and a fair amount of cumulative experience evaluating the actual earthquake response of designed structures, the complicated loadings—which are as much a function of the building design as they are of the environment in which structures are built—are idealized as a series of externally applied loads that are thought to mimic the loading effects of an earthquake. Complicated though the approach is, many buildings can be designed for earthquakes by engineers with little familiarity with dynamic behavior.

Our conventional approach to blast design is similar to that for seismic design, in two important ways: (1) both loadings clearly are dynamic and, hence, solutions are energy-based, and (2) the way we detail structural elements determines the effective loads for which structures must be designed (meaning, we limit the strength we need to supply by allowing post-elastic behavior to dissipate energy). However, blast loading, with its extremely fast rise time and usually short duration, is either dynamic or impulsive, depending on the nature of the explosive, its distance from the subject structure, and the level of confinement that the structure creates for the expanding hot gases (Mays and Smith 1995).

The impulsive form of the very fast load rise time and very short load duration normally associated with blast loading requires analytical approaches that generally demand direct solution of energy balance equations (U.S. Department of Defense 2008; Mays and Smith 1995).

1.4 STRUCTURE AS AN INFLUENCE ON BLAST LOADS

The pressure and duration of the impulse associated with a blast are influenced by reflections of the shock front (U.S. Department of Defense 2008). Reflection sources include the ground below the detonation point and building surfaces that have sufficient mass or ductility to remain largely in place for the duration of the impulse. When shock fronts are reflected, their pressures are magnified as a function of the proximity, robustness, and material characteristics of the impacted object (Bangash 1993). The more robust that object, the greater the reflected energy because less energy is dissipated by the response (such as ground cratering) of the surface. These variations often are neglected in conventional design.

For instance, facades normally are designed on the assumption that they are perfect reflectors of the shock front. Designers following common procedures are assuming that the facade components remain stationary for the duration of the impinging shock front, causing peak pressures and impulses sufficient to reverse the direction of the shock front. In practice, there can be some displacement of the facade during the loading cycle. This displacement reduces the effectiveness of the reflector, and correspondingly the impulse.

Analyses for interior explosions have additional complications, as designers attempt to deal with the multiple reflections of the shock front within the structure, and pressures that develop from containment of expanding hot gases (Mays and Smith 1995)–a phenomenon normally neglected for external explosions. Further, the geometry of the confining volume and the location of the explosion within the volume can substantially affect the pressures on surfaces (U.S. Department of Defense 2008). The science that describes the pressure history on interior surfaces is complex, and not generally considered rigorously in common blast-resistant design processes.

Providing blast venting through frangible components to mitigate the effects of interior explosions is even more complex, since the release time for the venting component is a key, but difficult to assess, factor in the determination of the magnitude of the pressure buildup. Approximations usually govern the analyses (U.S. Department of Defense 2008).

Clearly, there is interplay between the performances of building facades and frames. While in most cases the primary reason we enhance the performance of a facade is to protect occupants, we gain protection for the structure as well. Blast shock fronts that are not repelled by the facade will advance into a building, inducing pressures on interior surfaces of the structure and threatening interior columns, walls, and floor systems. Blast-related upward impulses on floor slabs can reverse force distributions in these structural elements. In systems that are not strong and ductile enough for these reversed forces, blast-induced deflections can fracture structural elements that are required to resist gravity loads. Hence, floor systems can fail after the direct effects of the blast pass and the slab falls back downward under the influence of gravity.

Of course, by designing the facade to resist the effects of an explosion, the designer is forcing the structure to become a support for the blast loads. Depending on the performance criteria, designers need to demonstrate that the framing system can support the applied loads, and that the structure as a whole will remain standing with an acceptable level of damage.

Building enclosures normally are designed to resist blast effects by inelastic flexural action, but it is possible to design facades to resist blast effects through catenary action as well. In particular, blast retrofits sometimes include new "catch systems" that are intended to reduce intrusion of blast pressures and creation of lethal missiles, by acting as a net inside the original exterior wall system.

In any case, the lateral displacement of the system often is large enough to open gaps between wall panels or between panels and floor slabs. When this happens, there is potential for leakage pressures to enter the building (U.S. Department of Defense 2002b), even when windows stay in place. This is particularly true in response to large, relatively distant explosions that have relatively long-duration impulses.

Pressure fronts that leak past facades that are damaged but remain in place normally are assumed to have insufficient energy to induce significant damage to interior structural components. However, these leakage pressures can cause personal injuries and damage to architectural and mechanical systems if they are not designed for resistance.

Add to the effects of leakage pressures the possibility that structural and architectural features on the inward-facing surfaces of facade components can become missiles when the facade sustains damage as it deforms, and there remains substantial risk to occupants inside blast-resistant buildings even with well-developed designs.

It is well established that breached fenestration leads to lethal missiles and internal pressurization (American Society of Civil Engineers 1999). Common design for blast resistance for malevolent attacks often is based on the premise that a significant fraction of the fenestration in a building will fail (General Services Administration 2003). This is due in part to the variability of the properties of glass, but also results from risk acceptance that employs the philosophy that an explosion is unlikely and that full, "guaranteed" protection is prohibitively difficult or expensive.

Hence, the effects of leakage pressures and missiles that are the product of building materials fracturing in response to a blast often can be destructive to the interiors of buildings, even when the facades of those buildings are designed to resist the effects of an explosion. Except when the most restrictive approaches to blast-resistant design are employed (e.g., with elastic response, so a building can remain functional), parties with standing in the design process need to understand that substantial interior damage and occupant injuries are possible should the design-base explosion occur.

1.5 STRUCTURAL RESPONSE

The shock front radiating from a detonation strikes a building component, it is instantaneously reflected. This impact with a structure imparts momentum to exterior components of the building. The associated kinetic energy of the moving components must be absorbed or dissipated in order for them to survive. Generally, this is achieved by converting the kinetic energy of the moving facade component to strain energy in resisting elements. Following the philosophy that blast events are unusual loading cases that can be allowed to impart potentially unrepairable damage to structures, efficiency in design is achieved through post-yield deformation of the resisting components, during which energy can be dissipated through inelastic strain.

Of course, this means that the components that need evaluation often are deformed far beyond limits normally established for other loading types, and many of the assumptions that form the basis for conventional design approaches might not be valid. For instance, recognition of the extreme damage state normally associated with dissipation of blast energy has led to debate about appropriate values of the strength reduction factors (Φ factors) to be used for design.

In conventional design (American Concrete Institute 2005; American Institute of Steel Construction 2005), the nominal strengths of structural elements are reduced by Φ factors to account for uncertainty in the actual strength of the elements, and for the consequences of failure. Their magnitudes for conventional design have been developed based on studies of structural responses that are commensurate with service performance of buildings and, for seismic design, responses that are anticipated to be sufficiently limited and ductile to allow elements to retain most of their original load-carrying capacity. Blast resistance, on the other hand, often takes structural elements far into the inelastic range, to where residual strengths might be reduced from their peaks, and alternative load-carrying mechanisms (e.g., catenary action) are engaged. Sometimes, designers anticipate complete failure of certain elements if they are subjected to the design-base event. In this environment, it is not at all clear that Φ factors developed for conventional, nonblast design are relevant.

Common blast-resistant design often takes the values of the Φ factors to be 1.0 (U.S. Department of Defense 2008). The bases for this approach range from the uncertainty about what the actual values ought to be to the observation that loads we assume for blast-resistant design are sufficiently uncertain that precision in the values for Φ is unjustified. It is further prudent to assume Φ =1.0 when performing "balanced design," in which each structural element in a load path is designed to resist the reactions associated with the preceding element loaded to its full strength. Using Φ =1.0 for determination of the full strength of the elements in the load path tends to add conservatism to the loads required for the design of the subsequent elements.

On the other side of the equation, designers often apply load factors equal to 1.0 to the blast effects (U.S. Department of Defense 2008). This follows from the lack of a probabilistic base from which to determine the design threat, and the rationale that conservatism can be achieved by directly increasing the design threat.

In any event, the absence of complete agreement on how to address strength reduction factors, and the valid observation that blast threats—particularly for malevolent explosions—generally are difficult to quantify, reduce our confidence in our ability to predict structural response with precision.

It is common in blast-resistant design to treat individual elements as single-degree-of-freedom nonlinear systems (U.S. Department of Defense 2008). Performance is judged by comparison of response to limiting ductility factors (i.e., the ratio of peak displacement to displacement at yield) or support rotations, with the response calculated as though the structural element were subjected to a pressure function while isolated from other structural influences. Of course, much more sophisticated approaches are pursued for critical structures and complicated structural systems. However, research on structural response for very high strain rates and very large deformations is limited, and results often are not widely disseminated. In many respects, the sophisticated software now available makes it possible to analyze with precision that exceeds our understanding of structural response.

Hence, the simplified, single-degree-of-freedom approach forms the basis for many designs. This approach usually is consistent with the precision with which we model the blast environment and our knowledge of element behavior, but it generally identifies the true level of damage only approximately. When considering elements as components of structural systems under the influence of blast, the response of the individual elements can differ significantly from that determined by analyses in isolation.

1.6 NONSTRUCTURAL ELEMENTS

Designers usually assume that the blast resistance of a structure is derived from the elements that they design for this purpose. While this clearly is true in large measure, in actual explosions, nonstructural elements—components disregarded in blast design—can act to reduce damage in a structure. It usually is conservative, and therefore prudent, to ignore these components because the designer cannot be certain about the reliability, or even the long-term existence, of building components that are not part of the structural design.

Nevertheless, elements with mass and ductility that stand between an explosion site and a target area can act to dissipate energy as they fail from the effects of the blast. In fact, designers sometimes do rely on specific sacrificial elements to reduce the blast effects on critical structural elements. The bases for this consideration are twofold: (1) through its failure, the sacrificial element dissipates energy that would otherwise be imparted to the structural element, and (2) for the brief time that the sacrificial element stays in place, it acts to reflect the shock front, thereby reducing the impulse felt by the protected structural element. For near-range conditions, when a bomb might otherwise be placed essentially in contact with a key structural element, a sacrificial element such as an architectural column enclosure can enhance survivability simply by inhibiting close placement of the explosive.

Of course, any shielding element that has inadequate strength, ductility, and connection to remain attached to resisting elements is likely to become a missile. Some of the energy these elements absorb is dissipated through strain, but the rest is retained as kinetic energy. The hazards created by these flying elements end only when that kinetic energy is brought to zero. Furthermore, care is needed in the evaluation of the value of shielding elements that are not positioned closely to the structure under consideration, since shock fronts reform beyond such elements, mitigating the protective value of the shield.

1.7 EFFECT OF MASS

The first influence of gravity comes to play when assessing the weights that the designer assumes are present in the structure at the time of an explosion. These weights, which are derived from the structure itself and its contents, act concurrently with the explosion-induced loadings. As a result, they "consume" some of the resisting capacity of the elements that are designed to resist the explosion. In addition, for the most part, they remain on the structure after the explosion and therefore must be supported by the damaged structure. The post-blast distribution of these weights often will be uncertain.

On the beneficial side, mass often augments the blast resistance of structural elements. Blast effects usually are impulsive, meaning that they impart velocity to objects through development of momentum. With momentum being proportional to the product of mass and velocity (Eq. 1-1), and kinetic energy being proportional to the product of mass and velocity squared, the larger the mass, the smaller the velocity and, hence, the smaller the energy that must be dissipated through strain (Eq. 1-2).

$$I = \int_{t_0}^{t_1} F(t)\,dt = MV \tag{1.1}$$

where: I = impulse
$F(t)$ = time-varying force
t = time
M = mass
V = velocity

$$E_k = \frac{1}{2}MV^2 = \frac{1}{2}\frac{I^2}{M} \tag{1.2}$$

where: E_k = kinetic energy

Gravity also must be considered when elements or overall structures deform. Vertical load-carrying elements often are designed to resist simultaneous vertical and lateral loads. Even when columns are not part of a structure's lateral load resisting system, it is common for them to be designed for an eccentricity of the vertical load to account for inevitable moments that will develop in use. Sometimes the magnitude of the eccentricity causing moment is assumed to be on the order of 3% to 10% of the element's cross section dimension (American Concrete Institute 2005). Response to blast often deforms vertical structural elements far more than limits assumed for conventional design. The designer needs to evaluate the ability to resist the resulting P-Δ effects, both for individual elements and for the structure overall.

Structures as a whole generally are not pushed over by a common explosion. The overall mass of a structure usually is large enough to keep the kinetic energy imparted to the structure as a whole small enough that it can be absorbed by the multiple elements that would need to fail before the building topples.

In many explosions that cause extensive destruction, the damage develops in two phases: (1) the energy released by the explosion degrades or destroys important structural elements, and (2) the damaged structure is unable to resist gravity and collapses beyond the area of initial damage. In some of the most devastating explosions, most of the structural damage has been caused by gravity (Federal Emergency Management Agency 1996, Hinman and Hammond 1997).

Normally, individual elements fail, necessitating the activation of alternative load paths within the structure to carry the gravity loads that remain after the direct effects of the blast pass. Studies that assess these alternate load paths need to consider the dynamic application of the redirected internal forces, as the sudden removal of load-carrying elements implies a change in potential energy, as portions of a structure begin to drop. This change in potential energy necessarily imparts kinetic energy that must be converted to strain energy for the falling mass to be brought to rest.

Hence, the evaluation of the full effect of a blast does not end with calculations of blast damage to individual elements or limited structural systems. Designers need to consider the ongoing effects in the damaged structure, under the influence of gravity.

1.8 SYSTEMS APPROACH

In our efforts to enhance the blast resistance of a facility, we need to remain cognizant about how our designs affect the performance and viability of the facility for nonblast events. As is always the case, there are competing goals and influences in the design of a facility, and those factors need to be balanced to achieve the most satisfactory end product.

Consider the conflicts between the structural performance preferred for seismic events and that preferred for explosions. One important goal in seismic design is to force failures to occur in beams before columns, so that the load-carrying capacity of columns is preserved even when the earthquake induces damage. This is accomplished by detailing connections between beams and columns so that plastic moments occur in the beams before the columns. This is the "strong column, weak beam" approach.

Consultants designing for blast often provide for the possibility that a column will be severely damaged by an explosion, in spite of our best efforts at prevention. When consultants assume that a column has lost its strength, they must develop alternative load paths to prevent a collapse from progressing from the initially damaged column through the structure. One form of alternative resistance involves making beams strong and ductile enough to span over the area of damage, thereby redistributing the load on the damaged column to adjacent columns. This requires strong beams which, if implemented without consideration of seismic response, can run counter to philosophies for robust seismic resistance.

Designers working to enhance blast resistance must also consider occupant egress and the needs of emergency responders. Blast resistance invariably includes fenestration with blast-resistant glass. By definition, such glass is difficult to break. Firefighters will need to use special tools and engage unusual tactics to fight a fire in a building that is difficult to enter and vent, and that has features that inhibit extraction of trapped occupants. Designers might need to compensate for blast-resistance features or enhance fire resistance.

Distance is the single most important asset to a structural engineer designing for blast resistance. The farther the explosion is from the structure, the lower are the effects that the structure must resist. Further, there often is merit to the construction of blast walls or line-of-sight barriers to add protection to a facility. However, the need to create an impenetrable perimeter, and the temptation to make it one that effectively hides the facility, can detract from the function of the facility.

First, there is the dilemma caused by features that are intended to keep aggressors away from a building, but that also block lines of sight to the building in the process. While such features add security, they also provide opportunities for the aggressors to effectively hide from observers in and around the building. A slowly developing assault may be more difficult to detect if the perimeter cannot be monitored effectively.

Next, there is the potential impact on the quality of life for occupants of buildings that have very robust defenses. Imposing perimeters and minimized fenestration display the robustness and the fortresslike design intent. While this might be perceived as an asset for what it says to the aggressor, it also communicates a sobering message to occupants and welcomed visitors. There has to be a balance between the means to provide the necessary resistance and the architectural and functional goals of the facility. Aesthetics need consideration for most facilities.

Overall security design needs to properly balance the efforts applied to the defense against a variety of threats. It is unsatisfactory to provide a very robust design to resist blast if the real threat to a facility is through the mechanical system. Clients will be unhappy if security protocols address perceived threats (e.g., outside aggressors detonating bombs near a building), but fail to prevent real threats (e.g., disgruntled employees intent on committing sabotage or violence inside the facility). Any overall security evaluation needs to consider all perceived threats and provide guidance that will allow clients to determine where best to apply their efforts to maximize their benefits. In many cases, a robust resistance to an explosion threat will not be the best expenditure of funds.

Given a security design developed for the spectrum of potential threats to a facility, owners sometimes face costs that exceed their means. When this occurs, and for facilities that risk assessments show to be at relatively low risk, owners must make decisions. Sometimes they instruct consultants to design to a particular cost, representing the amount that the owner can commit to the added security to be provided to the facility. In these instances, consultants must identify priorities that address the most likely threats and provide the greatest protection for the limited funds. When this happens, the consultants must explain to the owners the limitations of the options so that the owners can make educated decisions.

1.9 INFORMATION SENSITIVITY

When blast-resistant designs are for the security and safety of a facility in response to a threat of a malevolent attack, information about the assumed size and location of an explosion should be kept confidential. This information could be useful to an aggressor because it can reveal a strategy to overwhelm the designed defenses.

The common practice of specifying the design loads on drawings should not include a specific statement about the assumptions for blast loading when facility security is at issue. Potentially public communications among members of the design team and between the design team and the owner should avoid revelations about the design-base explosion.

In most cases, the design assumptions for accidental explosions are not sensitive. Precautions about security-related confidentiality usually do not apply, and customary processes for documenting the design bases may be followed. In addition, there might be legal requirements or other circumstances that dictate the documentation of otherwise sensitive information. As always, designers will need to comply with the law and to work with stakeholders in the design of a facility to contain the unnecessary dissemination of information that could potentially be misused.

1.10 SUMMARY

As consultants in the building design industry have been drawn into the matter of blast-resistant design, many have been handicapped by lack of familiarity with the blast environment, including not knowing how to determine loads for design, or with proper approaches for structural design. Consultants often anticipate that they will be able to provide effective designs by following approaches common in building design when blast is not an issue. Unfortunately, consultants expecting to apply their familiar approaches usually are proceeding along an improper path.

An explosion is a violent thermochemical event. It involves supersonic detonation of the explosive material, violently expanding hot gases, and radiation of a shock front that has peak pressures that are orders of magnitude higher than those that buildings normally experience under any other loadings. Designers hoping to solve the blast problem by designing for a quasi-static pressure are likely to be very conservative, at best, but more probably will simply be wrong.

Designers need to understand that the magnitudes of the pressures that an explosion imparts to a structure are highly dependent on the nature of the explosive material, the shape and casing of the device, the size and range of the explosion, the angle of incidence between the advancing shock front and the impacted surface, the presence of nearby surfaces that restrict the expansion of hot gases or that reflect pressure fronts, and the robustness of the impacted surface itself. Designers also need to understand that the durations of the pressures induced by an external explosion generally are extremely short compared to the durations of other loads and compared to natural periods of structures. Further, there is interplay between blast pressure magnitudes and durations, which is a function of distance from the detonation point, among other factors.

Designing for the very high peak pressures and short durations of blast loadings requires applications of principles of dynamic response. Accurate prediction of the peak response of a building will require the designer to analyze dynamic properties of the structure, and apply approaches that respect dynamic behavior. Further, most cost-efficient designs rely on deformation far beyond elastic limits to dissipate energy. Hence, many of the assumptions designers normally make when designing for loads other than blast do not apply when designing for blast resistance.

Consultants engaged in the design for blast resistance need to be qualified by education, training, and experience to properly determine the effects of an explosion on a structure. They must have specialized expertise in blast characterization, structural dynamics, nonlinear behavior, and numerical modeling of structures. Blast resistance designers must be licensed design professionals who are knowledgeable in the principles of structural dynamics and experienced with their proper application in predicting the response of elements and systems to the types of loadings that result from an explosion, or they must work under the direct supervision of licensed professionals with appropriate training and experience.

The present practice for blast-resistant design employs many approximations and, in many aspects, relies on incomplete understanding of the blast environment and structural behavior. While available approaches serve the public by increasing the ability of our structures to resist the effects of explosions, these conventional approaches generally are ill suited to provide a clear understanding of the post-blast condition of the structure. Consultants providing blast-resistant design need to understand the limitations of the tools they apply, and provide clients with appropriate explanations of the assumptions, risks, and expectations for the performance of blast-resistant structures. In many cases, those explanations need to make clear that the performance of the structure and the safety of individuals inside the protected spaces are not guaranteed.

REFERENCES

American Concrete Institute. 2005. *Building Code Provisions for Structural Concrete and Commentary* (ACI 318-05). Farmington Hills, MI: American Concrete Institute.

American Institute of Steel Construction. 2005. *Specifications for Structural Steel Buildings*. Chicago, IL: American Institute of Steel Construction, Inc.

American Society of Civil Engineers. 1999. *Structural Design for Physical Security: State of the Practice*. Reston, VA: American Society of Civil Engineers.

________. 2005. *Minimum Design Loads for Buildings and Other Structures* (ASCE/SEI 7-05). Reston, VA: American Society of Civil Engineers.

Bangash, M. Y. H. 1993. *Impact and Explosion: Analysis and Design*. Boca Raton, FL: CRC Press, Inc.

Center for Chemical Process Safety. 1996. *Guidelines for Evaluating Process Plant Buildings for External Explosions and Fires*. New York, NY: American Institute of Chemical Engineers Center for Chemical Process Safety.

Federal Emergency Management Agency. 1996. *The Oklahoma City Bombing: Improving Building Performance Through Multi-Hazard Mitigation* (FEMA 227). Washington, DC: Federal Emergency Management Agency.

________. 2003. *Reference Manual to Mitigate Potential Terrorist Attacks Against Buildings* (FEMA 426). Washington, DC: Federal Emergency Management Agency, Department of Homeland Security.

General Services Administration. 2003. *Facilities Standards for Public Buildings Service* (P100). Washington, DC: General Services Administration.

Hinman E. E. and D. J. Hammond. 1997. *Lessons from the Oklahoma City Bombing: Defensive Design Techniques*. Reston, VA: American Society of Civil Engineers Press.

Mays G. C. and P. D. Smith. 1995. *Blast Effects on Buildings*. London: Thomas Telford Publications.

National Research Council. 1995. *Protecting Buildings from Bomb Damage: Transfer of Blast-Effects Mitigation Technologies from Military to Civilian Applications*. Washington, DC: National Research Council, National Academy Press.

Smith P. D. and J. G. Hetherington. 1994. *Blast and Ballistic Loading of Structures*. Oxford: Butterworth Heinemann.

U.S. Department of Defense. 2002a. *DoD Minimum Antiterrorist Standards for Buildings* (UFC 4-010-01). Washington, DC: United States Department of Defense.

________. 2002b. *Design and Analysis of Hardened Structures to Conventional Weapons Effects* (DAHS-CWE) (UFC 3-340-01) (TM 5-855-1). Washington, DC: United States Department of Defense.

________. 2008. Structures to Resist the Effects of Accidental Explosions (UFC 3-340-02), Washington, DC: United States Department of Defense.

2 Design Considerations

Robert Ducibella and James Cunningham

2.1 INTRODUCTION

On April 19, 1995, a Ryder truck containing about 5,000 pounds of ammonium nitrate fertilizer and nitromethane exploded in front of the Alfred P. Murrah Federal Building in Oklahoma City, Oklahoma. The blast collapsed a third of the building, gouged out a crater 30 feet wide by 8 feet deep, and destroyed or damaged 324 buildings in a 16-block radius. The bomb claimed 168 confirmed dead.

On September 11, 2001, terrorists crashed two commercial jets into the World Trade Center twin towers. Another hijacked flight slammed into the Pentagon, while a fourth was forced down by passengers and crashed in a field near Shanksville, Pennsylvania. The total dead and missing numbered 2,992: 2,749 in New York City, 184 at the Pentagon, 40 in Pennsylvania, and the 19 hijackers.

More than any other acts of domestic or international terrorism, these two attacks have forever changed the American building-sciences community's relationship to the society it serves. Since the first skyscrapers appeared in the late nineteenth century, Americans have come to expect commercial structures of exceptional beauty and functionality. After Oklahoma City and September 11, many ordinary citizens also assume that new buildings are designed to protect people during explosions as well as other natural or man-made disasters. With few exceptions, however, significant movement toward achieving the recent imperative of both commercial utility and explosive blast resistance is a work in progress.

Design professionals have gained enormous experience with plans and models that anticipate structural responses to gravity, wind, and seismic loads. Preventing or curtailing random acts of terrorism by identifying their probability of occurrence and potential consequences, however, falls outside the general practice of structural engineering.

This chapter proposes an innovative and largely untapped approach to blast-resistant-building security design, a new paradigm in which senior individuals who have a breadth and depth of experience in the areas of site planning, civil, structural, mechanical, electrical, fire protection, and vertical transportation engineering; architecture; code and egress consulting; site planning; and security engineering collaborate on a total blast-resistant building security design. This team approach should take into consideration security and antiterrorist strategies that fundamentally affect site selection and building design. **In considering the design of blast-resistant buildings, the design professionals must partner to become *an effective security design team*.**

In more traditional building security design efforts, security professionals simply present the design team with the results of their risk assessment, and the security planning component is assumed to be largely finished. For reasons expanded upon hereinafter, the authors urge instead that the design team collaborate closely with security professionals and security engineers throughout the entire design process. In the iterative process of designing a blast-resistant facility, the architectural team should become a *security design team*, supported by homeland defense, intelligence, security, law enforcement, and blast consultant experts. Therefore, the term *security design team* as used in this chapter means the multidisciplinary building sciences/security/explosives experts group of professionals described above.

It is a concept capable of implementation and proven to work in a wide range of facility types and locations where occupants, assets, and business missions are deemed worthy of protection.

2.2 A NEW PARADIGM FOR DESIGNING BLAST-RESISTANT BUILDINGS, VENUES, AND SITES

The following paragraphs describe a structured framework for threat and vulnerability data gathering and for risk assessment. Security concepts such as design basis threat, consequence management, functional redundancy, building location, and critical-functions dispersal are explored. A brief checklist of security design considerations is presented, and the reader is introduced to the design principles and guidelines that are expanded upon in the handbook's subsequent chapters.

The suggested risk assessment model for blast protection has six parts:

1. A threat identification and rating, which is the security design team's analysis of what terrorists and criminals can do to the target.
2. An asset value assessment, which represents how much the project's people and physical assets are worth and what the responsible parties will do (and pay) to protect them.
3. A vulnerability assessment, which represents the attractiveness of the target, and areas of potential weakness and/or avenues of compromise.
4. A site-specific risk assessment, which is the product of these three studies. A credible site-specific risk assessment is the single most critical factor

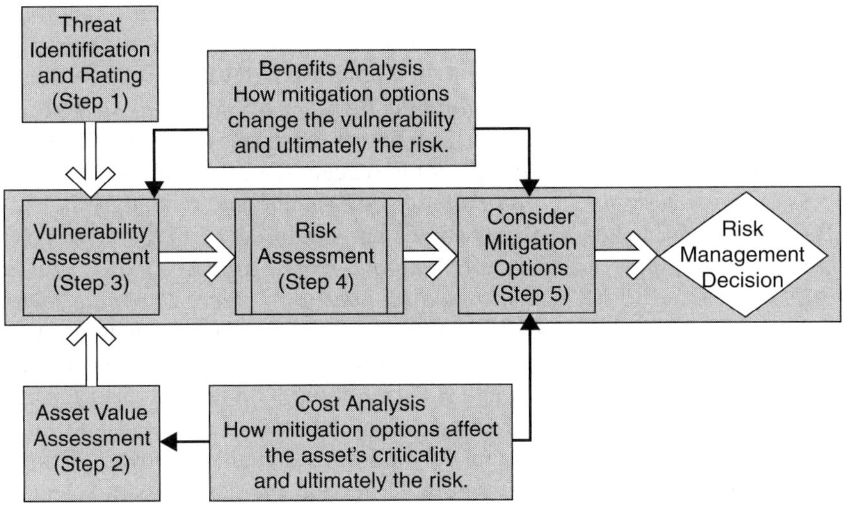

Figure 2.1 Risk Assessment Process (Adapted from FEMA 452)

of the security blast-design process; it is the basis and rationale for ensuring that the protective design strategies are incorporated in the multidisciplinary security/explosives/building sciences team approach that is stressed throughout this chapter.

5. Mitigation options, based on the risk assessment as a foundation.
6. Risk management decisions, driven by the mitigation options and informed by the available project resources.

Figure 2.1 shows the five steps leading to the risk management decisions, and their interaction with each other.

To be useful in influencing the design, the risk assessment should be completed early, preferably during the preliminary planning and conceptual design phase, but certainly no later than the completion of initial design documents.

Extensive unclassified information about explosive events is available from the FBI; Department of State; Department of Defense (DoD); Department of Homeland Security (DHS); Bureau of Alcohol, Tobacco, Firearms and Explosives (ATF); the U.S. Armed Forces; and other U.S. agencies. The authors urge the design team to obtain and use this information. Since this information is being constantly updated, we advocate that you use an Internet search engine such as Google, Mozilla Foxfire, or Ask.com to conduct your own research, thereby ensuring that your information is both current and relevant. For example, a quick search on the terms "Department of Defense explosive events" results in over 2 million hits while "FBI explosive events" returns over 600,000 citations, and "ATF explosive events" creates over 200,000 references.

There are numerous how-to guides that lay out systematic approaches to secure facility planning, design, construction, and operation. These various methodologies—the Department of Defense CARVER process, Sandia's RAMPART™ software tool, the *National Institute of Standards and Technology* (NIST)'s CET, and the Federal Emergency Management Agency (FEMA) building security series are among scores of candidates[1] —are all based on a process of threat assessment, vulnerability assessment, and risk analysis.

The authors have adapted and somewhat modified FEMA's *Risk Assessment: A How-To Guide to Mitigate Potential Terrorist Attacks Against Buildings* (FEMA 452[2]) and FEMA's *Reference Manual to Mitigate Potential Terrorist Attacks Against Buildings* (FEMA 426) as their risk assessment models to present the risk management process. Admittedly, FEMA 452 is a somewhat arbitrary choice of exemplar, although it is becoming a de facto industry standard among security professionals. There are many alternative methods—an Internet search of the term "risk assessment for buildings" resulted in over 3.7 million hits—but no matter which risk assessment version the design team selects, the methodology should be:

- Specifically written for the building sciences community of architects, engineers, and professionals who design not only high-security government facilities but private-sector structures as well.
- Intended to serve as a multi-hazard assessment tool of a building and its site, but readily adaptable to focusing closely on the explosive threat.
- Organized with numerous checklists, tables, and memory aids that will assist the design team in determining threats, risks, vulnerabilities, and mitigation options.
- Proven effective through years of extensive use in real-world threat and vulnerability assessments.

---

[1]The DoD's CARVER risk assessment process is a mnemonic rather than a model. First developed for the U.S. Special Forces in Viet Nam to target enemy installations, CARVER stands for *__C__riticality, __A__ccessibility, __R__ecognizability, __V__ulnerability, __E__ffect, and __R__ecoverability.* CARVER has recently become hard to find on the Internet but is widely available through the DoD, Homeland Security, ASIS International, or law enforcement agencies, among others. Sandia's *Risk Assessment Method—Property Analysis and Ranking Tool* (*RAMPART(tm)*) is a software-based methodology for assessing the potential risks of terrorism, natural disasters, and crime to buildings, particularly U.S. government facilities. Read more at http://ipal.sandia.gov/ip_details.php?ip=4420. NIST's *Cost-Effectiveness Tool for Capital Asset Protection (CET)* is a software-based risk assessment tool that building owners and managers can use to protect assets against terrorist threats. CET is available without cost on the NIST Web site. See http://www2.bfrl.nist.gov/software/CET/CET_4_0_UserManualNISTIR_7524.pdf.

[2]FEMA, located within the Department of Homeland Security, is the U.S. government agency tasked with disaster mitigation, preparedness response, and recovery planning. FEMA 452: *Risk Assessment: A How-To Guide to Mitigate Potential Terrorist Attacks* may be downloaded from www.fema.gov/plan/prevent/rms/rmsp452.shtm. FEMA 452 is a companion reference to the *Reference Manual to Mitigate Potential Attacks Against Buildings* (FEMA 426) and the *Building Design for Homeland Security Training Course* (FEMA E155). This document is also a useful companion to the *Primer for Design of Commercial Office Buildings to Mitigate Terrorist Attacks* (FEMA 427).

- Available in unclassified form and preferably at little or no cost. FEMA 452, FEMA 426, and their companion manuals, for example, are available for free from http://www.fema.gov/plan/prevent/rms/rmsp452.shtm.

Increasingly, the building sciences community is being challenged to incorporate high levels of security into the design of facilities, sites, and venues that do not yet exist. While the risk assessment tools spreading throughout the law enforcement, public safety, security, and building sciences communities can be extremely useful, almost all of them are geared to improving the security of sites and structures that are already standing. Consequently, it is essential that the risk assessment model that is selected has also been used to assess *future* buildings and venues, and to create security plans for virtual sites.

Traditionally, the building sciences community has prepared for natural disasters by following prescriptive building codes supported by well-established and tested reference standards, regulations, inspections, and assessment techniques. Many man-made hazards such as toxic industrial chemicals storage, and numerous societal goals such as life safety, have been similarly addressed. The building regulation system, however, has only just begun to deal with the terrorist threat. In the absence of high-quality regulatory guidance, the design team must fall back on its own resources and expertise—always remembering that the nature of the potential threat and the desired level of protection are equally important and inseparable design considerations.

2.3 A BRIEF HISTORY OF RECENT TERRORIST ATTACKS

Experts are quick to point out that terrorists almost invariably seek publicity and sometimes monetary reward or political gain as well. It should be emphasized, though, that terrorism has a powerful appeal to many of the marginalized people of the world. Throughout human history, asymmetric warfare—in this case, terrorism—has had an undeniable allure to some and the sympathy of many more. Because it can be spectacularly effective, terrorism will be around for the foreseeable future, hence the legitimate concern for designs to mitigate its effects.

As George Santayana famously said, "Those who cannot remember the past are condemned to repeat it." What follows, therefore, is a brief survey of broad trends in domestic and international terrorism, for the purpose of planning future security measures in site selection and facility design by learning from history (Santayana 1905, 284).

2.3.1 Terrorists' Use of Explosives

Explosives continue to be the terrorist's preferred weapon, since they are destructive, relatively easy to obtain or fabricate, and still comparatively easy to move surreptitiously on the ground and by sea. Terrorists are also well aware that explosives produce fear in the general population far beyond the geographical location of their intended target.

2.3.2 Vehicle-Borne Improvised Explosive Devices

The following case studies include just some of the vehicle-borne explosive devices that have been used in the past quarter-century.

April 18, 1983—The modern era of vehicle-borne improvised explosive devices dates from April 1983, when a massive truck bomb destroyed the U.S. Embassy in Beirut. The blast killed 63 people, including 17 Americans. The attack was carried out by a suicide bomber driving a van, reportedly stolen from the embassy in June 1982, carrying 2,000 pounds of explosives.

October 23, 1983—A Shiite suicide bomber crashed his truck into the lobby of the U.S. Marine headquarters building at the Beirut airport. The explosion, the equivalent of 12,000 pounds of trinitrotoluene (TNT) and alleged to be the largest truck bomb in history, leveled the four-story cinderblock building, killing 241 servicemen and injuring 60. Minutes later a second truck bomb killed 58 French paratroopers in their barracks in West Beirut.

September 20, 1984—A van loaded with an estimated 400 pounds of explosives swerved around several barricades and U.S. soldiers and penetrated the relocated U.S. Embassy annex compound in East Beirut. The suicide bomb exploded 30 feet from the building, killing 11, including 2 U.S. servicemen, and injuring 58.

February 26, 1993—World Trade Center's Tower One's underground parking garage was rocked by a powerful explosion. The blast killed 6 people and injured at least 1,040. The 1,310-pound bomb was made of urea nitrate pellets, nitroglycerin, sulfuric acid, aluminum azide, magnesium azide, and bottled hydrogen—all ordinary, commercially available materials. The device, delivered in a yellow Ryder rental van, tore a crater 100 feet wide through four sublevels of reinforced concrete.

April 19, 1995—Timothy McVeigh detonated an ammonium nitrate/fuel oil bomb in front of the Alfred P. Murrah Federal Building in downtown Oklahoma City, Oklahoma.

August 7, 1998—Two truck bombs exploded almost simultaneously at U.S. embassies in two East African capitals, killing 213 people in Nairobi and 11 more in Dar es Salaam. Some 4,500 individuals, principally Kenyans and Tanzanians, were injured. In May 2001, four men connected with al-Qaeda, two of whom had received training at al-Qaeda camps inside Afghanistan, were convicted of the killings and sentenced to life in prison. A federal grand jury has indicted 22 men, including Osama bin Laden, in connection with the attacks.

December 14, 1999—An apparent plot to bomb the Los Angeles airport was disrupted when Ahmed Ressam, a Canadian of Algerian background, was arrested at a United States–Canada vehicle border crossing in Washington State. Ressam had nitroglycerin and four timing devices concealed in his spare-tire well.

September 11, 2001—Al-Qaeda terrorists crashed two commercial jets into the World Trade Center twin towers, and another hijacked flight slammed into the Pentagon, while a fourth was forced down by passengers and crashed in a field near Shanksville, Pennsylvania.

June 14, 2002—A powerful fertilizer bomb blew a gaping hole in a wall outside the heavily guarded U.S. Consulate in Karachi, Pakistan. The truck bomb, driven by a suicide bomber, killed 12 and injured 51, all Pakistanis.

October 12, 2002—A bomb hidden in a backpack ripped through Paddy's Bar on the Indonesian island of Bali. The device was small and crude, but it killed the backpack owner, likely a suicide operative. The bar's occupants, some of them injured, immediately ran into the street. Fifteen seconds later, a second much more powerful bomb—estimated at slightly more than a ton of ammonium nitrate—concealed in a white Mitsubishi van was detonated by remote control in front of the Sari Club. This blast killed 202 and injured another 209.

May 12, 2003—Attackers shot security guards and forced their way into three housing compounds for foreigners in the Saudi capital of Riyadh. The terrorists then set off seven simultaneous car bombs, which killed 34 people, including 8 Americans and 9 Saudi suicide attackers, and wounded almost 200 more. The facades of four- and five-story buildings were sheared off. One explosion left a crater 20 feet across, while several cars and six or seven single-family homes within 50 yards of the blast were destroyed.

August 5, 2003—A powerful car bomb rocked the JW Marriott hotel in central Jakarta, Indonesia, killing 12 people and injuring 150. Police believe the suicide bomb, which severely damaged the American-run hotel, was concealed inside a Toyota car parked outside the hotel lobby. The terrorist group Jemaah Islamiyah is believed responsible.

August 19, 2003—Sergio Vieira de Mello, the U.N. special representative in Iraq, and at least 16 others died in a suicide truck-bomb explosion that ripped through the organization's Baghdad headquarters. The bomb-laden truck smashed through a wire fence and exploded beneath the windows of de Mello's office in the Canal Hotel in the late afternoon, destroying the building and shattering glass a half mile away. The concrete truck was said to have been chosen as a Trojan horse because of all the construction work going on in the area.

November 8, 2003—Seventeen people, including 5 children, were killed and more than 100 were wounded in an armed raid and suicide car-bomb attack on a residential compound in Riyadh, the Saudi Arabian capital. The huge blast from an explosives-laden vehicle came after gunmen tried to break into the compound and exchanged fire with security guards. The bomb could be heard across the capital, and Riyadh residents who lived miles away said they felt their buildings shake.

March 11, 2004—The routine of a morning rush hour in Madrid, Spain, was shattered when 10 simultaneous explosions occurred aboard four commuter trains. The attacks killed 191 and wounded more than 1,800. Thirteen improvised explosive devices concealed in backpacks were used, but three failed to detonate and were recovered by the authorities.

October 8, 2004—The Hilton hotel in the Red Sea resort of Taba, Egypt, was rocked by a large explosion that killed 29 people and injured at least 120 more. The powerful blast was caused by two separate suicide car bombs, each containing about 200 kilograms of explosives, which crashed into the lobby before detonating simultaneously.

March 2, 2006—A suicide car bomb killed 4 people and injured 50 others in a parking lot adjacent to the Marriott hotel in Karachi, Pakistan. The hotel is about 30 yards from the U.S. Consulate. Among the dead was David Foy, an American diplomat believed to have been the target of the attack. Police said they believed the bomber rammed a car packed with high explosives into the diplomat's vehicle. The force of the blast, the most powerful of its kind in Karachi to date, lifted the victim's vehicle into the air, hurled it over a 7-foot wall, and left a crater 6 feet deep.

June 2 and 3, 2006—Canadian security and intelligence services arrested 12 men and 5 youths on charges of plotting to bomb targets in southern Ontario, including Parliament. The terrorist ring had attempted to procure 3 tons of ammonium nitrate.

June 29 and 30, 2007—A car driven by suspected al-Qaeda sympathizers rammed into the main terminal of Glasgow International Airport, setting off a large blaze when the explosive devices caught fire instead of detonating. The previous evening in London's West End, an attempt to set off two car bombs was unsuccessful when the mobile phone triggering devices failed.

2.3.3 Person-Borne Improvised Explosive Devices

Person-borne improvised explosive devices (PBIEDs) are usually less powerful than vehicle-borne improvised explosive devices, but they are potentially more deadly because they are usually detonated very close to their intended victims.

In general, PBIEDs are targeted against civilian populations or at military personnel in civilian settings such as bus stops. The act usually has a political (not military) purpose, since the intention is to kill, maim, and terrorize a civilian population, and the target areas have few or no security measures in place to prevent bombings. For example, the so-called Second Intifada in Israel and the occupied territories, which occurred between October 2000 and October 2006, saw 267 clearly identified suicide bomber attacks, most of which involved person-borne explosives. More recent attacks in Iraq and Afghanistan continue this deadly trend.

PBIEDs are generally delivered to the target in the following ways.

- The explosives are transported in a package or backpack, and the bomber leaves the area before the explosion. The Madrid commuter attacks of 2004 and the Centennial Olympic Park bombing[3] in Atlanta, Georgia, during the 1996 Summer Olympics illustrate this technique.
- The PBIEDs are worn by the attacker, who expects to die in the explosion. (Suicide bombers, including female terrorists, are increasingly used to carry out such attacks.) The Amman, Jordan, hotel bombings of November 2005 were suicide attacks, as were the numerous Intifada-inspired explosive-vest attacks in Israel following the collapse of the Camp David II summit in 2000.
- The person-borne bomb may not always, in fact, be carried in the traditional sense of the word, but may instead be placed in a wheelchair, a wheeled computer case, or a small cart pulled by the aggressor. This delivery mechanism has the advantages of (1) depriving the security forces of the visual cues presented by a person with a bulky object under his or her clothes and (2) allowing the bomber to deliver more explosives. Consequently, its potential explosives design basis threat (DBT) value—a description of a specific explosive agent, a quantitative value, and a mode of attack that has a degree of likelihood of occurring—is significantly greater than that of body-worn devices. Mitigating the threat of a PBIED is essential in facility designs, as its successful delivery and detonation have the ability to invoke mass casualty, disproportionate damage, and progressive collapse.
- The person who delivers the bomb may be unaware of his or her role in the act of terrorism—for example, convicted Unabomber Ted Kaczynski's use of postal delivery workers. Express couriers, common freight carriers, or other similar second- and third-party delivery agencies such as FedEx or UPS might also be used. These bombs differ from the usual PBIED threat, as they arrive cloaked in the legitimacy associated with an expected delivery mechanism and deny security personnel the opportunity to profile the actual bomber. Using this delivery method, the explosive DBT value could be equal to the carrier's parcel weight and size management thresholds.

2.3.4 Locally Available Explosives

In assessing a threat, perhaps the most important question to answer is "Can the enemy obtain weapons?" The recent wave of terrorist bombings and bomb plots

[3] See http://en.wikipedia.org/wiki/Centennial_Olympic_Park_bombing for details.

around the world clearly answers that question, but the apparent ease with which powerful explosives are available is worth discussing.

Shortly after the Taliban retreated from Kabul in November 2001, a search of an al-Qaeda safe house uncovered a bomb-making training center. Among the documents seized were a detailed explosives instruction manual, a table of explosive mixtures classified by strength, and a table comparing the power of potential detonators such as acetone peroxide (Boettcher and Arnesen 2002). Shoe bomber Richard Reed used triacetone triperoxide (TATP) in his unsuccessful December 2001 attempt to down an American Airlines flight en route from Paris to Miami. In July 2005, however, the four London bus and subway bombers detonated homemade acetone peroxide explosives with deadly effect.

The al-Qaeda safe house also yielded handwritten lists of formulas that included instructions on how to make RDX and a version of C-4, the powerful military explosive used against the USS *Cole* in December 2000. The documents suggest that al-Qaeda was developing its own variant of C-4, known commercially as Semtex, for use in a wide variety of bombs. Creating explosives from the base chemicals, the training manuals noted, would avoid the security problems inherent in finding a supplier and in transporting the contraband into the target country.

The ability to buy explosives components locally "gives more latitude, more autonomy and possibly some degree of elusiveness," according to Tony Villa, an explosives expert who has worked extensively for the U.S. government (Boettcher and Arnesen 2002). Al-Qaeda is "not just a bunch of guys climbing along some jungle gym and going through tunnels and shooting their guns in the air," notes David Albright, a nuclear weapons design and proliferation expert (Boettcher and Arnesen 2002). "These are people who are thinking through problems in how to cause destruction."

Al-Qaeda videotapes, discovered in Afghanistan after the Taliban's defeat, also show a level of bomb-making skill that could allow terrorists to arrive unarmed in the target city and to easily buy and mix the ingredients of a high-explosive device. One of the captured videotapes, a detailed how-to guide, demonstrates the production from scratch of high-quality TNT and other bomb components using easily obtained materials. The videotaped demonstration reinforces the written bomb manuals but also makes the training much more effective.

"What we did see," Albright said, "is that when we compared this information on high explosives to the Internet that these [training manuals] are much more polished. They really did work with these formulas, tested these formulas, and developed a procedure of making these high explosives that led to effective high explosives in a safe manner (Boettcher and Arnesen 2002)."

"The overarching point here is that they can pick any venue or target city, arrive in that city and, based on the video tapes, construct a bomb using common materials," Villa explained (Robertson 2002).

2.3.5 Some Counterterrorism Considerations

Counterterrorism officials have drawn several lessons from these and many similar attacks:

1. Terrorists have always stuck to the operational maxim that what is simple is best, so the experts believe that increased security measures have pushed terrorists toward softer targets and easier-to-deliver but deadly vehicle-borne bombs.
2. The knowledge needed to create bombs and biological/chemical weapons is available in bookstores, in widely circulated training manuals, on CD-ROM, and on the Internet. Armed with this information, even an amateur terrorist can buy enough commercially available chemicals to make his or her contribution to global terrorism.
3. As Oklahoma City bomber Timothy McVeigh and the four July 2005 London transit jihadists have demonstrated, even novice terrorists can purchase bomb-making components and mix the ingredients to create a high-explosive device. For example, terrorists have combined acetone, hydrogen peroxide, and a strong acid such as hydrochloric or sulfuric acid to make triacetone triperoxide (TATP), a powerful and highly unstable explosive.
4. In the brief sample of bombing case studies above, bombers detonated their weapons simultaneously or nearly simultaneously 29 times, including the 7 synchronized blasts in Riyadh and 10 in Madrid.
5. Explosive charges greater than 2,000 pounds were used in nine of the bombings described above. Weapons weighing less than 1,000 pounds appear to have been used in eight cases, while the remaining seven explosive devices were described in terms such as "powerful" and "massive" without identifying the actual amount of explosives. This brief sample suggests that the continued use of large bombs carried in vans, SUVs, trucks, or passenger vehicles reinforced for this purpose is likely.
6. Suicide targets were often at hotels, in restaurants, and on public transportation—places where civilians had gathered.
7. Terrorists have planned phased explosions to target emergency responders, civilians who have not yet been evacuated, and bystanders.
8. Aggressive military, intelligence, and law enforcement measures since September 2001 have disrupted the terrorists' communications and logistical support. These worldwide activities have degraded the radical jihadists' ability to mount operations as sophisticated and coordinated as the hijackings of September 11 and the 1998 bombing of the embassies in East Africa. As the message of jihad reaches more disaffected young men, however, the bombings may be smaller in scale but more frequent.

9. The terrorists will opt for continual low-level bombings, therefore, as part of an ongoing campaign to inflict some "revenge"—less dramatic and less traumatic than the twin towers—on the enemy and to prove their relevancy.
10. The terrorists sought out and exploited gaps where security was either ineffective or lacking—most notably in the case of the U.S. commercial aviation industry on September 11, 2001.
11. Today's terrorists harbor no compunctions about killing or injuring innocent civilians.

2.4 COLLABORATING TO ANALYZE RISK

The first risk assessment probably occurred when a prehistoric couple heard strange noises outside their cave; in the intervening millennia, thousands of risk assessment formats have been developed. Today, in response to private-sector concerns, the security community's post-9/11 threat and vulnerability evaluations place far more emphasis on terrorism and criminal violence than they did even a decade ago.

A comprehensive risk assessment for blast-protection design involves close collaboration among city planners, architects, engineers, blast consultant subject matter experts, and law enforcement security professionals. In this group effort, collaborating professionals assess and select the security measures needed to detect, deter, prevent, defeat, mitigate, or recover from terrorists' bomb attacks.

This section introduces the analytical and decision-making steps that are widely used today to identify the critical assets requiring protection, determine the explosive threats, assess the vulnerabilities, and create mitigating measures. It begins with a thorough but flexible five-step methodology for preparing and executing a risk assessment.

2.4.1 Step 1—Threat Identification and Rating

The assessment process begins with a detailed understanding of the natural and man-made threats that could realistically occur in the geographic area in question and therefore pose a risk to the building or site being designed. To best prepare the design team to anticipate the broadest range of contingencies, the threat identification and rating step calls for the following four tasks.

Task 1-1. Identifying the Threats. In blast-protection terms, a threat is a potential event culminating in an explosion that damages or destroys a protected facility or its inhabitants.

Task 1-2. Collecting Information on Potential Attacks. Collecting information for an explosives-oriented threat assessment requires expertise in weapons and demolitions as well as in terrorist history, tactics, training, and targeting. An informed estimate of the aggressors' motivation and intentions against the future facility and its future occupants is also essential. The following list of questions is a quick overview of the kinds of information the team needs to gather:

- What domestic or international terrorist groups are the potential aggressors, and what are their capabilities?
- What has been expressed or can be inferred about their motivation and intentions against the clients?
- How skilled with demolitions are they? Are bomb-making materials available to them?
- Does the project have an economic, cultural, or symbolic significance that terrorists could exploit to justify their acts and to gain publicity?

Many informative checklists and memory aids are available for collecting these data, and several of the best have already been cited. The design team's security members should be well versed in at least one of these tools. They should also have information sources among state and federal law enforcement and homeland security agencies—if not by virtue of existing relationships, then certainly through contacts and their own credibility. Neighboring building owners, operators, and tenants can also be a valuable data source.

Task 1-3. Determining the Design Basis Threat for Explosives. The Department of Defense defines the design basis threat (DBT) as the hazard against which a building must be protected and upon which the protective system's design is therefore based.[4] In other words, the type and size of weapon the building and site must be designed to withstand is the explosive DBT.

To calculate an explosive DBT, the team must first estimate the potential for an attack by identifying the potential terrorists' tactics and the tools, weapons, and explosives they could employ. Terrorists have used explosives hidden in moving and stationary vehicles and have left behind explosives that were hand-carried to the target. Mail bombs and supply bombs—larger devices typically infiltrated through shipping departments—and explosive devices carried by suicide terrorists are also proven techniques. For reference purposes, large-scale truck bombs typically contain 10,000 pounds or more of TNT equivalent (all the following bomb weights are in terms of TNT equivalent). Bombs in vehicles ranging from small sedans to vans typically contain 500 to 4,000 pounds. Suicide bombers can unobtrusively deliver belts ranging in size from less than 10 pounds up to about 40 pounds. Hand-carried explosives are typically on the order of 5 to 10 pounds (Federal Emergency Management Agency 2005, 1–7).

Design documents do not normally specify the actual DBT explosive weights. Instead, the weights are expressed in terms such as W-1 (for example, a 10-pound bomb), W-2 (100 pounds), W-3 (1,000 pounds), and so forth. The actual values may vary from project to project as determined appropriate by the design team.

---

[4]This DoD definition of "threat" can be found at www.dtic.mil/doctrine/jel/doddict/data/d/01635.html.

If design documents containing specific DBT explosive weights fell into terrorist hands, they would become an extremely valuable aid to the enemy. Actual DBT weights are revealed only on a need-to-know basis.

The willingness of suicide terrorists to turn themselves into bombs precludes design professionals from adopting a design basis threat of "no explosives." Consequently, when the design team is developing protective design strategies for a site or building and its occupants, it must insist on a realistic cataloging of the project-specific explosive DBTs. This information mining is essential if structural and architectural facility designs are to provide credible protection.

The structural engineers' and architects' protection strategies are based on an estimation of the possible size and location of a detonation, coupled with a realistic estimate of the facility's ability to withstand an attack. Since the eventual design implementation concepts and components of a facility's structure; façade; evacuation, rescue, and recovery systems (ERR); and business continuity measures are built upon this analysis, establishing quantifiable DBTs is a fundamental first step in providing the design team with a framework for designing mitigation measures.

Task 1-4. Determining the Threat Rating. The team's threat ratings will evaluate the probability of various blasts occurring and the likely consequences if they do. The team should imbed *probability* and *consequences* in their consciousness; they are two extremely important concepts that security design planning teams far too often overlook.

Vehicle-borne bombs continue to be the terrorists' favorite tactic, primarily because trucks can carry potentially devastating explosive cargos. But terrorists will exploit the weakest links in a building's protective design; person-carried bombs have shown that relatively small weapons can cause great damage to exposed and structurally vulnerable building interiors. This reality appears to argue for defending buildings against every conceivable explosive attack, but total protection through design and engineering is impractical in terms of structural hardening and benefit-to-cost ratio. A potential solution for the "protect everything" dilemma is discussed in later paragraphs, when the authors address the complementary roles of building design, physical security, operational security, and continual situational awareness.

Facility designs outside North America are, of course, also vulnerable to explosives threats. For work outside of North America, it is especially important to develop a comprehensive risk assessment that is specific to the geographic area, the project site, and the facility.

In addition to the person- and vehicle-borne IED delivery methods, the use of standoff military ordnance, especially the use of shoulder-fired arms such as the ubiquitous rocket-propelled grenade (RPG), remains a significant concern. Structural engineers and architects should give careful consideration to these military-grade weapons as part of a competent protective design threat-mitigating strategy. Also, major headaches can be avoided by resisting the tendency to overestimate the man-made threat. Ninja-like terrorists in attack helicopters have not appeared yet, and the two-millennium history of terrorism clearly demonstrates that the enemy realizes that the best route to success is very often the easiest one.

The design team must provide project-specific information at the earliest stages in which conceptual designs are adequately advanced. The architects' and designers' timely participation in the initial risk assessment process will greatly influence the credibility—or lack of credibility—of the project design team's analysis of potential vehicle and pedestrian access routes. In short, the determination of the degree to which the project site, physical plant, and personnel are at risk from specific explosives must be informed by an interactive process of information exchange between the threat-assessment and design-team professionals.

The team, concluding its work on threat assessment, now produces a single threat rating for the entire project. Threat ratings are commonly expressed in a seven-level linguistic scale from very low probability to very high probability. "Very high" threats are considered credible and imminent, and "very low" threats are deemed negligible. Between those two extremes, the spectrum includes threats of high, medium high, medium, medium low, and low probability. The scale could be numerical instead, from 1 (very low) to 10 (very high). Some threat-ranking formats result in both linguistic and numeric ratings. This very-high-to-very-low scheme is also used in ratings of asset value and vulnerability.

Figure 2.2 illustrates the use of worksheets to consolidate the design team's knowledge in a single, agreed-upon format. In this example, the worksheet displays the probability of various improvised and military explosives being used in an attack and the anticipated damage as a consequence of the assault.

The authors encourage teams to use their own professional experience to question the raw scores produced by the threat-ranking scales. Applying a "Does this make sense?" filter is wise, because rating threats is more art than science. Deferring to the client's insistence that his facility is a likely terrorist target, for example, puts the assessment process at risk of being driven toward the "very high" end of the scale, resulting in overdesign. Of course, it is equally possible to underrate the danger, and the consequences of underdesigning could be grave.

This threat identification and rating exercise creates a prioritized team assessment of the potential dangers, especially of a terrorist bombing, to the project.

2.4.2 Step 2—The Asset Value Assessment

The team will quickly need to learn as much about their building, its site, and the surrounding area as they have discovered about the explosive threat. These assets can be tangible, including facilities, equipment, inventory, and, of course, people. Intangible assets include intellectual property, proprietary processes, reputation, corporate information, and image. A concerted effort should be made to become thoroughly acquainted with the client's facility and its adjacent environment. The team's knowledge about those unique tangible and intangible assets that require enhanced protection will be invaluable as the security design is developed.

In asset assessment, the first of four tasks is to identify the project's layers of defense.

Threat Determination Criteria								
Explosive Threats	Access to Explosives	Knowledge and Expertise	Threat History for Buildings, Tenants, and Environs	Asset Visibility and Symbolic Value	Asset Accessibility	Site Population and Capacity	Collateral Damage at Specified Distance	*Score*
Improvised and Millitary Standoff Explosive Devices								
1-lb. Mail Bomb	9	9						
5-lb. Pipe Bomb	9	9						
RPG or Similar Grenade/ Rocket Launcher	7	7						
20-lb. Cargo Bomb	9	9						
35-lb. Salchel Bomb/ Suicide Bomber	8	8						
500-lb. Car Bomb	6	8						
5,000-lb. Truck Bomb	4	8						
20,000-lb. Truck Bomb	2	6						
Natural Gas	2	8						

NOTE: The threat numbers are national and are used for illustrative purposes only.

Figure 2.2 Primary Explosive Threats to Notional Urban Building (Adapted from FEMA 426 and FEMA 452)

Task 2-1. Identifying the Layers of Defense. Layered facility protection is a time-tested security design engineering approach. Defense-in-depth creates concentric rings of security, each increasingly difficult to penetrate, which provide security and law enforcement officials with multiple opportunities for warning and reaction. A layered defense complicates the terrorists' operations, perhaps to the point that security forces are able to interrupt the aggressors' surveillance or intercept them early in their attack. The obstacles presented by a layered defense may cause the terrorists to attack a less well-defended target, or could allow building occupants time to relocate to predetermined defensive positions or find shelter in designated safe havens during an assault.

The first layer of defense in a typical urban setting is the surrounding area—the sidewalks, curb lanes, streets, other buildings, and the neighborhood itself. The structural composition of adjacent buildings can either enhance the project's security or increase its threat. Local law enforcement and traffic authorities are increasingly willing to support a security designer's street and curb-lane safety initiatives, particularly when realistic procedural measures precluding adjacent vehicle traffic lanes or parking are planned. Unobtrusive physical security devices—such as decorative anti-ram, crash-resistant street lights, benches, planters, and trash receptacles—are found often now on sidewalks and plazas. Embedded barricades, retractable bollards, and hydraulic barricades are also becoming more commonplace.

The exterior space between the line of vehicle or pedestrian interdiction and the project structure is the second layer of defense. Surveillance cameras and lighting must be part of any security design, but force protection engineered vehicle interdiction devices and pedestrian approach restrictions can be useful here too. Crime Prevention through Environmental Design (CPTED),[5] a multidisciplinary approach to deterring criminal behavior, may also create a disincentive to terrorist surveillance and operations. ASIS International,[6] the society for professional security officers, has extensive libraries of security equipment, concepts of operation, site civil and landscape counter-crime features, and operational procedures-related information to help in security design planning of this interstitial area between the protected building and the site perimeter.

The third layer of defense is the building itself, and it is here that security design measures can play a particularly effective role. Hardening structures, facades, and systems; providing a broad array of alarm and surveillance tools; and carefully designing and locating utilities and building systems, incorporating equipment that supports evacuation and recovery, are examples of critical elements in security design. Subsequent chapters of this handbook expand on topics such as the role of materials-performance criteria, structural systems design, facade protection, and survivability of ERR systems.

Task 2-2. Identifying the Critical Assets. The second task of the asset value assessment is to identify critical elements within each of the three defense layers—in other words, which resources must withstand an attack if the health and safety of the building's occupants are to be ensured. FEMA and ASIS, among others, publish extensive checklists that will greatly assist in pinpointing relevant issues.

Task 2-3. Identifying the Building Core Functions and Infrastructure. The security design team should next identify the building's, site's, or venue's core functions and infrastructure. The design team should use these data to inform designs that isolate, distribute, or replicate the equipment and processes that will

---

[5] A discussion of CPTED, including numerous references and external links, can be found at http://en.wikipedia.org/wiki/Crime_prevention_through_environmental_design.

[6] ASIS can be found at www.asisonline.org.

provide vital operations and services during and after an attack, or designs that use hardening strategies to improve the operational survivability of these operations and services.

One of the most successful strategies for the protection of building core functions and infrastructure in blast-protection design is dispersal and redundancy. If a bomb detonates adjacent to or within a facility, phones and telecommunications equipment, security and evacuation systems, fire suppression and smoke control systems, generators and uninterruptible power supplies, and other infrastructure elements deemed critical to evacuation, rescue, and recovery must continue to function. Here again, FEMA, ASIS, and others have developed extensive lists of questions to help assessment teams identify core security and life-safety functions and their interdependencies.

Task 2-4. Determining the Asset Value Ratings. The final task in Step 2 is to assign a value rating to each asset. The value rating is the security design team's judgment of the degree of debilitating impact that would be produced by the incapacitation or destruction of that asset. In other words, the consequences of damage or destruction are considered for each asset. Then a 1-to-10, very-high-to-very-low value is assigned to each asset. "Very high" would mean that the loss or damage of the building's assets would have exceptionally grave consequences, such as extensive loss of life, widespread severe injuries, or total loss of primary services, core processes, and functions. On the other hand, a "very low" ranking indicates that the loss or damage of the building's assets would have negligible consequences or impact. As was previously noted, there are five additional gradients between these two extremes (Federal Emergency Management Agency 2005, Table 2.5).

The asset value assessment lists the building's core functions and critical infrastructure elements. Next, the team assigns each element an intrinsic value based on the degree of debilitating impact that its incapacity or destruction would create. The product of the value asset rating exercise is a table that lists the various types of attacks against each of the facility's critical functions. The team then populates the table with high-to-low values expressing the consequences of loss or damage of each asset.

2.4.3 Step 3—The Vulnerability Assessment

Vulnerability is any weakness that aggressors can exploit to cause damage. To constructively influence design considerations, the vulnerability assessment must realistically appraise the building's projected functions, systems, and site features; it must pinpoint structural weaknesses and identify absences of necessary redundancy so that corrective actions and mitigations can be incorporated. Unlike threats, vulnerabilities are conditions and/or designs that people create themselves and over which they can therefore exercise some control. Close coordination among the assessment team, the building design professionals, and ownership is critical; it is essential to prevent a great deal of incompatible expectations and design deliverables down the road.

There are four tasks in assessing vulnerability:

Task 3-1. Organizing Resources to Prepare the Vulnerability Assessment. Task 3-1 has two subparts:

(a) Selecting the Assessment Team. Risk assessment is a complex undertaking with high stakes in terms of consequences and costs. Proper team selection requires some serious up-front time and attention, if only because, in unskilled hands, risk assessment tools can be blunt instruments. The best survey-tool authors stress that "teams created to assess a particular building should be senior individuals who have a breadth and depth of experience in the areas of civil, electrical, and mechanical engineering; architecture; site planning and security engineering; and how security and antiterrorist considerations affect site and building design" (Federal Emergency Management Agency 2005, ii).

(b) Determining the Level of the Assessment. The security community generally recognizes three increasingly detailed levels, or tiers, of risk assessment (explained below), so at this point a decision is required: how detailed must the evaluation be?

FEMA 452 guidelines for vulnerability assessment denote three assessment levels:

Tier 1—A study that identifies the most important vulnerabilities and their mitigation options—a "70 percent" assessment. A Tier 1 evaluation is typically conducted in approximately 2 days by one or two experienced assessment professionals working with the building ownership and key staff. It involves a quick look at the site perimeter, building, core functions, infrastructure, and any related drawings and specifications. A Tier 1 assessment is likely sufficient for the majority of commercial buildings and other noncritical facilities and infrastructure.

Tier 2—A full on-site evaluation by assessment specialists that provides a robust evaluation of proposed and/or existing systems' interdependencies, vulnerabilities, and mitigation options—a "90 percent" assessment solution. A Tier 2 assessment typically requires three to five assessment specialists. It can be completed in about 5 days and requires significant key building staff participation; access to all existing site buildings and areas, systems, and infrastructure; and/or an in-depth review of proposed building design documents, drawings, and specifications. A Tier 2 assessment should be sufficient for most high-risk buildings such as iconic commercial buildings, government facilities, cultural and educational institutions, hospitals, transportation infrastructure, and other high-value targets.

Tier 3—A detailed evaluation of the facility and site. A Tier 3 analysis uses blast and weapons-of-mass-destruction (WMD) modeling to assess the site and building's response, survivability, and recovery parameters. It involves engineering and scientific experts, requires detailed design information, and provides the maximum data for evaluating and developing mitigation options. Tier 3 modeling and analysis can often take several weeks or months for an existing facility and is a process involved at the Conceptual, Schematic, and Design Development phases of new facilities. It is typically performed only for high-value and critical structures. From 6 to 12 subject matter experts may be required, based on project complexity.

When the building and site are in the design phase and do not yet exist, the risk assessment data-gathering process described above must be modified considerably. A Tier 2 risk assessment of a *future* building would concentrate, for example, on conceptual design documents, drawings, and plans in an iterative process of discussion, impact assessment, and design or redesign. This situation is actually preferable, because it gives the design professionals an opportunity to build security into the structure's skeleton. The alternative—retrofitting protection into inappropriate or inadequate spaces—seldom offers totally satisfactory solutions.

The choice of the appropriate tier level depends on the building and site's location, its construction characteristics, its proposed use and profile, and its owners' or occupants' concerns about terrorism. The choice of tier level is also heavily influenced by the design team's initial threat and vulnerability appraisals; by local, state, and federal antiterrorism and public safety regulations and guidelines; and (inevitably) by benefit-versus-cost considerations.

Task 3-2. Evaluating the Site and Building. The authors have described how producing a robust building security design requires the team to identify potential threats and vulnerabilities, define the resources and people most at risk, and describe the consequences of potential losses. To bring order and predictability out of such complexity, the team's next task is to prepare a survey schedule and tentative agendas. The team needs, for example, to meet with stakeholders such as site or building owners, security and engineering personnel, IT directors, and emergency managers. The team will also need to inspect the area and facilities and review key information such as security plans and emergency procedures, if these documents exist. The key to timely execution and success is planning.

An important aid to planning is *threat mapping*, the clear and graphic representation of the site-wide explosive DBTs at vulnerable points, such as lobbies, mailrooms, and loading docks. Threat maps, which are based on architectural concept drawings, should encompass the site from its outermost perimeter to its most sensitive internal areas. This mapping can be done on existing blueprints or other site drawings. Each vulnerable area is labeled with a "W" number, according to the key that has already been created for designating explosive DBTs (W-1, W-2, W-3, etc.). Threat maps illustrate for design professionals how concentric layers of competently applied security measures can reduce, or in some cases largely eliminate, the facility's explosive threats.

Task 3-3. Preparing a Vulnerability Portfolio. It just makes good sense that the large volume of data that has been collected as part of the threat, risk, and vulnerability assessment in the previous steps should next be gathered into a well-organized security design document.

One of the most useful sets of assessment aids in this team portfolio can be the FEMA Building Vulnerability Assessment Checklists found at Appendix A of the previously mentioned FEMA publication 452, *Risk Assessment: A How-To Guide to Mitigate Potential Terrorist Attacks*. This series of tables prompts the team to adopt a consistent approach to evaluating security designs at their various stages of development. The checklist can be used as a screening tool for preliminary design vulnerability assessments and supports the preparation of each step in FEMA's *How-To Guide*. As the FEMA *Guide* explains:

> The checklist is organized into 13 sections: 1) site, 2) architectural, 3) structural systems, 4) building envelope, 5) utility systems, 6) mechanical systems, 7) plumbing and gas systems, 8) electrical systems, 9) fire alarm systems, 10) communications and information technology, 11) equipment operations and maintenance, 12) security systems, and 13) the security master plan. To conduct a vulnerability assessment or preliminary design, each section of the checklist should be assigned to an engineer, architect, or subject matter expert who is knowledgeable and qualified to perform an assessment of the assigned area. Each assessor should consider the questions and guidance provided to help identify vulnerabilities and document results in the observations column. If assessing an existing building, vulnerabilities can also be documented with photographs, if possible. The results of the 13 assessments should be integrated into a master vulnerability assessment and provide a basis for determining vulnerability rating during the assessment process. (Federal Emergency Management Agency 2005, Appendix A)

Task 3-4. Determining the Vulnerability Rating. The weaknesses the security design team has identified, which could be exploited during an explosives attack, are reflected in the vulnerability rating. As an example, the lack of critical-systems redundancy raises the vulnerability rating, of course, as do single points of failure among essential security and life-safety systems or structural systems incapable of withstanding the project explosive DBTs or an unforeseen event that could compromise a key primary structural element.

Like the previous threat and asset value ratings, vulnerability scores are commonly expressed in a 1-to-10 numerical scale or a very-low-to-very-high linguistic scale. A very high vulnerability rating is delivered from identifying one or more major weaknesses that make the facility extremely susceptible to an aggressor or hazard. A very low rating describes a building that integrates excellent physical security and comprehensive redundancies—in short, a building that would promise a high degree of physical and intellectual asset protection and be operational immediately after an attack.

To summarize where the discussion is now in the risk assessment process: In light of the identified potential threats (Step 1) and the team's categorization of the structure's critical assets (Step 2), the vulnerability assessment (Step 3) determines the opportunities to place at risk the project assets, and therefore sets the stage for (Step 4) the risk assessment, wherein the team and the client discuss and agree to the extent of acceptable risk.

The vulnerability assessment is a comprehensive listing of all weaknesses an aggressor could exploit to damage an asset. The product of Step 3 is another table, this one listing the building functions, each one with its high-to-low vulnerability rating, together with the associated vulnerabilities that the team has uncovered.

2.4.4 Step 4—The Risk Assessment

This is the step in which the team pulls together all the information it has compiled, all the tables it has completed, all the judgments it has made, and each of the threat, asset, and vulnerability ratings it has carefully assigned. This step is where the interdisciplinary teamwork pays off; the risk assessment, backed up by its demonstrable research, provides a credible foundation for security design decisions.

Task 4-1. Preparing the Risk Assessment Matrices. Focusing attention on explosive event protection naturally narrows the range of security-design considerations. If, however, the team wants to undertake an all-risk, all-hazard review, including chemical/biological/radiological attacks as well as cyberassaults, it can refer to FEMA 452 for a range of helpful checklists. Whether the analysis is broad or narrow, however, the way to estimate potential losses is to prepare a series of matrices, tailored to the individual task at hand and based on the analysis conducted during Steps 1, 2, and 3.

To estimate risk, the team must have analyzed a number of factors. First, it will have identified and ranked the threats that could harm the project. Next, it will have determined the value of the client's assets, including people who require protection. Finally, by using the results of the threat scores, asset value estimations, and vulnerability ratings, it will determine a vulnerability and asset rating system that identifies weaknesses an aggressor could exploit.

Task 4-2. Determining the Risk Rating. The design team next analyzes the threat ranking, asset value, and vulnerability rating developed in the previous steps and assigns a risk level for each critical asset. A specific risk such as a car-, truck-, or person-borne bomb can be expressed as the sum of the asset value times the threat rating times the vulnerability rating, or $R = AV \times TR \times VR$. This exercise also results in linguistic or numeric values, or both.

Task 4-3. Prioritizing Vulnerabilities with FEMA's Building Vulnerability Assessment Checklist. With a risk level now assigned to each critical asset, the team is ready to prioritize, or rank, the vulnerabilities it has identified. Prioritization is based on evaluating which vulnerabilities the aggressors will most likely exploit and which of them pose the greatest danger to lives, structures, and operations. The highest ranking is given, obviously, to the greatest vulnerabilities with the greatest consequences. Doing so makes it possible, in the next step, to determine the most effective mitigation measures.

2.4.5 Step 5—Considering Mitigation Options

After completing the risk assessment, stakeholders are frequently left with issues requiring mitigation, but they find their options limited by space constraints, esthetics, regulatory requirements, or costs. Decisions about allocating resources therefore must focus on the most practical mitigation options.

These are the fundamental mitigation questions:

1. What kind of action should be taken in response to the danger? There are four ways of responding to a threat:
   - *Acceptance* of a threat is a rational alternative that is often chosen when the threat has a low probability, low consequence, or both.
   - *Prevention* is the alteration of the target or its circumstances to diminish the risk that the terrorist or criminal will succeed.
   - *Interdiction* is any confrontation with, or influence exerted on, an attacker to eliminate or limit its movement toward causing harm.
   - *Mitigation* is preparation so that, in the event of an explosion, its consequences are reduced.
2. Does the response create new risks to the asset or others? The final step in analyzing the security program's efficacy is to be aware of new risks created by the prevention, mitigation, or interdiction of the threats under consideration.

Four tasks are involved in identifying mitigation alternatives:

Task 5-1. Identify Preliminary Mitigation Options. This exercise identifies the design team's available design options together with any relevant physical, technical, cyber, and procedural security alternatives. The goal is to create the strongest possible security design.

Task 5-2. Review Mitigation Options. Since every project has budget constraints as well as operations and maintenance limitations, it is seldom possible to totally eliminate a risk. Now is the time, therefore, to weigh mitigation options for their practicality and to match the most feasible mitigation opportunities with the most critical risks.

Task 5-3. Estimate the Cost. Some costs of protecting buildings are fixed, while others are variable. Funds for deploying protective equipment and creating a security zone-of-control are generally fixed; for example, the costs of anti-ram barricades to keep a vehicle away from the facility are comparable whether the same vehicle contains 50 pounds of TNT or 500. The blast-protection costs of facility construction, on the other hand, will vary according to the estimated threat level. Since the bomb size is outside building planners' control, the tailoring of blast protection to the building structure, for instance, is a function of the maximum credible explosive charge plus its estimated proximity when the bomb detonates.

Since cost is almost always a central issue, there are three mitigation approaches that can be used singly or in combination: (1) modify the design threat, (2) reduce the level of protection, or (3) accept the risk. Life-cycle costs need to be considered as well. For example, the recurring costs of maintaining and upgrading security and life-safety equipment can derail even the best protective programs.

Security design experience has repeatedly shown that attention paid to defending against man-made hazards during the preliminary planning and conceptual design phases significantly reduces the life-cycle costs of protection and increases the inherent security level provided to the building occupants.

In summary, the project security professionals' explosive threat assessment should figure prominently in the design of the facility facade, structural system, ERR systems, and other critical protective features. The design team should be directly involved in the risk assessment steps that identify which building design features and components can affect the type, size, and potential delivery method of an explosive device.

Task 5-4. Reviewing Mitigation, Cost, and the Layers of Defense. FEMA 452 includes a final series of tables containing examples of mitigation options and alternatives for each of the facility's security features. These security measures are organized by layers of defense and arrayed by degree of protection, cost, and effort. See Figure 2.3 below for FEMA 452 (January 2005) *Worksheet 5-1: Preliminary Mitigation Options.*

When the team has completed the risk assessment process, it will have identified and evaluated various mitigation options it can use as guidance when developing the building's protective design features and components.

2.4.6 The Continuing Role of Risk Management

The goal of stronger, safer structures begins with building security design and continues as security risk management throughout the building's life cycle. To support these initial and long-term goals, FEMA 452 asserts that the goal of the assessment process is to achieve the level of protection sought through implementation of mitigation measures in the building design. These measures may reduce risk by deterring, detecting, denying, or devaluing the potential threat element prior to or during execution of an enemy attack. The Department of

Prioritized Observations	Blast Regulatory Measures	CBR Regulatory Measures	Blast Repair and Strengthening of Existing Structures	CBR Repair and Strengthening of Existing Structures	Blast Protection and Control Measures	CBR Protection and Control Measures	*Worksheet 5-1 will help to identify your preliminary mitigation options.* *After you have priortized your observations/vulnerabilities (Task 4.3), proceed to rank them for impact during blast and CBR events. Using the first part of the Worksheet (Prioritized Observations) indicate if these observations merit a regulatory, rehabilitative, and/or protective measure and if they are directed at blast or CBR,*
Observation 1							*Using the second (Preliminary Mitigation Options for Blast) and third (Preliminary Mitigation Options for CBR) parts of the Worksheet, determine mitigation options that address the main concerns included in your observations and provided parameters.*
Observation 2							
Observation 3							
Observation 4							
Observation 5							
Observation 6							
Observation 7							
Preliminary Mitigation Options for Blast							
Mitigation 1							
Mitigation 2							
Mitigation 3							
Mitigation 4							
Preliminary Mitigation Options for CBR							
Mitigation 5							
Mitigation 6							
Mitigation 7							
Mitigation 8							

Figure 2.3 Worksheet 5-1: Preliminary Mitigation Options (From FEMA 452, Jan. 2005)

Homeland Security endorses the use of the following methodology (available in FEMA 452).to achieve this purpose.

Deter: The process of making the target inaccessible or difficult to defeat with the weapon or tactic selected.
Detect: The process of using intelligence sharing and security services response to monitor and identify the threat before it penetrates the site perimeter or building access point.

Deny: The process of minimizing or delaying the degree of site or building infrastructure damage or loss of life or protecting assets by designing or using infrastructure and equipment designed to withstand blast and chemical, biological, or radiological effects.
Devalue: The process of making the site or building of little or no value or consequence, from the terrorists' perspective, such that an attack on the facility would not yield their desired result (FEMA 452, 2005).

2.5 CONSEQUENCE MANAGEMENT

This section provides insight into one of the primary tools used in assisting design professionals and their clients in the development of decisions involving risk management. This section explores the value and process of evaluating consequences associated with explosive events and the major mitigating strategy categories used to reduce those consequences.

Consequential effects are, by their very nature, subjective evaluations and categorizations. The perception of what is and what is not valuable and/or essential is very much determined by the valuation process utilized by the institution or organization that is responsible for its codification. Consequently, institutional business mission statements, private organizational charters, government agency programs, stockholder expectations, legal proscriptions, and numerous other guidelines and/or mandates often identify different levels of expectation for post-event facility performance. These guiding documents and/or unwritten notions were generally not created for the purpose of assisting design professionals by clearly defining ownership's requirements for preserving institutional business-mission operations or the protection of people on premises. The absence of well-codified protection expectations leaves the design team with less than adequate information about the extent of consequential effects that ownership will find tolerable.

Design professionals must clearly understand the goals and expectations of the facility, its occupancy and the activities that will take place within the structure, site, and, in some cases, the campus or regional landscape or cityscape fabric they are creating for (and with) their client. This is an essential piece of early project information client data mining, which is often overlooked. Insufficient efforts and pursuit of a clear definition of client desires regarding consequence management are one of the most significant design process shortfalls. Mismanagement of this process will likely predestine the engineering efforts to fail to meet client needs or expectations.

To address this omission in the programming design phases, one of the first steps should be—during the earliest phases of project design programming and conceptualization—the creation of a brief security design mission statement. This document should clearly and concisely state the client's expectations for the facility's post-explosive event performance. These expectations are routinely overstated by clients. This is generally a result of their lack of sophistication or knowledge of the engineering complexities, architectural program compromises, post-construction institutional operational constraints, impact on project financials, aesthetic effects, and other factors better understood by the structural engineer and the balance of design team professionals. Consequently, this mission statement must be a consensus-developed document. It is reasonable to expect that it may take several iterations of authorship, editing, review, and subsequent approval.

Once agreed to, the consequence management statement should then become a part of the project's key design assumptions and design program objectives. It should be used as a means to guide design decisions, and to provide the basis for decisions regarding program conformance, as designs mature in the Schematic, Design Development, and Construction Document phases.

Once a clear concept of acceptable consequence management in terms of facility performance is understood and codified, there is a sound basis for developing the facility designs. This is essential to the process of achieving an acceptable balance between post-explosive event damage to the facility's physical and intellectual assets and the expectations of the client and society regarding life safety and institutional business-mission continuity.

No facility can be expected to provide a post-explosive event damage performance profile that ensures no injury, no loss of life, and no disruption to facility operations. **The challenge of consequence management is for the design professional and client team to establish reasonable goals and objectives. These need to be tempered by rational opportunities for success given the facility's design program, client occupancy and activities, project location and budget, law enforcement and security resources, and other economic and geopolitical factors, which are unique and specific to each design commission.**

Design professionals and their clients have opportunities to examine and implement strategies that can participate in achieving the level of consequence management identified in the project's key design assumptions and design program objectives, as stated above. These holistic design strategies usually include:

(a) Strategic site selection and facility location to reduce vulnerability to threats that are on-site, near-site, or regional.

(b) Within a specific building, the provision of equipment, personnel, and resource redundancies, so that the loss of a single critical system, facility management operational component, or structural element or system does not create a single point of unacceptable or catastrophic failure.

(c) Within a specific building, distribution of facility physical plant; intellectual and nonintellectual assets; operational programs; and evacuation, rescue, and recovery systems so that an explosive event in one location does not jeopardize the entire building, all of its occupants, and the facility-wide performance of ERR systems. (Note, however, that the use of diverse redundancy as a protective strategy should be examined during the design process in conjunction with the protective strategy of co-location of critical systems and/or resources housed within a well-protected environment. The intent is to foster survivability, thereby reducing consequential effects.)

2.5.1 Consequence Evaluation

It is universally understood by the architectural and engineering design community that explosive events can be responsible for significant damage to building structures, facades, and critical building systems. However, there is less awareness amongst the design profession during the initial planning phases of other extremely important and sometimes ethereal values such as an institution's public image and sense of solidarity. Without this awareness there is naturally less acceptance of design responsibility for the protection of these other, less tangible but extremely important client assets.

Additionally, other more essential issues associated with evacuation, rescue, and recovery operations generally are assumed by the design profession to be owner and operator responsibilities developed by those entities in response to the building's design. This approach is archaic.

Design professionals must address building designs so that the explosive event consequence management response is achieved through an evaluation of how evacuation, rescue, and recovery operations will occur after the explosive event. In this way, designs are informed to protect evacuation, rescue, and recovery building design features such as stairways, exit passageways, elevators, and building systems such as emergency power, emergency lighting, voice evacuation, fire detection, fire suppression, and smoke management. This consequence evaluation approach results in a building design that provides inherent tactical support and definitive criteria for an evacuation, rescue, and recovery concept of operations. This will always provide a superior product to an ERR strategy retrofitted to a design created with less inherent vulnerability.

The above design shortfalls can generally be attributed to the design team's focus on the intimate knowledge of and extensive involvement in developing all of the intricate details associated with each component of the building's architectural, structural, and support system infrastructure designs. Furthermore, the design community traditionally focuses on contract boundary limits as the extent of their responsibility for the ultimate scope of their work. **The range of effects from explosions rarely aligns with contract boundary limits of work. This fact must be taken into consideration as part of consequence evaluations.**

Facility occupancy after construction completion, with its complement of staff population assignments, support furnishings and equipment, and operational protocols usually is established with the greatest attention to detail after the building is commissioned. Consequently, these and other aspects of day-to-day facility management are logistically more distant entities and often are understood in less detail by the design professional team. **Design professionals should focus additional resources on more completely understanding post-construction occupancy and fit-out subtleties, in order for explosive event consequence evaluations to be more thoroughly responsive.**

Also, traditional contracts, governing the performance of the design team's scope of work, emphasize the most significant levels of effort up to and inclusive of building commissioning. They fall short of emphasizing the role of the design professionals in fully understanding a broader range of construction completion and subsequent facility occupancy concerns. These include all of the various intellectual and operational entities, exceptional conditions and crisis management, and facility restoration issues, which will be placed at risk and/or hindered by an explosive event once the structure is completed. **Understanding how the facility will operate under extraordinary conditions is essential to developing protective designs that address the consequential effects on occupants and operations, as well as on the physical plant.**

The structural engineering, architectural design, and balance of the design team professionals should ensure, at the commencement of the design process, that this disparate intellectual appropriation of design awareness is addressed. This should be accomplished through a process that focuses on the protection and post-explosive event performance of the structure itself in a manner that is carefully balanced against all of the other consequential explosive event effects, including day-to-day facility operations and the exceptional conditions associated with ERR.

A wide range of consequences must be evaluated. Understanding these consequences is critical to establishing the balance between (a) protective designs assigned to the facility's site perimeter protection program, facade, structure, evacuation, and rescue and recovery systems and (b) systems associated with life safety and business continuity. Mitigating each of the consequences that may occur if these areas of facility vulnerability are not adequately addressed will require a portion of the project's budget. **The project profile for security investments will in large part be determined by the consequence evaluation process. This process can be logically approached and informed by an evaluation of the following areas of explosive event influence:**

Site and Local Risk Management Context Although design professionals are traditionally charged with protective designs for the specific contracted-for client facility, explosive events routinely involve collateral damage to adjacent structures and/or public spaces and also place these additional lives and property at risk. This is a fact that cannot be ignored. The performance of the client facility must be considered within a regional context in order for the client and the design team to exercise the highest standard of industry care and to reduce the extent of design professional liability. The creation of a highly protected facility that exclusively focuses on protection of that specific facility's assets and does not take into consideration how its performance may affect adjacent structures, public spaces, and their populations cannot be considered the most prudent design solution.

Facility Structural System Protection of the facility's primary and secondary structure from progressive collapse and disproportionate damage associated with explosive events external or internal to the building is a baseline protective designconsideration in consequence evaluations. Catastrophic failures of an entire structure or a significant portion of it generally place large occupant populations at risk, both as a result of the immediate event effects and of disruption to ERR operations. Additionally, these structural failures invariably cripple critical business missions, sacrifice institutional operational viability, and adversely affect restoration of day-to-day operations.

Facility Facade System Explosive events, either internal to or external to the building, are likely to engage the facade in managing applied forces significantly in excess of the thresholds associated with conventional wind, seismic, and gravity loading. This routinely results in catastrophic glazing and glazing system failures, which are historically responsible for significant loss of life and injury to both occupants within and external to the targeted facility. **Airborne glass fragmentation continues to be identified as one of the most lethal aspects of building component responses to explosive events.**

Facility Evacuation, Rescue, and Recovery Systems While the immediate devastating effects of explosive events can be limited by protective design strategies, post-event evacuation functions and the subsequent activities of rescue and recovery are dependent upon the integrity of stairs, vertical transportation systems, power, lighting, voice communication, smoke management, and other systems. These systems can be placed out of service throughout the building or suffer disproportionate damage if they are not specifically designed in response to the project design basis explosive threats. **Maintenance of these critical ERR systems, outside of the immediate explosive event zone, is essential to reducing the extent of resulting lethality, injury, and facility disruption.** Design professionals, when evaluating explosive event consequences to ERR systems and operations, must take into consideration that these systems may be exposed to supplemental damage due to structural instability. Also, that extraordinary (terrorist) criminal event modalities have historically been strategically planned so that phased explosive events occur with the purpose of targeting emergency responders and those residual occupants on the scene who have not yet been evacuated.

HVAC Systems Explosive events are traditionally accompanied by the development of significant quantities of airborne particulate contaminants, including glazing fragmentation, toxic materials, and other byproducts of building component demolition. These are inevitably deposited within the HVAC system as a result of system breaches associated with the explosive event, or as a result of fans continuing to operate after the explosive event. This contamination can be of such significance as to warrant the demolition and entire replacement of the HVAC fan systems (including filters, coils, etc.), fan plenums, and distribution duct work.

Business Continuity The design professional community should naturally address issues of life safety as a first priority. However, considerations for institutional business continuity cannot be ignored as part of the initial consequence evaluation process, and may in fact be a client's most clearly stated objective. The extent of the business continuity protective design mandate will depend upon the client's mission statement and expectations. It will also be dependent upon the extent of resources available to provide building protective design strategies beyond those required to achieve the consensus of life-safety performance standards. **In general, clients' expectations regarding the financial benefits of business continuity are well stated. However, ownership routinely anticipates that the building codes and design professionals' experiences will adequately guide the design in the protection of occupants on site. This potential disparity in consequence management expectations must be identified and addressed accordingly.**

Restoration of Operations Explosive events are violent conflagrations with the potential to significantly disrupt facility physical plant, affect life safety, and violate viable occupancy. Every explosion will be initially treated by the law enforcement community as a crime scene, by the building department as a potential condemnation, by underwriters as a time for claim investigations, and by others as the site of numerous activities other than those originally intended as part of the building's use profile. There will be an inevitable period of investigation and severe facility occupant access restriction. **Design professionals should engage the client in post-explosive event consequence evaluations. These should very specifically focus on the extent of post-event facility demolition and restoration that is tolerable in order to meet client expectations regarding facility condemnation by the local building authority, the anticipated time and cost of reparations, and the extent of time required to reestablish baseline business-mission operations.**

Target Attractiveness To the extent that crimes involving the use of explosives can be deterred, a significant advantage can be achieved. While there are no specific predictive mechanisms that can be referenced as an engineering and design guideline for criminal deterrence, especially for extraordinary (terrorist) attacks, the concepts of target hardening and inaccessibility are universally accepted by the law enforcement and intelligence agencies as rational approach strategies for crime reduction. **Consequence evaluations should consider, at all stages of design, whether the facility will be an appealing target. It is especially important that target attractiveness be considered in the earliest phases of site selection and facility location within the site. Also, the extent of accessibility**

and opportunity for an aggressor to deliver the explosive to the target needs evaluation. Reduction in the perceived opportunity to achieve both of these potentials is a design goal worthy of substantial consideration.

2.5.2 Function Redundancy

This subsection identifies and provides insight into the benefits associated with the management of post-explosive event consequences through the application of redundancy in various building architectural elements, programmatic spaces, and/or building systems. Incumbent on the discussions regarding redundancy is the necessity to advance the concept of locational diversity, as adjacent redundancies are not efficient managers of consequences from an explosive event. Simply stated, adjacent assets are most likely to suffer the same damages.

This subsection text includes the identification of building systems involved in evacuation, rescue, and recovery, and discusses how redundant ERR systems may reduce the requirement for structural hardening or may otherwise inform structural concepts, while reducing consequential effects. This is essential as ERR systems have a primary role in enhancing facility life safety in response to an explosive event and the secondary consequences of such a circumstance.

This subsection also addresses business mission-critical systems and spaces and how redundancy and diversity of these systems may beneficially affect structural and facade designs from the standpoint of explosive event protection. Other aspects of redundancy associated with building components, systems, or program spaces are also discussed with a focus on how they may inform the structural designs in terms of acceptable post-explosive event effects.

The subject of structural redundancy, for primary and secondary structural elements, is not covered in this subsection, as it is the specific subject matter of other chapters in this handbook.

A primary facility performance goal for the design professional is to ensure that evacuation, rescue, and recovery systems remain operational outside of the specific area of influence associated with the project-specific explosive design basis threat event. This is a significant challenge since there are a number of systems that participate in ERR. Design professionals should anticipate that the following systems, at a minimum, are considered part of the ERR protection strategy:

1. Fire suppression systems, including their water supplies
2. Fire alarm, detection, exiting, and annunciation systems
3. Emergency voice communication systems
4. Emergency lighting systems
5. Video surveillance systems supporting ERR spaces and operations
6. Emergency way-finding and exiting systems (other than exit signage)
7. HVAC chemical, biological, and radiological (CBR) agent control

8. HVAC smoke control, purge, and pressurization systems
9. Elevators designated for Fire Department use
10. Access control systems engaged in emergency exiting
11. Fire and security command and control facilities
12. Emergency power systems supporting the systems listed above
13. Emergency exiting stairs and passageways

These are complex systems, with components distributed throughout a building. The opportunity for their disruption is high as a result of the high-energy devastation characteristic of an explosive event. Consequently, their survival outside of the immediate explosive event zone will depend upon significant investments in hardening of walls and slabs to protect both the front-end monitor and control portion of these systems and their extensive and pervasive distribution infrastructure. Consequently, this is rarely the most economical approach. It is also least likely to provide reliability equivalent to that of a design that implements redundant systems located with geographic diversity, whose extent of segregation is based upon informed explosive event calculations and modeling. **In summary, placing critical ERR system components outside of the blast effect zone is more often a more survivable and robust solution than attempting to provide a hardening strategy to overcome single points of vulnerability and consequent failures.**

Business-mission continuity depends upon the survival of critical systems, equipment, information, and intellectual resources. Consequently, they must be located within safe and secure spaces that are appropriately designed for human occupancy and provide conditions that do not exceed the systems' thresholds of environmental tolerance. These are fragile components, systems, and/or conditions that are as susceptible to disruption or compromise from an explosive event as the previously discussed ERR systems. Depending upon the type and extent of service and/or product provided by the institution, there may be a requirement for protecting a particular portion of or, alternatively, the entire facility. **Design professionals, in the Preliminary Planning and Conceptual Design phases, will need to partner with the client and determine the extent of the facility that must remain free of explosive event effects (as well as exterior support infrastructure) in order to achieve the client business continuity expectations.**

If a backup project site protective design approach is not identified as a viable option (see Section 2.5.4), then the strategy of individual physical and intellectual asset protection will need to be compared to a strategy employing redundant systems and personnel support within the project facility. These will need to be located at a dimensional separation established by informed explosive event calculations based upon the project explosive design basis threats and the locations where these threats can be delivered. These locations will be dependent upon the facility's security program. Where local hardening options exist, they should be carefully examined and a determination made as to whether they are the most cost-effective and reliable approach to a robust business continuity design concept. If so, then critical business operations and their support systems can be confined to a compact and unified occupancy program within a locally hardened portion of the structure.

External explosive events may place significant portions of a building's facade at risk, and the failure mode may involve large portions of the exterior envelope. Based on the explosive event specifics, including the threat assessment explosive DBTs, standoff, and architectural envelope design criteria, post-event facade response may involve significant inbound facade debris that could negatively impact life safety, ERR operations, and business continuity programs. **Design professionals, in the Preliminary Planning and Conceptual Design phases, should partner with the architectural team and develop preliminary ERR design strategies, informed by facade explosive event experience, so that determinations regarding facade hardening versus redundant diverse ERR systems are given due consideration and the proper balanced solution selected.**

While ERR components and operations and business continuity mission expectations traditionally receive the majority of design focus for protection from an explosive event, client requirements may dictate that other physical, intellectual, or operational assets survive the DBTs. To address this additional requirement, design professionals will need to clearly understand the type and locations of assets at risk. Subsequently, the post-explosive event architectural, structural, mechanical, electrical, and other related systems' performance requirements will need to be defined in a manner responsive to the asset protection profile. **Based on the complexity of the necessary support infrastructure, design professionals should conceptually investigate the option of individual system hardening versus redundant support systems at geographically diverse locations outside of the calculated explosive event effect zone.** Even if a redundant diverse support system is selected, it should be recognized that local hardening will likely be required at points of connectivity to the protected space(s).

In some cases, functional redundancy is not an option. The preservation of unique cultural artifacts, the protection of unique financial assets, co-located intellectual resources, critical unique industrial or financial processes, and other single-source assets do not allow the design professional the benefit of protection through diverse repetition. These will require competent hardening strategies specifically created in a manner responsive to each asset's vulnerability.

A discussion of functional redundancy and diversity, and mitigation strategies involving an analysis of hardening versus repetition and separation is not complete without consideration of effective strategies to preclude the introduction of explosives to the facility or to assets at risk, through the implementation of a competent security program. **Design professionals should vigorously pursue with ownership opportunities for threat reduction. The potential for implementing this approach will yield a more secure facility with greater**

functional capability for ERR and the least threat to business continuity or other single-source, high-value assets.

2.5.3 Building Location

This section explores and defines the opportunities that the structural and blast consultant engineers have available to assist the client in understanding the structural, architectural, site civil, constructability, cost, schedule, and operational impacts associated with explosive event protection during the site selection process.

Design professionals are generally given the opportunity to assist a client in site selection, or, alternatively, are asked to develop the best structural protective design response for a building on a particular client preselected site.

Site selection criteria will be influenced by the threat assessment as identified in Section 2.4 of this handbook.

When design professionals are engaged to work with the client on identifying the suitability of a site, or sites, and the placement of a structure on those site(s), the decision should be informed by a competent and thorough threat and risk assessment. This assessment must clearly address the qualitative explosive threats (vehicle-borne, man-portable/satchel, maritime vessel– delivered, airborne delivery, standoff military ordnance, etc.), the quantitative design basis threat values (amount of TNT equivalent for each of the threats identified, shaped or unshaped charge, etc.), and the extent of opportunity for establishing standoff between the threats and the protected facility. These are discussed in more detail in Section 2.4.

Similarly, when clients request that design professionals work within a given predetermined site (i.e., when informing the site selection process is not part of the design professionals' responsibilities), the limitations which that specific site will impose upon the structural engineers and other design professionals should be identified. This should be based upon the risk assessment results associated with an assessment performed for that specific site.

Site selection and building placement locational consequences can be addressed through a competent process of explosive event analysis. This analysis must be informed by the threat assessment. It should also be informed by standoff distance determinations, the evaluation of the resulting extraordinary loads imposed upon the facility by the explosive event threat, and the opportunities for mitigating strategy protective design concepts available to the design team.

Elimination of the explosive event threat remains the most attractive, competent, and cost-effective process in developing designs for facilities on any site. However, based on the threat assessment, it may be unreasonable to assume that the facility will not be exposed to explosive event effects. **Maintenance of standoff between the explosive device and the asset remains as one of the most powerful mitigating strategies available to the design team for the development of a structure that affords the best opportunities for risk and consequence management.** In the case of standoff military weaponry (such as shoulder-fired ordnance), should it be identified in the threat assessment as a design basis threat, standoff distance will likely be less important than obscuring line of fire, the provision of a predetonation screen, or the provision of other measures, as this form of attack can take place from a considerable distance.

In general, applied pressures and the duration of their application (impulse) to the structure drop off exponentially as the distance between the protected structure and the explosive event increases. The protection of the facility's facade; primary and secondary structure; evacuation, rescue, and recovery systems; business continuity–related intellectual assets and physical plant; and other preidentified entities worthy of protection will require a high level of design sophistication. The design process will need to balance budget allocations and architectural and structural design responses based upon the achievable standoff between the assets to be protected and the explosive event delivery location.

As part of the site selection process, the blast consultants, working with the threat assessment identified DBTs, should work closely with the client and design professional team. This process should establish the pressures and impulses that will be applied to the structure, its facade, inbound utilities, and other critical site civil features, based on the maximum achievable and realistic standoffs. This is essential so that an initial determination can be made on how these extraordinary loads (in excess of gravity, wind, and seismic) will affect the protective design capabilities of the following major building systems and/or afford protection to the following building components and occupants:

1. Major/critical inbound utilities, such as fire protection water service, utility electrical power, information technology and data services, domestic water, steam, and gas
2. Facility access control points for vehicular entry
3. Site-located critical utilities such as transformers, generators, cooling towers, fuel storage, domestic and fire service water storage, and communications infrastructure
4. Facility facade
5. Facility primary and secondary structural systems and elements
6. Vertical transportation systems, including stairs, escalators, and elevators internal to the building (these may be exposed to explosive event effects should the robustness of the facade design be less than the DBT event calculated pressures and impulse loads)
7. Evacuation, rescue, and recovery systems as identified previously, in Section 2.5.2, internal and external to the building (these may be exposed to explosive event effects should the robustness of the facade design be less than the DBT event calculated pressures and impulse loads)
8. Occupants within the facility
9. Business mission-critical systems and equipment located within and/or external to the facility

The opportunity to achieve maintenance of standoff between the explosive DBTs and the project assets, as noted above and as supplemented for each specific project, should be one of the primary focuses of the site selection process. This should involve a careful evaluation of how the explosive device delivery threat, as identified in the threat assessment, can be competently separated from the asset(s). Rational solutions involving vehicle anti-ram perimeters, pedestrian interdiction barriers, and other site civil and landscape features should be considered for their implementation feasibility. Inability to institute these protective features may have consequences on the building's design. Consequently, the following strategies should be considered:

1. To increase survivability of critical inbound utilities, and, therefore, the reliability of ERR systems and operations, business continuity, and the mitigation of secondary event effects (such as the ignition of a gas main as a result of the explosive event), these utilities should be located outside of an explosive event delivery zone. Alternatively, they should be provided with hardening protection and redundant diverse routing at dimensional separations identified by the blast engineer. **Utility exposure and explosive event survivability evaluation is an initial critical design determination, as it may inform substantial work and expenditure for utility hardening protection or diverse routing and/or establishing long lead time approvals and/or feasibility studies by local utilities.**
2. To increase protection for building occupants, ERR systems, critical business-mission operations, and other predetermined internal assets, the facade system for the building should be reviewed for its ability to be designed to reduce the threat of airborne fragmentation to both internal building occupants and those external to the building. This should include a determination as to whether a design can be implemented whose post-event performance will be adequately robust to preclude additional protective design measures internal to the building, or whether the failure mode requires supplemental protection of internal and external assets. **Facade robustness evaluation is a critical initial design determination, as it will influence the building's architectural aesthetic, internal space program assignments, and the dedication of additional expenditures for hardening and/or diverse redundancy of ERR and business mission-critical systems.**
3. Initial structural design concepts should be evaluated for their ability to increase the survivability of the primary and secondary structure, and, consequently provide significant enhancements in life safety and business-mission continuity, and reduction in both decontamination and restoration time to resume normal operations. These conceptual structural systems should be reviewed for their ability to be resistant to progressive collapse and disproportionate damage, based on the building location and the available standoffs. **Structural system survivability is a critical initial design**

determination, as the type of structural system for the building may, in fact, be determined by explosive event management, in lieu of being conceived to address traditional gravity, seismic, and wind loadings, in order to achieve the post-explosive event performance results. If a conventional structural system cannot satisfy the security performance requirements, then site selection and/or building placement on the site may have blast protection requirements that inform member sizing, connections, standoff, and other major portions of the design with consequential effects on construction schedule, budget, and the architectural program.

4. To ensure the most competent and effective means to facilitate evacuation, rescue, and recovery operations, the designers should evaluate egress, emergency responder vehicular access, and emergency responder entry. This should be performed through a process of establishing their exposure to the project explosive DBTs and the separations/standoffs that can be provided by the particular site. Vehicular access points to the site, approach configurations to the facility, and the conceptual location of exit stairs and elevators should all be reviewed for survivability. **The survivability of ERR facility components is a critical design determination, as it may inform major aspects of the site civil and landscape plan, and the location, quantity, and type of the facility's vertical transportation systems, as well as the hardening required for these systems.**

2.5.4 Building Dispersal/Distribution of Functional Programs

This subsection addresses the opportunities for consequence evaluation and management associated with the physical separation of internal building program spaces, critical ERR and business continuity support systems, and physical assets and occupants. It focuses on the relationships among facility vulnerability, facility diversity (geographically diverse facility entities and, also, redundancy and diversity of internal building assets at a single site), the blast analysis process and the associated analytic results, and how these factors and engineering analytics can inform whether physical separation and the resulting diversity are a more rational protective design solution than co-location and structural, architectural, and ERR system hardening.

The threat assessment process discussed earlier identifies credible threats appropriate for mitigation. If explosive event threats are subsequently identified, then both the site selection and facility design process will need to address their vulnerability to the specific explosive event types, design basis threat quantification, the means of delivery, and their adjacency to the protected asset.

Vulnerabilities will occur wherever the explosive event threat can be delivered at distances determined by the blast consultant to impose pressures, impulses, soil movements, and other explosive event effects in excess of the loads prescribed for management by the building code. Additionally, vulnerabilities may exist where these explosive events create conditions in excess of the criteria stipulated in the client's published design standards for project performance. **It is essential for a qualified blast consultant to inform the design team of these extraordinary loads, vibrations, soil displacements, and other consequential effects, and the distances at which they will be applied to the facility's structure, facade, and critical systems.** These values should not simply be obtained from resource and reference materials unless they are reviewed and subsequently modified and/or approved by a qualified explosives and structures subject matter expert with extensive experience and access to government-approved computer modeling software.

Subsequent to the identification of the anticipated explosive event effects, based on the type and delivery location opportunities for the explosive threat, conceptual building placements on the site and conceptual identification of ERR and business-mission or other critical assets should be identified by the design professionals working with the client team. Various options are likely to exist in the Conceptual and Schematic design phases, and each of these should be reviewed to identify explosive event vulnerability exposures. Based on the agreed-to, post-explosive-event facility design survivability performance criteria, the design team should consider:

1. The development of maximized enforceable standoffs.
2. The utilization of protective design facade, structure (primary and secondary), and internal partition hardening.
3. The addition of redundant ERR and other systems located diversely enough within the facility to ensure survivability to a single or multiple explosive events, as defined in the threat assessment. Note that this must take into account both external and internal explosive threat modalities.
4. The repetition of those portions of the facility at a diverse site (essentially another standalone clone of the assets at risk), at a distance ensured to be free from the explosive event effects at the project site as determined by the blast consultant.

Life-safety concerns for facility occupants and those external to the building are not specifically addressed by the concept of redundancy and/or diversity, other than through the benefit they receive from the protection afforded by survivable evacuation, rescue, and recovery systems. Those within the immediate blast effect zone (an area of influence that generally can be well estimated by the blast consultant) will suffer traumatically unless they are also protected by blast-hardened elements and free from exposure to airborne debris. However, the protection of those systems that aid ERR operations will significantly enhance the life safety of those who will need to exit the facility and provide rapid and efficient access by emergency responders immediately after the explosive event. These activities are essential in assisting those facility occupants affected by the event or those building occupants who fall within the ADA-defined profiles and require supplemental assistance for evacuation. In summary, diversity and redundancy are strategies best employed for the protection of ERR systems and building features, critical business-mission systems, equipment, and operations or other client-defined physical assets. **The initial phases of consequence evaluation and management should occur in the Conceptual and Schematic design phases and include a review of the opportunities for redundancy and diversity and the use of hardened facility design elements.**

The above notwithstanding, certain facility operations and client requirements may dictate the duplication of staff either within a facility or at a remote site. This will occur when individuals are considered to be of such importance that they are an integral component of business-mission continuity support. When this is a design parameter, the structural engineer, architect, and blast consultant must address the project's explosive design basis threats in a manner that identifies the extent of segregation required between the area where those threats may occur and the proposed location for this supplemental protected population.

This process must evolve from the development of initial "threat maps," which are based upon the available architectural concept drawings. These maps should be created to identify the location of explosive event delivery opportunities, both external and internal to the facility. This will require the parallel conceptual development of a facility security operations program. **Consequently, a security professional should be part of the design team's composition so that security systems and management operations are designed to limit the potential size, delivery strategy, and locational placement of explosives.**

Once there are reasonable key design assumptions regarding explosive event threat delivery locations, explosive event analytics should be performed to determine whether a co-located population of duplicate staff (and the infrastructure required to support them) can be adequately protected within a single building, or whether they need to be physically relocated to a remote site. Although this may appear to be a daunting and challenging process during the early stages of the project, where designs are truly conceptual, this will provide the best opportunity for informing the major structural and architectural aspects of the design. **The longer the process of determining how staff will be protected is delayed, the more difficult the protective design solutions will become. This will therefore generate a compromised set of strategies that will cost more and provide less competent protection.**

2.5.5 Disaster Recovery and Contingency Planning

Disaster recovery and contingency planning provide carefully conceived and tested alternatives for taking action when things go wrong. Since no private-sector structure can be built to survive every explosive attack, and since no security plan is perfect, every organization that believes there is a threat to its facility or proximate environs should create disaster recovery strategies, contingency procedures, and business recovery plans. Indeed, most boards of directors and insurance companies demand no less.

The purpose of contingency planning is to protect building occupants and speed business resumption, including regaining access to data, communications, workspace, building support systems, and other business processes. In short, the consequence management and the other security design concepts discussed in this handbook can and should support business continuity planning. Early in the conceptual design process, therefore, the design team should discuss post-blast recovery with ownership in the specific context of their contingency planning expectations. **Design professionals should insist on reviewing with the client the opportunity to incorporate contingency planning and business recovery concepts into their building design.**

From the occupants' perspective, the single most significant reason why contingency planning is important is that lives could hang in the balance. To the eventual building client institution, effective business recovery strategies are also critical in maintaining the public's confidence in the enterprise during a crisis.

2.6 THREAT REDUCTION

The design professional, security, and law enforcement communities acknowledge that facilities can be safer environments when those threats that place the institution and its occupants at risk are managed by a process of threat reduction and/or, preferably, elimination. When the law enforcement and intelligence agencies cannot ensure management of exposures to an explosive event, then institutions can achieve supplemental threat reduction by being placed as far from the threat as possible, so that the residual effects of the threat at the maximum achievable standoff distances are then addressed by the facility's site selection, facility placement within the site, and the facility's design. These relationships will be discussed in this section.

Explosive events can be categorized as those that occur:

1. As a result of the purposeful, beneficial, and legitimate use of explosives (e.g., rock excavation for foundations)
2. As a result of purposeful, willful, and malicious criminal acts, using explosive materials (e.g., a terrorist bombing)
3. As a result of a criminal act that does not initially use an explosive material, but that subsequently engages other resident agents possessing explosive capabilities (e.g., an arson event that then engages a compressed gas vessel)
4. As a result of an accidentally occurring event (e.g., a high-pressure steam line rupture)
5. As a result of a naturally occurring event (e.g., a lightning strike engaging an electrical transformer)

The law enforcement and intelligence communities and the security professionals responsible for securing a city, neighborhood, and/or facility site face an impossible challenge when tasked with the complete elimination of the explosive event threat. The prevalence of the suicide bomber creates exceptional opportunities for explosive event delivery that preclude providing the design professionals with a design basis threat of "no explosives." **Consequently, when design professionals are tasked with the development of protective design strategies for a site, building, and its occupants, they must insist on obtaining the specific explosive design basis threats, which are identified within a competent threat and risk assessment created with contemporary input from the law enforcement and intelligence agencies on explosive material types, delivery modalities, and any other related available specifics.** This information mining is essential if credible structural and architectural facility designs are to provide real-world protection.

Furthermore, it should be recognized that the law enforcement and intelligence communities cannot effectively identify the extent of explosive event threat reduction that a facility security program will accomplish, unless, perhaps, it is a government facility for which they have direct management control and involvement. In most cases, on-site and internal-to-building explosive event threat reduction must be identified by the client and their security professionals, or a highly qualified security consultant. This information must also inform the project threat and risk assessment so that quantifiable explosive design basis threats are represented on the previously discussed facility threat maps. This process will allow a clear and graphic representation of how explosive threat reductions can occur, starting at the facility site perimeter and migrating to the most sensitive internal2 areas requiring protection. **The intent of threat mapping is to clearly identify explosive quantity and locational placements to inform the design professionals on how the explosive event DBTs are reduced and/or eliminated as a result of being subjected to concentric layers of competently applied security measures.** This is the information which, as represented on the threat maps, will inform how threat reduction can be used by the design team to address malicious criminal acts involving the use of explosives.

To reduce the threat associated with naturally occurring events, the designers should gather relevant, project specific information and review the site and the facility designs at the earliest possible stages of design. The purpose of these investigations is to identify those sources of explosive event opportunity worthy of supplemental consideration and protection. This should involve the identification of other explosive event substances that contain inherent high-energy sources such as:

1. Compressed gas vessels and/or piping
2. Liquefied natural gas vessels and/or piping
3. Fuel storage and piping other than gaseous agents
4. High-pressure steam vessels and/or piping
5. Stored munitions and materials used in their fabrication

6. Electrical transformers
7. Combustible airborne particulates
8. Combustible chemicals and agents other than explosives
9. Chillers, boilers, and furnaces
10. Numerous others

These may occur either adjacent to or resident within the site and/or facility, or exposures may occur as a result of person and/or vehicle (motor vehicle, rail transit, aircraft, or maritime vessel) delivery within or adjacent to the facility.

2.6.1 Accidental Explosions

This subsection identifies potential accidental explosions that should be taken into consideration by the design professional team. The intent is to ensure that design professionals—especially the structural, mechanical, and electrical engineers and the architect—are aware of and provide protection for explosive events other than purposeful, malevolent criminal activity.

Explosive events occur as a result of the release of high-energy sources. While traditionally this condition is conceived of as the result of ignition of an accelerant, such as highly flammable liquids, compressed gases, or explosives, other high-energy sources—such as steam or water under extreme pressure, chillers, boilers, and other systems routinely found in facilities—are also candidates for imparting extraordinary pressures over extremely short durations.

The unabridged edition of the *Random House Dictionary of the English Language* defines an explosive event as "the act or an instance of exploding; a violent expansion or bursting. . ." and "a sudden, rapid, or great increase." **While the pressures, impulses, and plasma effects associated with an explosive event using extremely high-energy substances such as C4, Semtex, RDX, TNT, or their equivalents, are regarded as upper-level design thresholds for protection, design professionals should carefully review the entire project design with the intention of identifying other building systems and/or programmatic spaces allocated to the storage and/or distribution of substances and materials that could participate in an explosive event.** These systems and areas of programmatic use are unlikely to be identified by a traditional security threat and risk assessment, as these activities usually focus upon the purposeful introduction of high-energy explosives as the anticipated design basis threat.

At the earliest possible design phase, the structural engineer, mechanical and electrical engineers, and architect should review the facility design and identify whether any of the systems, equipment, or environmental conditions identified previously are resident within the project scope. This process may involve making specific inquiries of the client, based on the sophistication of their occupancy program, regarding what explosive exposures they know exist, as a result of their detailed familiarity with the facility operations that will occur in the project. **The design team should list, as part of the protective design strategy's key design**

assumptions, the potentially explosive operations, systems, and/or materials and substances that will be present as part of the building's use.

A list of potential explosive event conditions should be created based on the aforementioned plan review and client inquiry. Individual mitigating strategies should be proposed by the design team to address these exposures. This may include the following:

1. Elimination and/or reduction of explosive sources and/or quantities from the project
2. The introduction of detection systems to identify the conditions associated with the development of an explosive event condition prior to its detonation and/or release point
3. The introduction of suppression systems capable of reducing the post-explosive-event effects
4. The introduction of systems that extract the explosive event agent, upon its detection, and prior to its detonation/ignition
5. The elimination of ignition sources
6. Management of access control, for both persons and vehicles, to preclude entry and accidental initiation of the explosive event
7. The introduction of explosive event pressure relief features that direct the effects of the explosive event in a manner that provides acceptable post-event effects
8. Informed placement of potential explosive systems and/or environments so that their post-event effects meet the project's security performance criteria
9. The use of appropriate signage and graphics to prevent occupancies, activities, and/or behavior that could trigger the event
10. The use of hardening to preclude unacceptable and/or disproportionate damage from the event

These and other mitigating strategies should be considered for implementation by the design team to enhance the facility protection profile.

2.6.2 Intentional Explosions

This subsection addresses purposeful extraordinary criminal activity associated with the use of explosives as a means of attacking persons and property. Refer to Section 2.3, "A Brief History of Recent Terrorist Attacks," which identifies sources of related explosive event history and demographic information. While engineering design guideline publications can assist the design professional by providing pertinent related historical information, ownership and design professionals must anticipate that history will repeat itself and that the design and client community have a responsibility to engage in predictive security planning.

This handbook advocates the use of the extensive explosive event information available from numerous agencies, such as the FBI, the Department of State, the Department of Defense, the Department of Homeland Security, the U.S. Armed Forces, and others, as a means to reinforce confidence in providing—well beyond the publication date of this manual—protection against explosive events.

While the primary use of explosives was in the military arena, criminals and terrorists have adopted the use of explosive devices as the most commonly used tactic to inflict devastating results. **Contemporary with the publication of this handbook, law enforcement agencies, the intelligence community, terrorism experts, and security professionals continue to emphasize that the man-portable improvised explosive device (MPIED) and vehicle-borne improvised explosive device (VBIED) are the weapons of choice for the criminal to use when contemplating an attack against a population and/or facility, with the intention of achieving mass loss of life, disproportionate property damage, and the inevitable publicity this generates.**

Facility designs outside of North America are also subject to the use of explosives as part of the criminal/terrorist act modality. In addition to the MPIED and VBIED delivery methods, the use of standoff military ordnance, especially the use of shoulder-fired arms (such as the rocket-propelled grenade [RPG]) remains a significant concern and should be given careful consideration by a structural engineer and architect as part of a competent protective design mitigating strategy. **It is especially important, for work outside of North America, that design professionals insist upon the use of a competent threat and risk assessment process that is specific to the project facility, in order to determine whether standoff military weaponry is to be considered as part of the mitigation criteria.**

It is also essential for the design professional team, especially the structural engineer and architect, to understand the relationship among procurement of explosives, their delivery, placement, and subsequent detonation, in order to clearly understand the rational and design consequences associated with the identification of design basis threats. The protection strategies and eventual design implementation concepts and components of a facility's structure, facade, ERR systems, and business-mission continuity are founded upon mitigation of either a predetermined explosive event type, size, and detonation location, or a predetermined pressure, impulse, and post-event facility performance standard. **Consequently, the establishing of quantifiable design basis threats is a fundamental first step in providing the necessary guidance to the design team, as this will be the basis for subsequent analysis and design mitigation development.**

Explosive design basis threats (DBTs) must be provided as part of the threat and risk assessment by a competent security professional. The project-specific and identified DBTs will be based upon the opportunity for the aggressor to procure the explosive, the ability for it to be delivered to the point of attack, and the opportunity to achieve a detonation sequence that will achieve the full value of the DBT value. Threat and risk assessment professionals, working with the law enforcement and intelligence communities, have resources to assist in the development of DBT values based on these generic parameters. However, it is essential for the design professional team to participate in the risk assessment process by the provision of project-specific information, at the earliest stages when conceptual designs are adequately advanced. This is necessary as it will inform the ability for explosive delivery mechanisms to be considered credible or not credible, based on project design features that affect the mobility of vehicles and persons intent on delivering the VBIED and MPIED, respectively. **While the threat and risk assessment authors can provide quality information regional to the site, the extent to which the project site and facility physical plant are at risk from specific explosive event DBTs must be informed by an interactive process of information exchange between the threat assessment and design team professionals.**

Large quantities of explosives are readily delivered by vehicles. These values are considered public sector information-sensitive. However, there is a trend toward this information becoming common knowledge within the design community as more explosive event protection mitigation is practiced. However, this handbook continues to respect the need for this information to remain confidential. Consequently, the explosive delivery capabilities for passenger-operated vehicles (cars); passenger vans, sport utility vehicles, and pickup trucks; small box trucks; large box trucks and trucks with vessels carrying liquids; and the larger classes of vehicles such as tractor trailers have all been assigned explosive event design basis threats by the law enforcement community. **Quality sources of VBIED DBT information are the ATF, TSWG (Technical Support Working Group), and other DHS agencies. Design professionals should request this information from the project's security professionals engaged in the project threat and risk assessment**.

MPIED DBTs have been credibly established as a result of their extensive use in numerous criminal acts. Again, there is a range of explosive event DBT sizing based on the delivery mechanism. Individuals may deliver explosives either by transporting them directly as a body-worn device or as a separately carried parcel or satchel. The increased use of well-packaged high explosives allows for competent concealment and provides the law enforcement and security professional with significant challenges for rapid identification through the routine use of informed visual observations. There are emerging programs that employ behavioral identification, parcel recognition, and other predictive strategies employed under the nomenclature "situational awareness." These are supplemental procedures only; they are helpful in explosive device identification, but should not be considered substitutes for physical protective designs.

The body-worn explosive device DBT is not insignificant and is capable of inflicting structural damage. The extent of this damage can be dramatically increased through the use of high-energy explosives and specific charge shaping.

The satchel may be carried or packaged in a wheeled parcel pulled by the aggressor. Wheeled parcels often are harder to detect because the ease with which they can be transported reveals less through body language. Consequently, the MPIED has a potentially large value range. It is significantly in excess of the body-worn DBT value.

Explosives can also be purposefully delivered using the U.S. Postal Service, express couriers, common freight carriers, or other similar second- and third-party delivery agencies. These differ from the MPIED threat, as they arrive cloaked by the legitimacy associated with an expected delivery mechanism and devoid of the opportunity for profiling the carrier. The range of DBT value is entirely dependent upon the parcel/freight weight and size management threshold criteria established by the carrier. Suffice it to say that the range of explosives that can be delivered by a carrier generally exceeds that which an aggressor could deliver personally. With the advent of GPS and other handheld intelligent geospatial and initiating technologies, criminals are capable of remotely detonating shipped explosives at explicit latitudes, longitudes, and elevations, allowing targeting by building, by floor, and by room, assuming satellite signal transmission strength is adequate.

In summary, design professionals should use the security consultant's estimates of the VBIED and MPIED threats in the design of protection strategies for the structure, facade, ERR systems, and so forth. They should insist upon active participation in those portions of the threat and risk assessment that involve the identification of building design features and components that can affect the type, size, and location of a bomb. Also, a competent security systems and operations professional should work with the threat and risk assessment and design team to advise on the method to limit the delivery of explosives.

2.7 VULNERABILITY REDUCTION

This section addresses the critical topic of how vulnerability of an institutional facility, including its physical plant (buildings), occupants, and assets can be secured—both from the standpoint of safety and of business-mission continuity—by the application of rational and justifiable mitigating strategies. Facility vulnerability can be defined as the capability of, or susceptibility to, being compromised, damaged, destroyed, or otherwise violated in a manner that is detrimental to human safety and institutional viability.

It is unreasonable to expect that government agencies can entirely eliminate the threat of a bomb. As mentioned previously and as so aptly demonstrated during the Second World War, the kamikaze attack or the actions of the suicide bomber generally are successful, as they defy the basic concepts of military engagement, law enforcement, and protective designs, which assume that even the enemy or the criminal exercises the intent of self-preservation. **In an environment in which criminal threats cannot be dismissed, only diminished, institutions remain vulnerable, and protective designs that address those vulnerabilities are necessary.**

Previous portions of this handbook also identified the potential for accidental explosive events, which also place structures and their occupants at risk. While these may be significantly diminished by competent planning and protective design strategies, there always remains the opportunity for exposures resulting from protective designs that are overwhelmed by larger threats and/or combinations of events that were not anticipated in the risk assessment.

Factors of safety (FOS) are routinely employed in the designs executed by engineering professionals, especially structural engineering, which benefits from the values that have established industry standards for seismic, wind, and gravity loads. Unfortunately, there are no specific FOSs for intentional explosions, and for accidental explosions there are very limited published guidelines (such as those published by electrical utility companies for transformers, by steam utility providers for high-pressure steam distribution systems, etc.) identifying the quantified extent of additional protection that should be designed for by the structural engineer. **Consequently, the identification of facility vulnerability to an explosive event, which in large part depends on the value of the explosive threat to be mitigated, remains an inexact science.** While the military has developed practical approaches to address military attack ordnance of particular sizes and types, this information is not explicitly helpful in identifying the vulnerability of conventional structures to explosions associated with nonmilitary ordnance.

The process of vulnerability reduction must therefore occur within a design environment in which there is a presumption that the threat cannot be completely eliminated, that its quantification is an informed but still subjective value, and that there are no preestablished industry factors of safety. **Irrespective of the difficulty in specifically quantifying explosive events, there remain viable opportunities and design considerations for use by design professionals to address facility vulnerability. They are addressed in the following sections and can be summarized as standoff distance, physical security and protective design, operational security, and structural design responses.**

2.7.1 Standoff Distance

Standoff distance is the amount of space denied to an aggressor between his threat device and the area being protected; for example, ram-resistant vehicle barriers set 100 feet from the building facade create a 100-foot standoff. Since the destructive effects of an explosion decrease rapidly over distance, space is the most effective and cost-efficient safety measure of all the blast-mitigation strategies available to building designers and engineers. (See the later chapters on blast phenomena and blast loading for detail.)

The DoD and other federal entities[7] provide uniform tables of minimum standoff distances that the design team can use to establish and enforce their

[7]See the National Counterterrorism Center (NCTC) Bomb Threat Standoff Distances table at www.nctc.gov/docs/2006_calendar_bomb_stand_chart.pdf.

agreed-upon security and safety levels. **Determining the correct standoff margin with precision is challenging, however, since estimating the required standoff distance involves an accurate approximation of the bomb's composition and explosive weight. Because the terrorists hold the initiative, such calculations are often simply educated guesses and hence become dependent on the accuracy of the threat assessment.**

Employing standoff design principles to protect the facility's facade and primary and secondary structure, ERR systems, business-continuity-related intellectual property, physical plant, and other protected assets requires sophisticated planning, additional budget allocations, and probably architectural innovation or restraint. **Putting space between the explosive device and the assets remains one of the design team's best mitigating strategies for risk and consequence management, so realizing the positive benefits of standoff distance should be a primary focus of the site selection process**. Consequently, vehicle anti-ram perimeters, pedestrian interdiction barriers, and other site civil and landscape features should be given serious consideration.

2.7.2 Physical Security

Physical security equipment and operational security measures such as personnel, policies, and procedures are proven defenses that can keep explosives away from the facility. Although security programs are a clear and observable protective measure—think of controlled underground parking and the badges now required in most office buildings—design teams often neglect to build competent and enforceable physical security equipment and operational security program spaces into the facility's infrastructure.

2.7.3 Operational Security

Architects' and engineers' failures to anticipate security program needs seem rooted in their clients' tendency to institute these measures only after building occupants begin arriving. But any facility constructed to resist explosive attacks needs—to cite only a few examples—controlled access areas, delineated zones of control, and places where vehicular and pedestrian entry is strictly managed. **Design professionals should vigorously pursue with ownership the opportunity for positioning the required protective equipment and security operations spaces into their planning cycle in a manner compatible with the intended facility operation's plan.** The final design will yield a more secure facility with greater functional capability for detection, deterrence, prevention, and recovery.

2.7.4 Structural Design

This subsection provides a nontechnical discussion of how structural design should be considered as one of the major mitigating strategies for reducing facility vulnerability. Since the analytical and engineering specifics of this topic are addressed in much greater detail in other sections later on in this handbook, this subsection is intended to introduce and validate the role of explosive-event-responsive structural designs. This section also identifies the significance associated with making the determination that structural robustness is to be part of the facility's protective design strategy, along with the accompanying effects on the design process, product, construction cost, and risk acceptance.

The decision to enhance a facility's structural design to resist the effects of explosive events and thereby reduce its vulnerability will have project management design implications. These include:

1. Owner decisions regarding risk tolerance and acceptance
2. Project cost allocations
3. The effects on the role of structure in architectural design
4. Design space implications for architectural use programs
5. Design implications for mechanical, electrical, and plumbing (MEP) and other building infrastructure systems
6. Implications associated with construction quality control for both structural shop drawing submittals and during construction field inspections
7. The requirements imposed on document production and management to maintain confidentiality of the security-sensitive information included in the construction document packages

Assuming that standoff distances are optimized through a competent security program, which achieves maximum available separation between the explosive DBT and the structure, and that a combination of physical and operational security strategies achieves credible reductions to agreed-to explosive DBTs, remaining vulnerabilities from the explosive threat will need to be managed by the facility's structure. **Facility protective designs are dependent upon reducing project-specific vulnerabilities, and the extent of mitigation that must be performed by the structure is the residual threat not capable of being addressed by separation from the threat and the extent of threat that is not capable of being detected and precluded from access to the site and the facility.**

Even small explosive events create extraordinary load magnitudes in excess of wind loads on facility facades. These applied loads to the facade must be transferred back to and managed by the building's structural frame. Additionally, explosive events, based on the project-assigned DBT values, will very likely apply significant loads directly to both primary and secondary structural members and elements, such as columns, girders, trusses, shear walls, connections, and slabs. This may come about as a result of larger DBT values at some distance from the structure (externally), or these larger DBTs may have proximal/adjacent access to the structure itself. Structural exposure to and the consequent vulnerability to these threats can occur from either the VBIED or MPIED, as noted previously.

Additionally, these same VPIED and/or MPIED threats may internally threaten the structure, depending on the project occupancies and the extent and efficacy of the facility's security programs in managing the migration of the threat. **In summary, the structural frame/load-carrying system of the building will likely provide the most significant protective value in reducing vulnerability to an explosive event, as the loss of primary and secondary structure will usually create circumstances that place significant numbers of building occupants; evacuation, rescue, and recovery systems; and business missions at severe risk.**

However, with the decision to engage the structure in facility vulnerability reduction, one of the most expensive building components is now subject to modification in order to provide performance beyond that required by most building codes and conventional wind, gravity, and seismic loading. While European building codes (such as those in the United Kingdom) currently embrace the requirement for design professionals to address the removal of a single column (for unspecified reasons, essentially requiring the designers to address a "threat-independent" vulnerability), this requirement has not yet been uniformly adopted by building codes throughout North America. And, when this requirement has been included in a particular regional building code, or as a programmatic design conformance requirement, the removal of a single column may not provide the extent of protection required to preclude progressive collapse and disproportionate damage from threats of specific sizes as identified within the project threat and risk assessment DBTs. These DBTs (used in "threat-specific" or "threat-dependent" designs) may be of such a size and capable of achieving adjacency to the structure so that they affect the load-carrying capability of more than a single primary element. These circumstances, therefore, require a greater extent of mitigation and consequent vulnerability reduction than would otherwise be provided by the use of the threat-independent criteria only. **Consequently, a fundamental risk management decision must be made by project ownership and codified in the key programmatic design and facility performance assumptions to be used by the design professionals: specifically, whether the structural system for the building will provide protection against the actual project threat assessment identified explosive DBTs ("threat-dependent" design approach), or the extent of protection afforded by addressing the removal of a single primary structural element ("threat-independent" design approach), or, potentially, both criteria. The client-defined extent of post-explosive-event facility performance must be clearly understood and agreed to in order for there to be a clear expectation regarding structural performance in vulnerability reduction. This may have a significant project cost impact, and, during the initial Conceptual phases of the project's design, magnitudes of probable cost should be assigned to this aspect of vulnerability reduction and a determination made regarding funding viability and assignment.** Inability to fund a competent protective structural design response will inevitably lead to a commensurate increase in vulnerability and an increase in the owner's risk exposure.

Based on the project explosive event DBT values and the locations where these threats can migrate within the project site and structure, a wide range of structural protective design strategies may emerge. This may involve the introduction of additional structural members, increasing the size of structural elements, the introduction of new structural framing and load-carrying techniques such as trusses and outrigger systems, addition and diversity of building materials, connection detailing, and other schemes, all of which may affect the overall appearance of the structural framing systems. Based on the aesthetic role of structure in the particular project's architectural vocabulary, there may be a direct visual translation of these protective design influences on the architect's visual design intent. **During the initial Conceptual phases of the project's design, blast consultants and structural engineers should participate in design charrettes with the project architects. These charrettes should address the symbiotic relationship between the structural scheme options and architectural design intent to achieve the required blast protection and the desired structural component of the architectural concept. Failure to address this vulnerability reduction opportunity at this early phase within the project may prejudice the opportunity to develop the most architecturally acceptable, security-competent, and cost-effective structural scheme.**

While visual aesthetics often dominate design decisions, structural designs responding to the requirement for blast mitigation may involve potentially significant impacts on programmatic space use, size, and arrangement. Increased member sizes for blast protection, the addition of structural members or systems, and/or the introduction of physically enforced standoff to improve structural survivability will, of necessity, reduce space available for other functional purposes. Also, large structural systems such as trusses, additional shear walls, compression and tension rings, outrigger systems, and other similar systems may be of such size and/or space invasiveness as to dictate relocation of architectural space programs and/or MEP systems to floors other than where these structural systems need to occur to provide the required levels of protection. These space constraints may not be the only motivation to reassign architectural use programs or MEP systems to alternate locations within the system. Based on the facility security program, the introduction of the threat and risk assessment identified DBTs may place these critical structural systems at an enhanced level of risk. This occurs because it may not be possible to preclude (due to the space utilization and occupant activity programs) the MPIED or VBIED from having access to the desired locations where these structural system enhancements optimally would occur. **Consequently, during the initial Conceptual phases of the project's design, the blast consultant, structural engineers, and security professionals should participate in design charrettes with the project architects so that there is agreement upon the ability to feasibly implement a security program that precludes the explosive DBTs from having access to critical portions of the structure. Also, these sessions should resolve the identification of structural impositions on MEP systems' design and deployment, and the anticipated space impacts, both in terms of space availability and occupancy use**

programs. Failure to address these structural vulnerability reduction opportunities, at this early phase within the project, may prejudice the opportunity to develop the most competent and cost-effective structural schemes, architectural designs, space utilizations, and MEP systems' placement.

Value engineering has become an inherent project management process throughout the design and construction profession. It has been shown to provide beneficial results by optimizing the engineers' design intent through a process involving the application of previous experiences through a peer review by other design professionals, contractors, and/or material fabricators. As long as this process continues, opportunities exist to modify structural systems in a manner that may or may not reduce their ability to provide the enhanced protection they must provide when they are used as part of an explosive event facility vulnerability reduction strategy. The science of designing structural systems to resist explosive event loading is a unique and less well understood art than the management of wind, gravity, and seismic loading. It is extremely likely that those involved in the value engineering process may not fully understand the extent of blast load structural system investment included by the original design engineers, and recommendations for materials and/or structural systems' design optimization may be based upon inadequate knowledge or experience. **Any value engineering and/or other similar "design optimization" process that involves review and recommendations regarding structural systems engaged in blast vulnerability management should include the blast consultant and the structural engineers responsible for designing the systems under evaluation.**

Furthermore, during the construction phase of activity, the structural engineers should remain involved in the shop drawing approval process to ensure that the materials' fabrication, erection, and field installation techniques and details continue to respond to the analytically proven designs that were developed as part of the original blast mitigation work to achieve the required project vulnerability reductions to the threat-assessment-identified explosive DBTs. **Value engineering and quality assurance involvement during the construction phase by the blast consultant is essential. As a result of the blast consultant's analysis and recommendations to protect against an explosive event, the structural materials' selection and the systems' design and detailing may well be, of necessity, more robust than that otherwise required for traditional gravity, wind, and seismic loadings. Without the benefit of this blast analysis, the rationale for unconventional material oversizing, encasement, overly robust connections, and other structural enhancements may appear to be simply overly conservative structural design and detailing, and may thereby create an intuitive but false opportunity for invalidated optimization by the contracting community.**

Structural designs providing explosive event protection are codified within the drawings and specifications used for both procurement and directing contractor construction-related activities. Of necessity, details associated with member sizing, connections, reinforcing detailing, plating, and other blast protection techniques must appear in adequate detail to inform bidding, shop drawing development, and field work. **The structural engineering professionals should develop, with the project architect and ownership team, the means to ensure the nondisclosure of, and maintenance of the confidentiality of, structural enhancements intended for the purpose of reducing vulnerability to the project explosive DBTs. This process should be in place prior to the development of any documentation, whether Conceptual designs, blast analytics, or more detailed documentation normally associated with the Schematic, Design Development, and Construction Document phases of design work.** This should include a series of published policies and procedures to manage electronic and hardcopy design information.

2.8 RISK ACCEPTANCE

It is universally acknowledged within and outside of the design profession that no facility can actually be designed, constructed, and operated in such a manner as to be totally risk-free. There are simply too many variables and unknowns that can create exposures and vulnerabilities that cannot be foreseen or completely mitigated. The challenge is to reach a rational and justifiable position of balance among available resources to be applied to facility design, construction, fit-out, and subsequent operations, and to do so based on the best contemporary knowledge available at the time of design and on reasonable foreseeable future vision. These resources include finances, design talent, design time and schedule, site opportunities and constraints, materials availability, contracting skill, construction schedules, post-construction operational staffing, intellectual resources to manage operations and create and enforce viable policies and procedures, and last but not least, client risk tolerance and commitment.

This section discusses the implications of designing facilities to meet explosive threats effectively with less emphasis on resource implications and with the primary goal of achieving lower levels of risk. This approach is generally considered to be "designing to address the threat" or, more simply, "design to threat." The emphasis being that vulnerability reduction and threat mitigation are the leading programmatic requirements for consequence management and are intended to dominate the protective design decision process.

This section also discusses the implications of implementing designs that include mitigating strategies as they can be afforded by available resources, and the consequent potential increases in risk exposure if these resources are limited. This approach is generally considered to be "designing within budget" or, more simply, "design to budget." This remains as one of the more significant challenges to the design professional and client team, as there are almost always limited resources available, and this then requires an eventual trade-off between resource allocation and protective design implementation. Just as it is unreasonable to assume that no design and facility operation will be devoid of threats, vulnerabilities, and risks, so is it unrealistic to assume that there are unlimited resources available to achieve complete risk elimination. **The process of adjudicating this tension between resource consumption and facility security enhancements will require the design professionals to clearly identify optional levels of protective enhancement so that client risk management and the risk tolerance allocation process have a basis for rational decision making.**

2.8.1 Design to Threat

This subsection focuses on the process of assigning and allocating resource dedications adequate to fully address the threats identified in the threat and risk assessment process as referenced in Section 2.4 and the associated advantages in terms of increased opportunities for substantial risk reduction. Consequently, the success of the design team in achieving the extent of risk acceptance identified by ownership will, in part, be based upon a clear understanding of the project-specific threats identified in the threat and risk assessment. As mentioned previously, for explosive events, this will require a clear and concise definition of the details associated with the types of events. This may include purposeful criminal, man-made devices or naturally occurring opportunities for there to be an explosion that places the facility at risk. For the criminal act, this will include the type of explosive to be used for analytical modeling (usually, this is expressed in terms of TNT equivalent), the size of the design basis threats (usually, this is expressed in terms of a specific weight of TNT equivalent), the delivery mechanisms (MPIED, VBIED, etc.), and other specific characteristics (such as charge shaping, etc.). It is important to note that a project may well have multiple DBTs for a variety of different threat scenarios. For other explosive event opportunities, well-defined criteria need to be established, including the source of the explosive agent, its containment or delivery vessel or system, quantities, and other relevant specific data necessary to achieve the appropriate level of analytical modeling.

Once the threat data are defined and the post-explosive-event facility performance criteria are fully understood and agreed to with ownership, the design to threat approach requires the design professionals to engage in the process of clearly identifying facility vulnerabilities. This will, of necessity, involve the services of security professionals so that the potential delivery of threats can be clearly defined based on the architectural design, planned electronic and operational security countermeasures, and other explosive threat-reduction program elements. This will allow a clear understanding of remaining areas of explosive vulnerability to both criminal and noncriminal explosive threats. Since the resource investments required to substantially reduce risk from explosive events are normally substantial, addressing the specific areas of greatest vulnerability and then identifying remaining areas of lesser vulnerability is prudent. This process of prioritizing vulnerability from most significant to least significant is positively dependent upon the design professionals clearly identifying, and agreeing with ownership on, where an explosive threat can occur. **The development of a rational and known to be competent security program, which identifies with confidence limited areas of threat potential, is the most economical means to assign resources to limited areas of vulnerability mitigation. This will achieve the best risk reduction with the least resource allocation.**

While reducing the quantity of vulnerabilities has substantial virtues, those remaining vulnerabilities may be significant and challenging. The design to threat approach, quite simply, requires the design professional to develop the appropriate mitigating strategies during the design process, so that the post-event performance criteria meet the standards originally established during the Conceptual Design phase, as codified in the project key design assumptions. In anticipation that required resource allocations will not be trivial, the design professional team may find it most beneficial to develop two or three alternate schemes of structural system design, so that there is an inherent and preestablished response to the formal or informal value engineering process, which will inevitably occur as part of any well-managed project.

For the structural engineer, the design to threat process must assume that threat reduction opportunities have been fully exploited. Simply stated, opportunities for reducing the DBTs have been taken into consideration by law enforcement agencies and by other project mitigation strategies developed by ownership and the balance of the design professionals. As an example, this may include the provision of security personnel, equipment, and policies and procedures, such as the use of explosive detection devices and screening operations as a means to reach an agreed-to explosive DBT level at various points throughout the site and facility. **Assumptions involving the definitive reduction of explosive DBTs and their potential delivery locations should be specifically validated by a well-orchestrated and documented series of design charrettes involving ownership, the architect, the security professional, and the structural engineer.**

The design to threat approach may, in fact, inform ownership that structural system protection, hardening, or alternate load transfer path designs create substantial resource allocation hardships. **Irrespective of the development and presentation of various structural design options intended to address the explosive threat, it may be necessary to carefully examine the means to reduce the threat, reduce the number of vulnerabilities, or have ownership participate in greater risk acceptance.** This may occur as a result of the unattractive and/or unmanageable financial or programmatic implications of the structural protective design options. It is for this reason that addressing the opportunity to employ the design to threat approach during the earliest Conceptual and Schematic design phases is essential. The viability of designing to specifically achieve the required threat reduction through a process of structural design mitigation may identify the need to impose and implement the other threat mitigation resources. This may include the development of protective architectural and site civil design features, deployment of security systems, equipment, and manpower, and recurring capital expenditures for operations. It may also involve increased restrictions associated with space-use programs and/or redesign of the systems associated with mitigating the effects of accidental explosive events.

2.8.2 Design to Budget

This subsection discusses the counterpoint to the design to threat process. Specifically, the design to budget process can be invoked by ownership when they determine that finite resources are available and a set quantity is established; the extent of protection provided is then tailored to achieve the best risk reduction benefit within those strictly defined resource boundaries. The term "budget" in this section of the handbook refers to the budget of overall resources, a definition purposefully intended to be more than just project funding. However, more often than not, the project resources that will have the most confining effect on protection strategies intended to mitigate the explosive event will be the project funds available to spend on security. Occasionally, other resource issues, such as lack of specific protective design materials and/or the skilled labor to install them, will also create resource constraints and may direct the design professional to utilize an overabundance of those protective strategies that are the most available. As an example, the inability to include a competent electronic security system as part of the project, whether due to lack of skilled labor, system components, or the ability for that system to be reliably maintained and serviced, may dictate that explosive event facility vulnerabilities require greater investments in physical hardening to make up for the inability to implement early warning detection and control of access.

Once there is agreement on the project definition of budget resources and the extent of funding available for an agreed-to definition of "project security," the design professionals should conduct a series of design charrettes. At a minimum, these should include the project architect, structural engineer, blast consultant, MEP and vertical transportation consultants, site civil and landscape designers, and security professionals. Other consultants, including traffic engineers, lighting consultants, geotechnical engineers, and others may need to participate as well, based on the project's location, occupancy program, and complexity. **The intent in sequestering the design teams' resources is to create a series of consensus-developed options (at least two but preferably three) wherein each design discipline inherits a small portion of the responsibility for mitigating the facility's vulnerability to explosives. This has repeatedly been shown to be the least expensive and most effective approach to resource allocation and will provide the best opportunity to meet both protective design and budget objectives.** In short, this is the most expeditious path toward "designing to budget."

Unfortunately, these design charrettes, while yielding the most comprehensive and effective solutions at the least cost, often still identify funding expenditures in excess of client resources. This is usually a product of design professional conservatism, wherein each professional advances protective designs that have the absolute least opportunity to fail at that professional's particular mission. Ownership's misallocation of funding (underfunding), due to their extremely limited experience and understanding of the investments required to create truly credible explosive event consequence reduction designs, is also a factor. Sometimes, however, there are simply insufficient overall project funds to create great architecture that includes spaces that are intended to support continued critical operations in environments of considerable threat potential.

When there is a disparity between the ownership's available resource pool and the resource requirement dedications to meet the originally agreed-to post-explosive-event performance, the design to budget process approach does not need to be abandoned. However, the design approaches that remain will all involve an extent of risk acceptance by ownership. Quite simply, at this point, there is an admission by the design and ownership teams that vulnerabilities will continue to exist in the project that may not be adequately addressed. In order to achieve the best result, the project vulnerabilities to the explosive event DBTs identified in the threat and risk assessment will need to be listed and prioritized based on the anticipated consequential effects of an explosive event. This can be most successfully accomplished if there is a comprehensive and prioritized list of consequential effects. A thorough understanding of the prioritized vulnerabilities and prioritized consequential effects will allow the available project resources to be applied in a manner that provides the best extent of protection and risk management. **Once a determination has been made about how limited project resources are to be allocated, a concise written protective design narrative should be authored by the design team. This narrative should be subsequently reviewed and approved by ownership so that there is a legacy document that clearly explains the rationale behind resource allocation and the extent of explosive event protection provided.**

It is essential for the design professionals to understand that client bracketing of funds for project security will need to cover investments other than facade, structural systems, ERR systems, and sensitive space hardening. Security strategies to reduce the explosive event threat will also require expenditures for site civil vehicle approach interdiction features, electronic security systems, both personnel and vehicular screening equipment, and the costs for security staffing and the build-out of their programmatic support spaces. Based on the project cost-estimating techniques and/or models used, redundant diverse ERR systems and architectural features, and redundant diverse business continuity systems may be lumped into the overall security budget. **In order for the design to budget process to be effective—and in order to avoid mismanagement of the extent of funding available for actual explosive event security mitigating building features, components, and systems—the design professionals must engage in a reconciliation process with ownership. This partnership should address and agree upon what project features and components constitute an explosive event protective design strategy and the budget available to achieve it.** This activity should take place at the time when the project budget is in the process of being established.

2.9 SOME RECENT EXAMPLES OF SECURITY DESIGN "BEST PRACTICES"

The authors' combined 80-plus years of security experience have acquainted them with a broad range of effective and workable protective building-design concepts and mitigation solutions, summarized below, that are in daily use around the world. (Because of the sensitivities surrounding post-9/11 security, venues are not specified.[8]) Allocating space for these measures, and incorporating utility requirements for them, are considerations that security design teams need to address at the earliest design stages.

1. *Defense in Depth*—Creating a layered zone defense built upon concentric rings of interlocking physical security, security systems and equipment, personnel, and policies and procedures.
2. *Security Zones of Control*—Enforcing a distinction between different physical spaces and the level of security assurance that each space must provide. Some locations are designated as sterile; that is, everyone in them has been screened for explosives (as well as chemical/biological/radiological materials and other hazardous or prohibited items). Other spaces are identified as secure; that is, only people with proper access privileges are permitted to enter. Other areas may be open to the general public and without restrictions. Additional zones of control may also exist based on the project type, size, and location.
3. *Redundancy and Diversity*—Designing for the redundancy and diversity of critical assets, including security, ERR, and MEP systems, as well as client-specific physical and intellectual assets.
4. *Exceptionally Protected Zones*—Specifically designing for the preservation of unique cultural artifacts and the protection of unique financial assets, intellectual resources, critical unique industrial or financial processes, and other single-source assets that do not allow for protection through diverse repetition. To protect one-of-a-kind assets, hardening solutions have to be created in a manner responsive specifically to each asset's vulnerability.
5. *Personnel Screening*—Identifying and screening all individuals and their personal items for explosives and other proscribed or hazardous materials.

[8]These sites include the U.S. Capitol, the Pentagon, Central Intelligence Agency (CIA) Headquarters, Boston's Logan International Airport, the London Ring of Steel, London's Canary Wharf commercial office complex, and the Eurotunnel between the United Kingdom and France. This best practices section also includes observations from the Sears Tower in Chicago and the Statue of Liberty Visitor Screening Center. Relevant security measures at a worldwide insurance company headquartered in lower Manhattan and a number of comparable private sector enterprises, as well as several U.S. border inspection stations, are also incorporated.

6. *Vehicle Screening*—Identifying and screening all vehicles to prevent vehicle-borne explosive (and other) threats from entering managed street segments, on-site parking, subgrade parking, and cargo offloading areas.
7. *Cargo and Mail Screening*—Ensuring that qualified and trained personnel screen for explosives all cargo deliveries destined for the buildings' core interior areas in accordance with the facility-specific DBT. For example, security personnel would inspect all overnight and freight couriers, and other nonpostal/mail services for explosives and other threats. In some very high-security venues, all delivery vehicles, mail, and cargo may need to be inspected off-site before delivery to the protected premises.
8. *Secondary Screening Capability*—Enforcing a clearly defined separation between the primary inspection process for screening people and vehicles and detecting anomalies, and the secondary inspection process, which resolves any anomalies through detailed search, investigation, and testing. At least two different detection technologies should be used to complement each other—for example, using explosive-detection canines, explosive vapor sampling, trace detection, or detailed visual inspections to conduct secondary investigations when high-energy imaging appears to detect explosive materials.
9. *Space and Provisioning for Security Control Centers*—Ensuring that the facility command, control, communications, and intelligence (C3I) center that monitors and controls security systems and operations is afforded the appropriate space, utilities, security, environmental protection (from smoke and chemical/biological agents, as appropriate), and backup systems.
10. *Segregation and Protection of Building Mechanical, Security, and ERR Equipment*—Isolating critical security, ERR, and MEP systems and areas from tampering and disabling by unauthorized individuals.
11. *Crime Prevention through Environmental Design (CPTED)*—Designing CPTED—a concept that incorporates environmental design features into the overall security strategy for deterring crime and unwanted behavior—into an area or facility.

2.10 RELATED PHENOMENA

Designing to comprehensively mitigate the explosive event requires an understanding of the consequential effects associated with the release of significant amounts of energy over very short durations. Design professionals must be cognizant that the explosive event alone is not the only threat consequence requiring design attention and subsequent mitigation. Post-explosive-event effects may be numerous based on the type of explosive event, location internal or external to the facility, the type of building construction, internal contents, occupancy use program, and numerous other contributing factors such as the flame, smoke, and fuel contribution of the structure itself and its contents.

No discussion of "related phenomena" would be complete without the acknowledgment that there will be the potential for severe injuries and fatalities; damage to critical facility assets, both intellectual and physical; disruption of business missions; and the need for implementing evacuation, rescue, and recovery operations. Additional effects will include crime scene management, post-event facility evaluations, insurance claim investigations, and reconstruction and restoration of physical plant, institutional operational viability, and public image.

However, design professionals have the opportunity to influence a limited venue of related explosive event phenomena. With appropriate assistance from blast and security consultants, engineers can design to contain and manage fires which may follow some explosions. Also, limitations to the extent of post-event damage and structural system collapse can be analyzed and responded to, so that the structural performance of the facility is engaged in a limited participation in life-safety, ERR, and business continuity exposures. Well-developed facade designs can reduce injuries and fatalities while also protecting interior spaces, the ERR, and critical business continuity systems. Design professionals can also identify the intended means of evacuating building populations and engage in the planning and providing for the subsequent arrival of emergency responders so that effective ERR can enhance the opportunity for occupant safety. In doing so, they can clearly identify those building features, systems, and components that require an enhanced level of explosive event protection as it is extremely likely that they will be severely compromised by such an event unless they receive extraordinary protection.

Of the numerous explosive-event-related phenomena, the following handbook sections emphasize the need for the design team to pay particular attention to and dedicate both design and building resources to the mitigation of progressive collapse, protection of ERR, and management of attendant fires.

2.10.1 Progressive Collapse

As mentioned earlier in this chapter of the handbook, the losses of primary and secondary components of the building's structure are potential results of an explosive event, whether a purposeful and malicious act or an accidental or naturally occurring incident. The extent of damage, as previously noted, will depend upon the size and type of event, its location, and the extent to which the structural system in the building has been prepared for managing the extraordinary applied loads and any significant impacts due to airborne debris.

Since life safety is generally the paramount criterion, the definition of disproportionate damage used to define the project's post-explosive-event security performance criteria may be dependent upon the size and location of the building's occupant population. High-rise office buildings with commercial occupancies traditionally allocate 150 to 200 square feet per person as a means to estimate population densities. With common floor plate sizes ranging from 25,000 to 45,000 square feet, there are opportunities for approximately 150 to

300 people per floor. The loss of multiple floors, especially throughout the entire height of the building, which may be as many as 70 or 80 floors, defines catastrophic loss of life circumstances. Hence, progressive collapse from one floor to the next is deemed unacceptable. However, the definition of disproportionate damage may be more conservative than the prevention of damage outside of the area of explicit and direct explosive event effects. **Design professionals should, at the earliest possible design phase in the project, review with the architect the anticipated occupancy profiles for the building's interior and exterior spaces. They should examine the extent of failure associated with the threat-assessment-identified explosive event DBTs or other possible explosive sources within the facility, should no supplemental structural robustness be added for explosive event management. At this point, a determination should be made whether the consequent resulting structural damage invokes too high a casualty count for the owner's and design professionals' risk tolerance.** If that is in fact the case, then while the damage may not be structurally disproportionate, the life-safety consequences may be unacceptable. This may then inform a more conservative definition of allowable disproportionate damage. **Since the design professionals, from a structural standpoint, generally are responsible for designing to preclude progressive collapse and to achieve nondisproportionate damage, clear definitions of these terms, especially disproportionate damage, should be reviewed with ownership and codified as part of the project's key design assumptions.** The significance of these definitions becomes even more evident for low-rise facilities that have significant occupancies concentrated in relatively small areas, such as performance venues or transportation structures. In these environments, there may be, for estimating purposes, one person for every 4 to 6 square feet, which may translate into thousands of persons within a relatively small area. Again, a traditional structural definition of disproportionate damage may define acceptable structural damage to four column bays as a result of the loss of a single column. However, the loss of these bays may translate to, using a 40-foot-square bay module, 6,400 square feet of occupancy. In this example, at a density of 6 square feet per person, there are potential casualties of over 1,000 people. This is an unacceptable result. Therefore, each project must be individually evaluated and the progressive collapse designs developed accordingly.

In order to competently evaluate either preexisting structures intended for renovation and/or expansion or the design of new structures, the services of a qualified blast consultant working in partnership with the structural engineer are essential. The determination of the extent of anticipated damage and the opportunities to identify the extent of potential mitigation will be based upon their interactive use of structural analysis, blast effects science and software, historically informed engineering judgment, and the validation of results conclusively drawn from these experiences and analytical tools through actual live testing of structural systems with actual explosive events. **Failure to properly anticipate and define the circumstances and designs that will not prevent progressive**

collapse and that will not prevent disproportionate damage will preclude the elimination of potentially fatal flaws in the design and place both ownership and the design professionals at considerable risk.

2.10.2 Disruption of Evacuation, Rescue, and Recovery Systems

Careful study of post-explosive-event conditions clearly identify the value provided by building designs that provide a safe and secure means of egress for internal occupants and equally safe and secure means of reentry for emergency responders. **Time-motion egress analytics, which model the amount of time required for a building to engage in a complete evacuation of all its occupants, routinely show dramatic improvements when all of the intended means of egress are available and unimpaired.** Or, stated another way, the loss of egress capacity associated with an explosive event can be expected to dramatically increase the time required for people to exit a facility at risk and to substantially diminish emergency responder efficacy at extracting the base population of handicapped individuals and those incapacitated by the explosive event itself. This can be expected to translate into higher injury and casualty counts, as every moment is crucial during facility occupant extraction.

Structural integrity of stairs and elevators (and sometimes escalators, as these may also be used as an egress mechanism in some facilities) may not translate to a tenable ERR environment. As exemplified by the 1993 use of explosives at the World Trade Center by terrorists, stairwells became filled with smoke, even though they were structurally adequate to continue to function as viable means of egress. This smoke condition hindered exiting operations and was allegedly responsible for smoke inhalation, and a significant number of related adverse health consequences were claimed. Structural integrity maybe inadequate if not supported by survivable smoke control/pressurization systems combined with interlocked and functionally survivable smoke detection.

At the time of authorship of this handbook, there is a distinct and evolving trend for building codes and other legislative documents that guide and/or direct the design community to require the survivability of evacuation, rescue, and recovery systems after an explosive event. Considerable attention has been focused on this subject as a result of the extensive studies carried out by the National Institute of Standards and Technology (NIST) of the World Trade Center attacks of September 11, 2001. Numerous building safety recommendations were made by the NIST committee, many of which focus on a need for additional legislation and design responsiveness to the enhancement and preservation of ERR systems, building features, and design components. It is reasonable to expect, and this handbook anticipates, that these NIST recommendations will become the foundation for new language in national and international building codes. It is reasonable to assume that such significance has been focused on ERR survivability that these codes will call for increased stairwell widths; more robust stairwell construction; the use of elevators for ERR operations, and consequently the construction of more robust and water-resistant hoistways and lifts; enhanced fireproofing ratings, adhesiveness and impact resistance; improved fire detection and reporting; more survivable and more competent fire suppression systems; more effective emergency voice communication systems; and other related support infrastructure improvements. **With these known to be valuable and soon to be codified ERR building features, design professionals should acknowledge that one of the most significant explosive-event-related phenomena requiring attention is the preservation of facility ERR operations**.

In Section 2.5.2, a comprehensive list of ERR systems was provided. While it is not reasonable to expect that these systems will survive in the immediate zone of the most powerful explosive event effects, design professionals should carefully analyze the location where the threat-assessment-identified explosive event DBTs can occur. They should then design ERR systems so that they are appropriately protected to achieve survivability and continue to provide functionality to the balance of the facility. This may involve protective hardening of the spaces where they occur or the design of redundant systems with adequate geographical diversity so that they are not disabled by the DBT events. To accomplish this effectively is no small design endeavor. A quick review of the ERR building components will intuitively identify that they are likely to be spread throughout the entire facility. This may therefore require a detailed review of the entire facility and all of the attendant systems and analytical modeling of the protective design features in response to all of the various explosive DBTs. **Consequently, the design professionals should carefully review the client's expectations, code requirements, and good engineering practice and develop a clearly stated standard with respect to the desired performance of ERR systems and building features in response to the project's explosive DBTs.** To do this competently, it must be remembered that many systems that support ERR are functionally dependent upon each other. As an example, stair pressurization, smoke purge, and smoke control fans, necessary to maintain egress and firefighting, are integrally linked to the building management system (BMS), fire command systems, and emergency power, and the compromise of any one of those by the explosive event will render that system inoperative. Consequently, all of these systems must be systematically analyzed and protected. As another example, fire suppression systems, necessary to manage potential attendant fires, are dependent upon the integrity of fire suppression piping, pumps and control panels, emergency power, and fire alarm systems. Again, a compromise of one of these components may place the entire system at risk.

There are simply too many permutations of ERR systems integration that could be placed at risk by explosive events to competently note and review each of them individually in this handbook for the purpose of developing uniquely responsive mitigating strategies. The appropriate process to ensure that ERR systems have the greatest probability of post-event survivability is further discussed elsewhere in Chapter 2. as part of consequence management, and is founded upon the strategy of overlaying ERR systems on the project threat maps and having their designs and/or hardening protection informed by individual explosive event modeling using the project DBTs.

2.10.3 Attendant Fires

Not all explosive events are accompanied by an immediately ensuing fire development. The rapid expansion of gases at hypersonic speeds and the instantaneous consumption of oxygen as part of the explosive event chemistry often deny fire development. To that end, the oil industry has used oxygen starvation as a means of suppressing oil well fires through the use of explosives. However, certain types of explosive events, such as those caused by high-pressure gas mains, often have an initial burst of high energy but then continue to burn at extreme temperatures and have the opportunity to subsequently engage combustible materials or continue to generate adequate heat from their own fuel source to subject structural elements to ductile yielding and reduction in load-carrying capacity. Mitigating strategies often involve the opportunity to eliminate the participation of secondary fuel sources ignited by an explosive event. **The design professionals should carefully review the storage and distribution systems for flammable agents and their exposures to the project explosive DBTs and include in their design pump shutdowns, automatic and manual breach control valves, and/or other strategies so that these fuel sources have limited burn times. Additionally, the design of robust encasement around fuel storage and/or distribution systems should be considered.** Examples include the installation of generator fuel storage reserves in what are essentially concrete bunkers providing both explosive event ignition protection and elimination of secondary-source fuel supply.

Attendant fires pose a range of problems. As mentioned previously, they may challenge the structural integrity of the facility as a result of extreme temperatures, especially applied in an environment where fireproofing may be damaged and/or absent as a result of the initial blast pressures and abrasion/impact of airborne debris associated with the explosive event. Although it is anticipated that building code legislation will likely require increased fire ratings for structural members and the specification of more physically resilient fireproofing methods, it is unlikely that these material specifications will provide adequate post-explosive-event performance reliability. **The structural engineer and architect should partner with the project blast consultant and security professional and determine the opportunity to receive enhanced benefits from the specification of robust fireproofing strategies.** This may include the use of concrete or masonry encasement. Reasonable solutions for structural system protection from fire will continue to include fireproofing, fire suppression systems, and alternate load path designs. **However, the most effective mitigating strategy remains the reduction and/or elimination of the explosive threat, and the reduction of building materials and contents that have the opportunity to participate as fuel sources in the fire development triad of oxygen, fuel load, and an ignition source.**

The National Fire Protection AssociationStandards has quantified the parameters by which building occupants are affected by fire events. Temperature, air toxicity levels, and visual obscurescence are all factors that affect either health or the ability to extricate oneself from the fire event. Explosive events that are accompanied by attendant fires are likely to be characterized by significant quantities of airborne debris, irrespective of the smoke contribution associated with the fire. Air quality may rapidly approach lethal thresholds as a result of escalating air temperatures (140°F is the upper-level threshold for human exposure). Air quality may also become untenable as a result of toxic air particle concentration from the explosive demolition effects and/or from the fire's oxygen consumption or toxic gas generation. These conditions combined with visual obscurescence create a potentially deadly environment. This situation and the previous concern regarding structural integrity are exacerbated by the likely failure of sprinkler or gaseous suppression systems in the immediate area of the event. **However, containing the event-based fire to a smaller radius of effect will do much to improve life safety for those outside of the explosive event zone. Consequently, the design of sprinkler systems with smaller fire control zones, increased water application density, fast response sprinkler heads, and survivable water supplies provides a very reasonable and competent means to both improve life safety and reduce property damage.** As part of any explosive event protection strategy, the blast consultant and the structural engineer should partner with the fire suppression engineer for the purpose of optimizing the sprinkler system's response to an explosive event. This should include sprinkler coverage zone sizing; the use of automatic and/or manual breach control valves, in conjunction with alternate main and/or pipe riser cross connections, to ensure system performance outside of the explosive event zone; increased hazard classifications to improve water availability and discharge density; head selection; and other pertinent aspects of suppression systems' design with the intention of fire containment to a limited area. This will have the dual benefit of reducing loss of life and property and reducing the extent of structure exposed to extreme heat.

There may be less opportunity to influence the selection of the actual building construction materials and the interior fit-out of furniture, fixtures, and equipment, to reduce the building's internal fuel load, but, as an explosive event design mitigation strategy, it should not be summarily overlooked. The less there is to burn, the shorter the duration and overall exposure of the building's occupants and structural elements to the deleterious effects of a fire born out of an explosive event. Similar design logic follows in the selection of building construction materials. The specification of noncombustible components will reduce fire damage. Generally, structural systems and building contents are regulated by the Building and Fire Codes with the intent to reduce combustibility. **However, whenever the project threat assessment identifies explosive events as a threat to be mitigated, building and interior fit-out components should be specified with an eye toward reducing the facility fuel load.**

2.11 SECURITY DESIGN CONSIDERATION GUIDELINES

The following bullets summarize major points in the building security design considerations process in a condensed format.

1. During the project's design's initial Conceptual phases, the blast consultant and structural engineers should participate in design charrettes with the project architects so that the symbiotic relationship between the structural scheme options and the architectural design concepts to achieve the required explosive event protective designs are considered for potential implementation. Failure to address this vulnerability reduction opportunity at this early phase may prejudice the opportunity to develop the most architecturally acceptable, security-competent, and cost-effective structural scheme.
2. The entire security design team should participate in the creation of a brief design mission statement, which clearly and concisely states the client's expectations for the facility and site's post-explosive-event performance. Since these expectations are routinely overstated by clients—as a result of their lack of sophistication or knowledge of the engineering complexities, architectural program compromises, post-construction institutional operational constraints, impact on project financials, aesthetic effects, and other factors better understood by the structural engineer and the balance of the design team professionals—this mission statement must be a consensus-developed document.
3. On projects where explosive event consequence management is a programmatic design response requirement, design professionals should insist that the threat assessment and site-specific DBT be provided to them at the earliest possible date. This is essential if the specific site and facility vulnerabilities are to be identified and mitigating concepts developed for potential implementation.
4. During the site selection process, the security and blast consultant explosive event engineering professionals should work closely with the client as full members of the building security design team. Based on the maximum realistic standoffs, the pressures and impulses that will be applied to the structure, its facade, inbound utilities, and other critical site civil features should be factored into the site selection decision. This is essential so that an initial determination can be made on how these extraordinary loads (in excess of gravity, wind, and seismic) will affect the protective design capabilities of the major building systems.
5. The initial phases of consequence evaluation and management should occur in the Conceptual and Schematic Design phases and include a review

of the opportunities for redundancy and diversity and the use of hardened facility structural, facade, and ERR design elements.

6. Insufficient efforts to clearly define the client's consequence management expectations are one of the most significant design process shortfalls and are a major reason why the engineering efforts fail to meet client needs or expectations.
7. Deliberate or accidental explosive events rarely impact only the areas defined within contract boundary limits of work, a fact that must be taken into consideration as part of consequence evaluations. During the initial stages of site planning and facility design blocking, stacking, and facade selection, consequence evaluations should consider the aggressors' opportunity to deliver the explosive to the target and the extent of damage a detonation external to the building or site could cause.
8. Design professionals must address building designs so that the explosive event consequence management response is based on an evaluation of how ERR will occur. In this way, designs are informed to protect ERR building design features and provide inherent tactical support and definitive criteria for an ERR concept of operations.
9. While the pressures, impulses, and plasma effects associated with an explosive event using extremely high-energy substances such as plastic explosives are regarded as upper-level protection design thresholds, the security design team should carefully review the entire project design to identify other building systems or storage spaces that could participate in an explosive event.
10. Any value engineering or other similar "design optimization" process that involves review and recommendations regarding structural systems engaged in blast vulnerability management should include the blast consultant and the structural engineers responsible for the systems under evaluation.

2.12 CONCLUSION

It is the authors' hope that this discussion of security design considerations has given the reader an increased awareness of how a competent and well-organized security design approach, initiated at the beginning of the design process, can significantly influence the eventual success or failure of the facility's explosive event protective design strategies. Managing the design and security interdependencies and complexities of today's blast-resistant construction, in an uncertain and ever changing threat environment, requires the contributions of security, explosives, and building sciences domain experts working together with ownership risk managers, forming a group that can then appropriately be called the *security design team*.

REFERENCES

Boettcher, Mike and Ingrid Arnesen. 2002. Al-Qaeda Documents Outline Serious Weapons Program. CNN.com (January 25, 2002), available at http://www.isis-online.org/publications/terrorism/cnnstory.html.

Federal Emergency Management Agency. 2005. *Risk Assessment: A How-To Guide to Mitigate Potential Terrorist Attacks* (FEMA 452). Washington, DC: Federal Emergency Management Agency, Department of Homeland Security. Available at http://www.fema.gov/plan/prevent/rms/rmsp452.shtm.

Robertson, Nic. 2002. Bomb-making video reveals scope of al Qaeda threat. CNN.com (August 21, 2002), available at http://archives.cnn.com/2002/US/08/21/terror.tape.main/index.html.

Santayana, George. 1905. *The Life of Reason, or the Phases of Human Progress: Introduction and Reason in Common Sense*. New York: Charles Scribner's Sons.

U.S. Department of Defense. *DOD Dictionary of Military Terms*. http://www.dtic.mil/doctrine/jel/doddict/.

3 Performance Criteria for Blast-Resistant Structural Components

Charles J. Oswald

3.1 INTRODUCTION

Performance criteria specify quantitative limits on the response of structural components that are intended to achieve the blast design objective or strategy for the building design, such as those summarized in Table 3.1. A structural component is designed by calculating the blast load on the component, using a dynamic analysis to determine the maximum component response, and then checking the response against quantitative limits that are consistent with the overall building design objective.

In typical static design, the stress level of component response is limited to prevent failure. However, this approach cannot be used for blast design, since controlled, ductile yielding is usually part of the design intent. Therefore, response limitations are typically placed on the maximum dynamic deflection of the component to prevent component failure, where this deflection accounts in an approximate manner for the amount of acceptable damage or plastic strain.

The amount of conservatism in the design depends on the allowable component deflection limits. Usually, the allowable component deflection is limited so that either repairable, or unrepairable, damage occurs, but not component failure. In other cases, when component failure from a worst-case explosion scenario is almost impossible to prevent, response limitations may be placed on the post-failure velocity of the components and the amount of overall building collapse.

Table 3.2 shows different areas of blast design and typical performance goals for each area. The performance goals are affected primarily by the type of blast threat for each area. For example, terrorist attacks involve many uncertainties related to the explosive threat, and consequently this area of blast design has more basic design performance goals. Explosive safety, on the other hand, involves explosions of known stored or manufactured materials, and its performance goals have more defined levels of protection. Hardened military structures typically contain mission-critical equipment and personnel (i.e., command and control

Table 3.1 Summary of Blast Design Strategies

Performance Goals	Design Strategy	Comments
Prevent component failure	Limit response of components protecting personnel and equipment to levels below those causing failure. A safety factor against failure is provided by minimizing expected damage levels.	This is the primary approach used to protect building occupants against injuries from an explosion.
Limit structural collapse	Prescribe more conservative allowable response criteria for load-bearing components. Also, design buildings to withstand localized failure of a primary or load-bearing component (i.e., design against progressive collapse).	UFC 4-023-03 (UFC 2005) and GSA (GSA 2003) have specific criteria to design against progressive collapse. Building collapse will almost always cause a very high percentage of fatalities to building occupants.
Maintain building envelope	Design against failure of building cladding elements, including windows and doors.	Blast overpressure entering buildings can cause significant injuries and extensive damage, but not usually many fatalities.
Minimize flying debris	Design cladding components using an expected worst-case explosive threat to fail without becoming hazardous projectiles to the occupied space. This may be used for a limited building area receiving the worst-case blast loads.	Mitigating the extent and severity of injuries to occupants in building areas without overall structure collapse is very dependent on minimizing flying debris.
Prevent cascading explosion events	Design structures around stored or manufactured explosives to resist projectiles and blast pressures from a nearby explosion without failing catastrophically and detonating explosives within structure. Also, design building components to protect personnel and control equipment necessary for shutting down potentially explosive surrounding industrial processes in an explosion.	This includes proscriptive requirements based on testing in DDESB 6055.9 (DDESB 2004) and blast design requirements to protect equipment in UFC 3-340-02 (UFC 2008) for cases involving high explosives and ASCE (ASCE 1997b) for cases involving industrial explosions.

centers) that must be protected against a wide array of enemy threats, but these threats are relatively well defined.

3.2 BUILDING AND COMPONENT PERFORMANCE CRITERIA

Blast design and analysis are primarily component-based, whereby the applied blast load and response of each component in the building are determined

Table 3.2 Summary of Blast Design Areas

Design Area	Performance Goals	Comments
Explosive safety	Protect personnel Protect equipment, supplies, and stored explosives Prevent cascading explosive events	Design against accidental explosion where the amount and location of potential explosives are relatively well quantified. All inhabited areas are designed to prevent injuries. This type of blast design generally has the most design conservatism.
Hardened military structures	Preserve mission-critical functions	Buildings designed to resist attacks from specified weapons with a near-miss or direct hit such that given operations can still be performed.
Antiterrorism	Prevent mass casualties Minimize flying debris Limit structural collapse	There are uncertainties in the amount and location of potential explosions caused by terrorists. The usual intent is to build in a given amount of blast resistance using a level of protection against a given defined threat. A maximum standoff distance to the protected building is established with barriers and inspections for explosives.
Process safety	Protect personnel Prevent cascading events Minimize financial losses	Explosive events can be modeled relatively well, but initial conditions of the accident causing explosion must be assumed. Neutral risk philosophy is employed such that people inside buildings should have the same protection as those outside buildings. Ability to shut down operations is often important to prevent cascading events.

individually. However, performance goals are usually set in terms of life safety, functionality, and reusability for the entire building. Therefore, damage or response levels for individual building components must be established to achieve the overall building performance goal. A framework for this process has been developed by the American Society of Civil Engineers (ASCE, forthcoming) based on a similar approach developed by the U.S. Army Corps of Engineers, Protective Design Center (PDC) (Protective Design Center 2008b). In this approach, a building Level of Protection is selected based on desired performance goals, as shown in Table 3.3. Table 3.4 is then used to determine the associated damage levels for each type of component in the building. Therefore, this approach determines component damage levels that are consistent with an overall building performance goal. It can also be used to assess the Level of Protection provided by an existing building, based on calculated component damage levels.

Table 3.4 refers to three types of components: primary, secondary, and nonstructural. Primary components are members whose loss would affect a

Table 3.3 Building Levels of Protection

Level of Protection	Building Performance Goals	Overall Building Damage
I (Very low)	*Collapse prevention:* Surviving occupants will likely be able to evacuate, but the building is not reusable; contents may not remain intact.	Damage is expected, up to the onset of total collapse, but progressive collapse is unlikely.
II (Low)	*Life safety:* Surviving occupants will likely be able to evacuate and then return only temporarily; contents will likely remain intact for retrieval.	Damage is expected, such that the building is not likely to be economically repairable, but progressive collapse is unlikely.
III (Medium)	*Property preservation:* Surviving occupants may have to evacuate temporarily, but will likely be able to return after cleanup and repairs to resume operations; contents will likely remain at least partially functional, but may be impaired for a time.	Damage is expected, but building is expected to be economically repairable, and progressive collapse is unlikely.
IV (High)	*Continuous occupancy:* All occupants will likely be able to stay and maintain operations without interruption; contents will likely remain fully functional.	Only superficial damage is expected.

number of other supported components and whose loss could potentially affect the overall structural stability of a building area. Examples of primary structural components include columns, girders, and any load-bearing structural components such as walls. Secondary components generally are supported by a primary framing component and can fail without creating widespread structural damage.

Table 3.4 Expected Component Damage for Each Level of Protection

	Component Damage Levels		
Level of Protection	Primary Structural Components	Secondary Structural Components	Nonstructural Components
I (Very low)	Heavy	Hazardous	Hazardous
II (Low)	Moderate	Heavy	Heavy
III (Medium)	Superficial	Moderate	Moderate
IV (High)	Superficial	Superficial	Superficial

Hazardous damage—The element is likely to fail and produce debris.
Heavy damage—The element is unlikely to fail, but will have significant permanent deflections such that it is not repairable.
Moderate damage—The element is unlikely to fail, but will probably have some permanent deflection such that it is repairable, although replacement may be preferable for economic or aesthetic reasons.
Superficial damage—The element is unlikely to exhibit any visible permanent damage.

Examples of secondary structural components include non-load-bearing infill masonry walls, metal panels, and secondary steel framing members such as studs, girts, purlins, and joists. Nonstructural components include interior non-load-bearing walls, and architectural items attached to building structural components.

Nonstructural components are not usually designed specifically against design blast loads. In general, their performance has received very little attention, since testing and research for blast-resistant design has focused on structural components. However, when appropriate, depending on the level of protection, these nonstructural components can be attached to structural members with strengthened connections in manner similar to that used in earthquake-resistant design.

3.3 RESPONSE PARAMETERS

Component blast damage levels, such as those in Table 3.4, have been correlated to the maximum dynamic deflection of the component with two nondimensional parameters: the ductility ratio (μ) and support rotation (θ). The reasons for basing the correlations on these simplified parameters are discussed in this section, as well as the limitations of this approach. The support rotation is defined in Figure 3.1:

$$\theta = \tan^{-1}\left(\frac{2x_m}{L_{\min}}\right)$$

where: L_{min} = is the shortest distance from a point of maximum deflection to a support

It is based on the shortest distance from a support to the point of maximum component deflection ($L_{\min}$), where the point of maximum deflection is determined from yield line theory. $L_{\min}$ is equal to one-half the span length of one-way spanning components, except for cantilevers where $L_{\min}$ is the whole span length. The yield line pattern for two-way spanning components can be determined from charts in UFC 3-340-02 (Unified Facilities Criteria Program 2008), which is an update of TM 5-1300 (DOANAF 1990).

The ductility ratio is defined as the ratio of the maximum component deflection to the component yield deflection, as shown in Equation 3.1. The yield

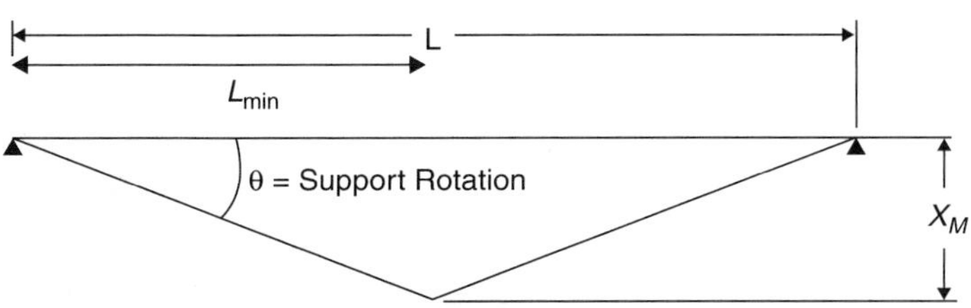

Figure 3.1 Support Rotation Angle

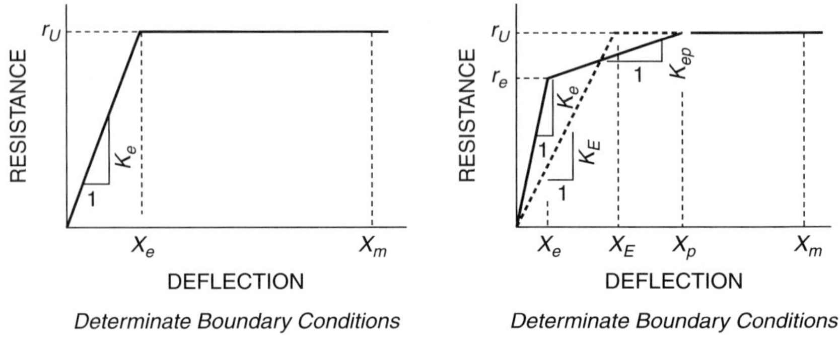

Figure 3.2 Resistance-Deflection Curves for Flexural Response

deflection for determinate components is the deflection when the stress in the maximum moment region equals the yield strength. The yield deflection for indeterminate components is usually equal to the equivalent yield deflection, X_E, in Figure 3.2. This is an average ductility ratio, which is less than the ductility ratio at the initial yield location and greater than the ductility ratio at the final yield location. It is calculated as shown in Equation 3.2, such that the area under the resistance-deflection graphs for the simplified elastic-perfectly plastic resistance-deflection curve yielding at X_E (i.e., dashed curve in Figure 3.2) and the actual, multi-stiffness resistance-deflection curve (i.e., solid curve in Figure 3.2) are equal at deflections greater than the deflection at which the component becomes a mechanism (i.e., X_p in Figure 3.2).

$$\mu = \frac{X_m}{X_y} \tag{3.1}$$

where: μ = ductility ratio
X_m = maximum component deflection
X_y = deflection causing yield of component
= X_e for determinate components (see Figure 3.2)
= X_E for indeterminate components (see Figure 3.2)

$$\begin{aligned} X_E &= X_e + X_p\left(1 - \frac{r_e}{r_u}\right) \\ K_E &= \frac{r_u}{X_E} \end{aligned} \tag{3.2}$$

Given values of ductility ratio and support rotation have traditionally been used as response criteria in blast-resistant design, where these criteria are the quantitative limits that define the upper bounds of each response level or damage level for blast-loaded components. These two parameters are used as the basis for response criteria because they are related to the amount of component damage, as discussed below, and because of their design-level simplicity. They are both based on the maximum dynamic component deflection, which is relatively easy to measure in component blast tests and to calculate with simplified dynamic analysis methods used for blast design.

Damage occurs in blast-loaded components as plastic strains approach the material failure strain at some location in the component. The ductility ratio is an approximate measure of component plastic strain based on the assumption that the curvature in the maximum moment regions increases proportionally with deflection after yielding, and plane cross sections remain plane. In this case, the ductility ratio is a measure of the ratio of the total strain to the yield strain in the maximum fibers. This approach is applicable to cases where the strains causing the component damage occur at the same location where initial yielding occurs, such as the extreme fiber of the maximum moment region of a steel beam. In reinforced concrete, on the other hand, damage is primarily controlled by compression strains as they approach a given limit value (i.e., the concrete crushing strain), whereas initial yielding occurs in the tension steel at a deflection that usually causes relatively low compression strains in the maximum moment region. Therefore, the ductility ratio does not correlate well with flexural damage to reinforced concrete members. Also, very ductile steel members typically develop tension membrane response at high ductility levels, which can causes high strains and failure around the connections, rather than at midspan, where initial yielding occurs.

The support rotation correlates better to component blast damage than the ductility ratio in these cases. In the case of reinforced concrete and masonry components, the support rotation is related to the cross section rotation at maximum moment regions, and therefore to the compression face strain, which primarily causes damage and failure to underreinforced components. This is the case when the component responds as a ductile mechanism after yielding of the reinforcing steel. Very little damage occurs prior to yielding of the reinforcing steel.

In the case of tension membrane response, the support rotation increases proportionally with the maximum deflection-to-span ratio and therefore increases with the amount of tension membrane response. Support rotation limits help ensure that the amount of tension membrane force in a ductile component that can undergo large deflections is limited to values that do not allow connection failures. Large deflections of primary components can also cause failures at connections to supported components. Many blast tests and field investigations at accidental explosion sites have shown that steel components connected to their supports with screws, bolts, and welds typically fail at the connection, rather than at the points of maximum strain in the component (i.e., yield locations). Therefore, ductile steel components typically have separate response criteria in terms of ductility ratio, to limit plastic strains at the maximum moment region, and in terms of support rotation, to limit strains from tension membrane response near connections or other critical areas (i.e., utility cutouts in well-attached metal studs).

The ductility ratio is usually most useful for determining damage levels with low ductility ratios in the range of 1.0 to approximately 3.0. This applies to low damage levels for ductile components and low to high damage levels of relatively brittle components, such as wood components. Damage from brittle, nonflexural response modes, such as shear, buckling, or connection failure, can also be correlated to the ductility ratio if the yield deflection, X_y, in these cases equals the maximum component deflection causing a critical shear, buckling, or connection strain in the component and these strains increase proportionally with increased component deflection.

Limit support rotations and ductility ratios that establish component damage levels generally cannot be derived from theoretical considerations alone. These parameters are simplified for practical reasons and failure modes are complex and influenced significantly from actions, such as combined bending, shear, and axial load, which are difficult if not impossible to quantify and characterize universally. Also, applicable damage levels typically have qualitative, rather than quantitative, descriptions. Therefore, limit values for these parameters are based primarily on empirical correlations between observed damage levels to components from blast tests and corresponding values of support rotations and ductility ratios calculated from the component's measured maximum dynamic deflection. This is discussed more in the next section.

A limitation in this approach is that the support rotation and ductility ratio terms do not capture all the parameters that are actually related to the observed component blast damage. For example, an analysis of data from fifteen one-way-spanning reinforced masonry walls tested in a shock tube showed that the relationship between the measured support rotations and the observed damage levels was strongly dependent on the wall reinforcement factor, which compares the reinforcing steel ratio to the balanced steel ratio, as well as to the support rotation (Oswald and Nebuda 2006). The test walls with a low reinforcement factor sustained significantly less damage at the same support rotation than walls with a higher factor. Also, a statistical study of several hundred static reinforced concrete beam and slab tests showed that the support rotation at failure was not constant, as generally assumed for blast design, but was a function of several parameters, including the reinforcing steel ratio, the shear span to depth ratio, and the ratio of the steel reinforcement yield strength to concrete compression strength (Panagiotakos and Fardis 2001). Therefore, if two concrete panels with different spans, thicknesses, and reinforcement ratios are both subject to blast loads causing the same given support rotation (e.g., 2 degrees), it is possible that they will not have the same observed damage level. A similar situation is possible for other component types. Fortunately, this is alleviated to some extent by the qualitative nature of damage levels, which encompass relatively broad ranges of damage.

In spite of the limitations of the support rotation and ductility ratio as definitive measures of component damage, it is the judgment and experience of engineers who have developed response criteria relating these parameters to component damage that they are adequate and generally conservative for blast design and analysis purposes. More accurate terms that include more properties associated with a component's blast damage level, and have acceptable simplicity for design-based methodologies, may be developed in the future.

3.4 EMPIRICAL CORRELATIONS BETWEEN RESPONSE PARAMETERS AND COMPONENT DAMAGE

Support rotations and/or ductility ratios that define the damage levels of blast-loaded components have been established using correlations between these two parameters calculated from available test data and observed damage levels in the test components. The amount of available test data varies with component type, since blast testing is typically conducted by various government agencies and organizations to address specific blast design considerations, and the data include varying amounts of measured material properties, blast load information, and response values. Also, there is an almost complete lack of blast test data for some component types and damage levels. Deflection-controlled static tests that provide information on component performance at post-yield deflections are considered for component types with limited available blast test data. These static data are generally conservative because each strain and deformation level exists for a sustained period of time, and can therefore result in more damage than the same strain level that will only exist from a blast load for a few tenths of a second, or less.

The most comprehensive published study that has correlated blast test data with component damage levels is part of the *Component Explosive Damage Assessment Workbook* (CEDAW) methodology (Protective Design Center 2008a). CEDAW is a fast-running methodology to assess component blast damage based on pressure-impulse (P-I) diagrams that was developed to be as consistent as possible with available blast test data. In this study, data were collected from over 300 blast tests on 10 different component types and used to calculate the support rotation and ductility ratio for each test, using single-degree-of-freedom (SDOF) dynamic analyses. Then, this information was used to determine the rotations and ductility ratios that bounded each observed damage level for each component type, using scaled P-I diagrams.

This effort is illustrated in Figure 3.3 and Figure 3.4, which are based on damage levels defined in Table 3.5. Pbar and Ibar in the graphs are the peak positive phase pressure and impulse, respectively, scaled (i.e., divided) by relevant component dynamic response parameters to create dimensionless terms that are directly related to component response. Therefore, components with very different blast loads and structural properties, but the same Pbar and Ibar values, should have the same ductility ratio or support rotation values (i.e., a strong component with a high blast load and a weaker component with a lower blast load that both have response with the same ductility ratio or support rotation would have the same Pbar and Ibar values). This should be true in both blast tests and dynamic response calculations. The Ibar terms for component response in terms of ductility ratio are different from those for component response in terms of support rotation.

Figure 3.3 shows a scaled P-I diagram with curves of scaled blast loads calculated with SDOF analyses causing ductility ratios (μ) of 1, 3, 6, and 12 for a corrugated steel panel. These curves are plotted with data points of scaled blast loads causing damage levels ranging from Superficial to Hazardous Failure to

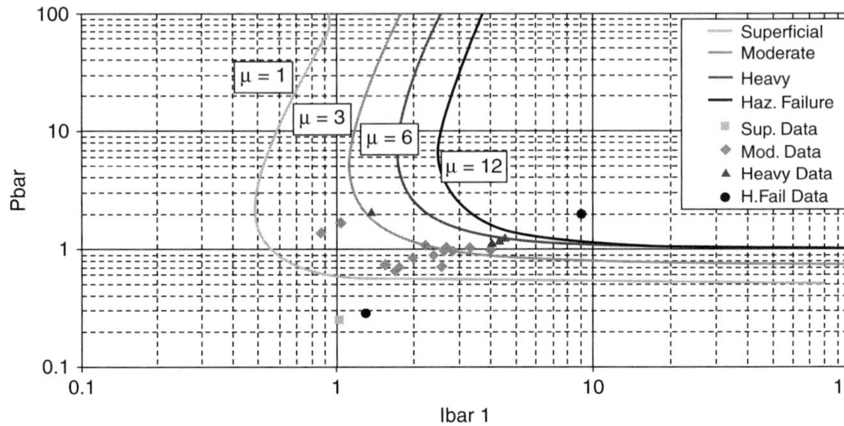

Figure 3.3 Scaled P-I Curves in Terms of Ductility Ratio vs. Scaled Data for Flexural Response of Corrugated Steel Panels

corrugated steel panels in blast tests. Figure 3.4 is a similar scaled P-I diagram with curves of scaled blast loads calculated with SDOF analyses causing support rotations (θ) of 3, 6, and 10 degrees plotted against the same data that are scaled by Pbar and Ibar terms based on component support rotation. In both of these figures, the curves with the lowest support rotation or ductility ratio correspond to the lowest damage level (i.e., Superficial), and curves with higher support rotations and ductility ratios correspond to higher damage levels up to Hazardous Failure.

Figure 3.3 and Figure 3.4 were used to determine that the ductility ratios and support rotations shown on the figures bounded each of the damage

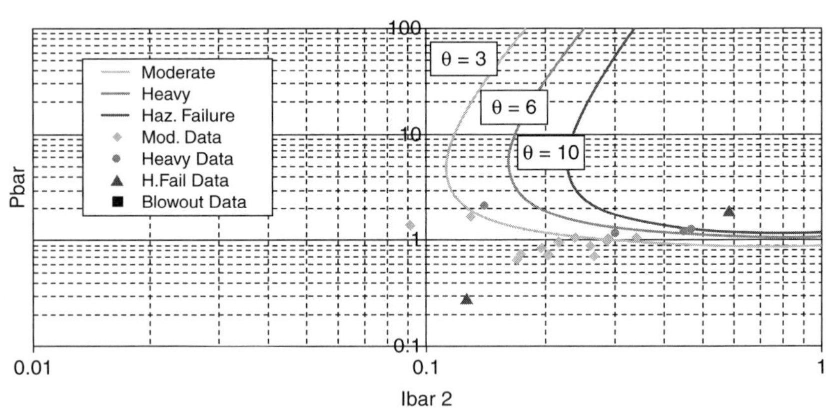

Figure 3.4 Scaled P-I Curves in Terms of Support Rotation (Degrees) vs. Scaled Data for Flexural Response of Corrugated Steel Panels

Table 3.5 Component Damage-Level Descriptions

Damage Level	Component Damage
Blowout	Component is overwhelmed by the blast load, causing debris with significant velocities.
Hazardous Failure	Component has failed, and debris velocities range from insignificant to very significant.
Heavy	Component has not failed, but it has significant permanent deflections, causing it to be unrepairable.
Moderate	Component has some permanent deflection. It is generally repairable, if necessary, although replacement may be more economical and aesthetic.
Superficial	No visible permanent damage.

levels, and therefore could be used as response criteria for corrugated steel panels (Protective Design Center 2008a). Similar curves were generated for all other component types where there were available test data. The CEDAW *Workbook* (Protective Design Center 2008a) has figures similar to Figure 3.3 and Figure 3.4 for each component type and a full description of the test data. Table 3.6 shows a summary of the response criteria determined for each component type that was developed in this manner. No criteria are shown for Superficial Damage, since this damage level was correlated to a ductility ratio of 1 by definition for the CEDAW study. No test data were available to develop response criteria for component types not shown in Table 3.6, including hot-rolled steel beams and primary framing components.

PDC TR-08-07 (Protective Design Center 2008a) and Oswald and Nebuda (2006) have a detailed discussion of the Pbar and Ibar terms used to develop the empirical correlations between component performance levels and support rotations and ductility ratios. These terms were developed for component response in terms of support rotation and ductility ratio and for different types of response modes (i.e., flexure, tension membrane, etc.) based on conservation of energy equations between the work energy and kinetic energy from the blast load and the strain energy of the component in the applicable response mode. The parameters included in the Pbar and Ibar terms vary based on the type of response mode for the component and the type of response parameter representing the component response (i.e., ductility ratio or support rotation). Also, the Pbar and Ibar terms were developed to account for both positive and negative phase blast loads since both types of blast loads generally affect component response and damage levels. Extensive checks were performed on the validity of the Pbar and Ibar scaling terms, as summarized in PDC TR-08-07 (Protective Design Center 2008a). The scaling terms are not exact since they were derived based on some simplifying approximations, but numerous comparisons showed they caused the SDOF-based scaled P-I diagrams to be within 20% of more exact solutions in almost all cases.

Table 3.6 Response Criteria for Upper Bound of Each Damage Level for Each Component Type

	Ductility Ratio			Support Rotation			Support Rotation w/ Tension Membrane		
Component	Moderate	Heavy	Haz Failure	Moderate	Heavy	Haz Failure	Moderate	Heavy	Haz Failure
One-way corrugated metal panel	3	6	12	3	6	10			
Cold-formed girt and purlins							4	12	20
Open-web steel joist				3	6	10	3	6	10
One-way or two-way reinforced concrete slab				2	5	10			
Reinforced concrete beam				2	5	10			
One-way reinforced masonry				2	8	15			
One-way or two-way unreinforced masonry[1]				1.5	4	From data w/o SDOF			
Wood stud wall	2	3	4						
Reinforced concrete column (shear failure)		6							
Steel column (connection failure)		1							

[1]Response criteria are only valid for SDOF analyses with brittle flexural response and axial load arching.

3.5 RESPONSE CRITERIA DEVELOPMENT

The information in Table 3.6 represents the most extensive and formalized use of blast test data to develop response criteria. This information was an important part of the development of several of the current blast response criteria, However, it was not the only consideration. Each organization that has developed response criteria defining given performance levels considers a range of factors, such as the exact definitions of its performance levels relative to the available data, how to address component types where the data are inadequate, whether to discount some data or data sets as not applicable, and the level of conservatism to be included in the response criteria. There is very limited published information on these considerations or on the compilations of the data that were considered.

Table 3.7 shows a summary of response criteria that have been developed for each area of blast design. Ideally, there would be one general set of response limits on component response applicable for all the performance levels and design strategies in Table 3.1. There is strong movement in this direction, but currently there are separate sets of response limits associated with performance levels established by the different organizations associated with each area of blast design. These differences are due in part to the fact that these organizations have different missions, design objectives, and types of explosive threats.

The following subsections provide some detail on the criteria in Table 3.7. In all cases the response criteria are stated in terms of either ductility ratio, support rotation, or both terms, as discussed in Section 3.2. In cases where response criteria for both terms are provided, the overall component damage level or performance level is controlled by the response criterion that causes more severe damage or lower performance. In addition to the response criteria discussed below, the engineer should also always consider any design-case-specific requirements on component deflection and limitations of response criteria, as discussed in Section 3.6. It should also be noted that all the response limits stated in this document may be updated in the future. Historically, response criteria have been updated as more information on component response to blast loads has become available and as new component and material types are used for blast-resistant design.

3.5.1 Explosive Safety Criteria

Blast-resistant design criteria are included in UFC 3-340-02 (Unified Facilities Criteria Program 2008), which has been developed by the explosive safety design community under direction of the U.S. Department of Defense Explosive Safety Board (DDESB). Explosive safety refers to safe design and operation of manufacturing and storage areas for high explosives that can accidently explode. This primarily includes military and military contractor facilities with explosives, but it also includes industries such as fireworks and commercial explosives manufacturers.

Most blast-resistant design prior to 1990 was for explosive safety and critical military structures, and the majority of the initial blast testing of structural

Table 3.7 Summary of Response Criteria Publications

Organization	Publication	Type of Criteria	Comments
Department of Defense	UFC 3-340-02 (Unified Facilities Criteria Program 2008)	Design criteria for explosive safety.	Published in draft form in 1984 and finalized in 1990 as the tri-service manual TM 5-1300, NAVFAC P397, and AFM 88 22. This document generally has the most conservative criteria. It has been updated recently with some revisions as UFC 3-340-02.
Task Committee on Blast Resistant Design of the Petrochemical Committee of the Energy Division of ASCE	Design of Blast Resistant Buildings in Petrochemical Facilities (American Society of Civil Engineers 1997b)	Performance criteria for three component damage levels. These criteria are commonly used for blast design for industry related to process safety.	Initial performance criteria for blast design from 1997. These criteria will be updated in a new edition of this publication in 2009.
U.S. Army Corps of Engineers, Protective Design Center (PDC)	Single-Degree-of-Freedom Structural Response Limits for Antiterrorism Design, PDC TR-06-08 (Protective Design Center 2008b).	Performance criteria for four component damage levels associated with building levels of protection. These criteria are commonly used for design against terrorist threats.	Performance criteria developed in 2006 that are significantly more comprehensive than previous criteria. Also, they are based more explicitly on available blast test data.
ASCE Blast Protection of Buildings Standards Committee	Blast Protection of Buildings (American Society of Civil Engineers, forthcoming)	Performance criteria for four component damage levels. These criteria can be used for a wide range of blast design except explosive safety.	Based very strongly on criteria in PDC TR-06-08 from U.S. Army Corps of Engineers.

components against conventional (i.e. non-nuclear) explosives was conducted in support of explosive safety design. Since that time, an increase in terrorism and industrial safety requirements by OSHA have considerably increased the amount of blast-resistant design.

Response criteria for blast resistant design related to explosive safety for the U.S. Department of Defense are defined in UFC 3-340-02 for the component types shown in Table 3.8. Inhabited buildings subject to blast pressures from

Table 3.8 Response Criteria for Explosive Safety Design from UFC 3-340-02

	Protection Category[a]			
	1		2	
Element Type	μ_{MAX}	Θ_{MAX}	μ_{MAX}	Θ_{MAX}
Reinforced concrete[b]				
Conventional slabs	—	1°[c]	—	8°
Beams with stirrups (closed ties)	—	2°	—	8°
Exterior columns	3	—	—	—
Precast concrete				
Prestressed	1	2	—	—
Non-prestressed	3	2	—	—
Compression members	1.3	—	—	—
Reinforced masonry[b]				
One-way[d]	—	1°	—	
Two-way[d]	—	2°	—	
Structural steel				
Beams, purlins, spandrels, girts	10	2°	20	12°
Frame structures	—	2°[e]	—	—
Plates	10	2°	20	12°
Open-web steel joists				
Controlled by maximum end reaction	1	1°	—	—
Otherwise	4	2°	—	—
Cold-formed steel floor and wall panels				
Without tension membrane action	1.75	1.25°	—	—
With tension membrane action	6	4°	—	—

Note: Where a dash (—) is shown, the corresponding parameter is not applicable as a response limit.
[a]See Table 3.9 for definitions of Protection Categories.
[b]Reinforced with conventional steel rebar.
[c]May be increased to 2° with either shear reinforcement or tensile membrane action.
[d]Non-reusable walls.
[e]Relative sidesway deflection between stories is limited to H/25.

accidental explosions must resist the blast loads with response criteria for Protection Category 1 in Table 3.9. Buildings protecting critical equipment must resist the blast loads with response criteria for Protection Category 2. The component damage associated with each Protection Category is not well defined in UFC 3-340-02, but Protection Category 1 is considered to provide a factor of safety against failure of at least two.

The performance criteria in Table 3.8 were first published in 1984 in a draft version of TM 5-1300 and then finalized in 1990. They are unchanged in the recent conversion of TM 5-1300 into UFC 3-340-02. These criteria are more conservative than more recent criteria discussed in the next sections, which are based in part on many recent blast tests on structural components that were not available at the time the response criteria for TM 5-1300 were developed. It

Table 3.9 Protection Category Descriptions for Structural Components from UFC 3-340-02

Protection Category	Description for Structural Components
Category 1	Attenuate blast pressures and structural motion to a level consistent with personnel tolerances
Category 2	Protect equipment, supplies, and stored explosives from fragment impact, blast pressures, and structural response

is also the intent of TM 5-1300/ UFC 3-340-02 to provide design criteria with a significant level of conservatism, since it is used to help protect workers at explosive production and storage facilities, who spend a significant amount of time exposed to the risk of accidental explosions.

3.5.2 Response Criteria for Antiterrorism

The Protective Design Center (PDC) of the U.S. Army Corps of Engineers published *Single-Degree-of-Freedom Structural Response Limits for Antiterrorism Design* in 2006 (i.e. DoD Antiterrorism response criteria) and updated it in 2008 (Protective Design Center 2008b). These response criteria have also been adopted into the forthcoming ASCE document "Blast Protection of Buildings" (American Society of Civil Engineers, forthcoming). The PDC response criteria define limit values for ductility ratio and support rotation corresponding to four different damage levels that apply to twelve different common structural component types with a total of 38 different subcategories based on component characteristics, such as response mode and types of connections. They also consider the cases where components are primary components (i.e., major framing components) or secondary components (i.e., cladding components). These response criteria are used to design DoD facilities against explosive terrorist threats and can also be used for non-DoD government buildings, such as those that provide blast protection required by the Interagency Security Committee (ISC) Security Criteria (Interagency Security Committee 2004).

The DoD Antiterrorism response criteria are shown in Table 3.10. Some of the criteria are based on the response limits developed from blast test data acquired to develop CEDAW (shown in Table 3.6) for corresponding component types. The response criteria for other component types are based primarily on static test data, as summarized in Table 3.11. The response criteria based on static test data are generally more conservative than those based on blast test data, as discussed previously. For example, there are limited blast test data that indicate that conventional metal stud wall systems can resist significantly higher loads than those calculated with SDOF analysis using the response limits in Table 3.10 (Grumbach et al. 2007).

Table 3.10 DoD Response Criteria for Antiterrorism Design

Element Type	Expected Component Damage							
	Superficial		Moderate		Heavy		Hazardous	
	μ_{MAX}	Θ_{MAX}	μ_{MAX}	Θ_{MAX}	μ_{MAX}	Θ_{MAX}	μ_{MAX}	Θ_{MAX}
Reinforced concrete[a]								
Single-reinforced slab or beam	1	—	—	2°	—	5°	—	10°
Double-reinforced slab or beam without shear reinforcement[b,c]	1	—	—	2°	—	5°	—	10°
Double-reinforced slab or beam with shear reinforcement[b]	1	—	—	4°	—	6°	—	10°
Prestressed concrete[d]								
Slab or beam with $\omega_p > 0.30$	0.7	—	0.8	—	0.9	—	1	—
Slab or beam with $0.15 \leq \omega_p \leq 0.30$	0.8	—	$0.25/\omega_p$	1°	$0.29/\omega_p$	1.5°	$0.33/\omega_p$	2°
Slab or beam with $\omega_p < 0.15$ and without shear reinforcement[b,c]	0.8	—	$0.25/\omega_p$	1°	$0.29/\omega_p$	1.5°	$0.33/\omega_p$	2°
Slab or beam with $\omega_p < 0.15$ and shear reinforcement[b]	1	—	—	1°	—	2°	—	3°
Masonry								
Unreinforced[c,e]	1	—	—	1.5°	—	4°	—	8°
Reinforced[a]	1	—	—	2°	—	8°	—	15°
Structural steel (hot-rolled)								
Beam with compact section[f]	1	—	3	3°	12	10°	25	20°
Beam with noncompact section[c,f]	0.7	—	0.85		1	—	1.2	—
Plate bent about weak axis	4	1°	8	2°	20	6°	40	12°
Open-web steel joist								
Downward loading[g]	1	—	—	3°	—	6°	—	10°
Upward loading[h]	1	—	1.5	—	2	—	3	—
Shear response[i]	0.7	—	0.8	—	0.9	—	1	—
Cold-formed steel								
Girt or purlin	1	—	—	3°	—	10°	—	20°
Stud with sliding connection at top	0.5	—	0.8	—	0.9	—	1	—

(Continued)

Table 3.10 (*Continued*)

Element Type	Expected Component Damage							
	Superficial		Moderate		Heavy		Hazardous	
	μ_{MAX}	Θ_{MAX}	μ_{MAX}	Θ_{MAX}	μ_{MAX}	Θ_{MAX}	μ_{MAX}	Θ_{MAX}
Stud connected at top and bottom[j]	0.5	—	1	—	2	—	3	—
Stud with tension membrane[k]	0.5	—	1	0.5°	2	2°	5	5°
Corrugated panel (1-way)with full tension membrane[l]	1	—	3	3°	6	6°	10	12°
Corrugated panel (1-way) with some tension membrane[m]	1	—	—	1°	—	4°	—	8°
Corrugated panel (1-way) with limited tension membrane[n]	1	—	1.8	1.3°	3	2°	6	4°
Wood[o]	1	—	2	—	3	—	4	—
Blast doors								
Built-up (composite plate & stiffeners)	3	1	10	6	20	12	—	—
Plate (solid)	3	1	20	6	40	12	—	—

Note: Where a dash (—) is shown, the corresponding parameter is not applicable as a response limit.
[a]Reinforced with conventional steel rebar.
[b]Stirrups or ties that satisfy sections 11.5.5 and 11.5.6 of ACI 318 and enclose both layers of flexural reinforcement throughout the span length.
[c]The ASCE" Blast Protection of Buildings" document states that these response limits are applicable for flexural evaluation of existing elements that satisfy the design requirements of Chapters 6 through 8 but do not satisfy the detailing requirements in Chapter 9 of the document, and shall not be used for design of new elements. This note does not appear in the DoD Response Criteria for Antiterrorism Design.
[d]Span to thickness ratio greater than 5. Also, reinforcement index $\omega_p = (A_{ps}/bd)(f_{ps}/f'_c)$. Also see Equation 3.3.
[e]Values assume wall resistance controlled by brittle flexural response or axial load arching with no plastic deformation; for load-bearing walls, use superficial or moderate damage limits to preclude collapse.
[f]Limiting width-to-thickness ratios for compact and noncompact sections are defined in ANSI/AISC 360.
[g]Values assume tension yielding of bottom chord with adequate bracing of top chord to prevent lateral buckling.
[h]Values assume adequate anchorage to prevent pull-out failure and adequate bracing of bottom chord to prevent lateral buckling.
[i]Applicable when element capacity is controlled by web members, web connections, or support connections; ductility ratio for shear is equal to peak shear force divided by shear capacity.
[j]Also applicable when studs are continuous across a support.
[k]Requires structural plate-and-angle bolted connections at top, bottom, and any intermediate supports.
[l]Panel has adequate connections to yield cross section fully.
[m]Typically applicable for simple-fixed span conditions.
[n]Limited to connector capacity; includes all standing seam metal roof systems.
[o]Values shown are based on very limited testing data; use specific test data if available.

Table 3.11 Bases for DoD Antiterrorism Response Criteria

Component Type	Primary Basis for Response Criteria
Reinforced concrete slabs	Response limits from CEDAW study based on several blast test series on simply supported wall panels where damage occurred primarily in maximum moment region.
Reinforced concrete beams	Blast test data for slabs assumed to apply to reinforced concrete beams.
Hot-rolled steel beams	Static test data.
Cold-formed steel panels and girts/purlins	Response limits from CEDAW study based on several blast test series on wall systems with cold-formed beams and corrugated steel panels where lower damage levels occurred primarily in maximum moment region, characterized by local buckling of compression flange, and higher damage levels including failure occurred at the connections due primarily to in-plane reaction forces from tension membrane response that occurs at large deflections.
Cold-formed metal studs	Static test data.
Open-web steel joists	Downward response limits from CEDAW study based on one test series.
Wood	Response limits from CEDAW study on two test programs on wood stud walls. One static and one dynamic test series both indicated that typical plywood panels did not create a fully composite section that added significantly to ultimate resistance of stud wall system.
Prestresssed concrete components	Static test data.
Reinforced masonry	Several test programs with shock tube and high-explosive tests.
Unreinforced masonry	Several test programs with shock tube and high-explosive tests documented in PDC TR-08-07 (2008a) and Wesevich et al. (2002).

3.5.3 Response Criteria for Blast-Resistant Design of Petrochemical Facilities

The Task Committee on Blast Resistant Design of the Petrochemical Committee of the Energy Division of ASCE published the manual *Design of Blast Resistant Buildings in Petrochemical Facilities* in 1997 (American Society of Civil Engineers 1997b), and an update will be finalized in 2009. This manual has three levels of component response to blast loads, as shown in Table 3.12, and provides response criteria in terms of maximum ductility ratio and/or support rotation criteria for each response level. Separate criteria are defined for different component types (i.e., steel beams, concrete slabs, etc.), and the criteria address primary and secondary components, frame components with and without significant axial load, and shear controlled response for reinforced concrete components.

Table 3.12 Component Damage Level Descriptions

Damage Level	Description
High	Component has not failed, but it has significant permanent deflections, causing it to be unrepairable.
Medium	Component has some permanent deflection. It is generally repairable, if necessary, although replacement may be more economical and aesthetic.
Low	Component has none to slight visible permanent damage.

In general, the response limit values are quite low for components in which shear or compression is significant compared to flexural response. Where adequate shear capacity is provided, much larger response limit values are associated with each response or damage level.

The updated version contains new response criteria that are based on a combination of criteria developed for the original publication and the response criteria in PDC TR-06-08. The intent for the new response criteria was to expand the existing criteria to consider a wider range of component types and subcategories relevant for typical industrial buildings and to replace or modify existing response criteria with values from PDC TR-06-08 where there was a strong case to be made based on the CEDAW data (Oswald 2008). Also, differences in damage level definitions used by the PDC and by the ASCE committee (see Table 3.12) were considered.

Table 3.13 shows the updated response limits for steel components, and Table 3.14 shows the updated response criteria for reinforced concrete and masonry components. The criteria values for Medium and High response of cold-formed steel girts and purlins, reinforced concrete and masonry, and prestressed concrete components in Table 3.13 and Table 3.14 are based on similar values for Moderate and Heavy component damage levels, respectively, in PDC TR-06-08. The criteria for Superficial damage in PDC TR-06-08 were not used in the ASCE response criteria because this damage level was considered more conservative than the Low damage level in Table 3.12. Therefore, values for Low response of these component types are approximately one-half the values for Medium response. The reinforcement index that is used as part of the response criteria for prestressed concrete components in Table 3.14 is defined in Equation 3.3.

The response criteria for all other component types not listed in the previous paragraph are the same, or very similar, to values developed for the original publication of the manual. In some cases this is because the values in PDC TR-06-08 are similar to those in the original publication of the manual or there were not enough supportive blast test data to justify changes.

$$w_p = \frac{A_{ps}}{bd_p}\frac{f_{ps}}{f_c'} \tag{3.3}$$

Table 3.13 Response Criteria for Steel Components

Component[1]	Low Response		Medium Response		High Response	
	μ_{MAX}	θ_{MAX}	μ_{MAX}	θ_{MAX}	μ_{MAX}	θ_{MAX}
Hot-rolled steel compact secondary members (beams, girts, purlins)[2]	3	2	10	6	20	12
Steel primary frame members (with significant compression)[2,3,4]	1.5	1	2	1.5	3	2
Steel primary frame members (without significant compression)[2,3,4]	1.5	1	3	2	6	4
Steel plates[5]	5	3	10	6	20	12
Open-web steel joists	1	1	2	3	4	6
Cold-formed light gage[6] steel panels (with secured ends)[7]	1.75	1.25	3	2	6	4
Cold-formed light gage[6] steel panels (with unsecured ends)[8]	1.0	—	1.8	1.3	3	2
Cold-formed light gage[6] steel beams, girts, purlins, and noncompact secondary hot-rolled members	2	1.5	3	3	12	10

[1]Response limits are for components responding primarily in flexure unless otherwise noted. Flexure controls when shear resistance is at least 120% of flexural capacity.

[2]Primary members are components whose loss would affect a number of other components supported by that member and whose loss could potentially affect the overall structural stability of the building in the area of loss. Secondary members are those supported by primary framing components.

[3]Significant compression occurs when the axial compressive load, P, is more than 20% of the dynamic axial capacity of the member. P should be based on the ultimate resistance of the supported members exposed to the blast pressure. See PDC TR-06-08 for detailed examples of calculation for P.

[4]Sidesway limits for moment-resisting structural steel frames: low = (height)/50, medium = (height)/35, high = (height)/25.

[5]Steel plate criteria can also be applied to corrugated (crimped) plates if local buckling and other response modes are accounted for in the analysis.

[6]Light gage refers to material which is less than 0.125 inches (3 mm) thick.

[7]Panels must be attached on both ends with screws or spot welds.

[8]Panels are not attached on both ends (for example standing seam roof panels).

where: A_{ps} = area of prestressed reinforcement in tension zone
b = the member width
d_p = the depth from compression face to center of prestressing steel
f_{ps} = calculated stress in prestressing steel at design load
f'_c = the concrete compressive strength

3.5.4 Blast Resistant Doors

Table 3.15 shows recommended response criteria for design of structural components in blast doors based on a desired door performance category from *Design*

Table 3.14 Response Criteria for Reinforced Concrete (R/C) and Masonry (R/M) Components

	Low Response		Medium Response		High Response	
Component[1]	μ_{MAX}	θ_{MAX}	μ_{MAX}	θ_{MAX}	μ_{MAX}	θ_{MAX}
R/C beams, slabs, and wall panels (no shear reinforcement)		1		2		5
R/C beams, slabs, and wall panels (compression face steel reinforcement and shear reinforcement in maximum moment areas)		2		4		6
Reinforced masonry		1		2		5
R/C walls, slabs, and columns (in flexure and axial compression load)[3]		1		2[2]		2[2]
R/C and R/M shear walls & diaphragms	3		3		3	
R/C and R/M components (shear control, without shear reinforcement)	1.3		1.3		1.3	
R/C and R/M components (shear control, with shear reinforcement)	1.6		1.6		1.6	
Prestressed concrete ($w_p \leq 0.15$)[4]	1			1		2
Prestressed concrete ($0.15 < w_p < 0.3$)[4]	1		$0.25/w_p$	1	$0.29/w_p$	1.5

[1]Response limits are for components reinforced with conventional rebar or prestressing strands responding primarily in flexure unless otherwise noted.
[2]A support rotation of 4 degrees is allowed for R/C components that have compression face steel reinforcement and shear reinforcement in maximum moment areas.
[3]Applicable when the axial compressive load, P, is more than 20% of the dynamic axial capacity of the member. P should be based on the ultimate resistance of the supported members exposed to the blast pressure. Refer to PDC TR-06-08 for detailed examples of calculation for P.
[4]The reinforcement index, w_p, is defined in Equation 3.3.

of Blast Resistant Buildings in Petrochemical Facilities (American Society of Civil Engineers 1997b). Table 3.10 shows similar design information from PDC TR-06-08. Blast doors can be designed using SDOF analyses of door components responding in flexure such that the responses of all components comply with the applicable response criteria in these tables. Both inbound and rebound response of blast doors must typically be considered during blast door design. The structural performance of metal doors and frames and their restraining hardware (such as latches and hinges) at the load equal to the maximum calculated resistance from the SDOF analyses can be verified by applying an equivalent static pressure to blast doors in accordance with ASTM F2247 (ASTM 2003b).

Blast-resistant doors are typically designed by vendors according to specifications from the design engineer and architect. Blast door performance can also be demonstrated by blast tests. There is an ASTM committee that has developed a draft specification for blast testing of doors (i.e., ASTM WK1902).

Table 3.15 Recommended Response Criteria for Blast Door Components

Performance Category	Description	Response Criteria	
		μ_{MAX}	θ_{MAX}
I	The door is to be operable after the loading event, and preestablished design criteria for stress, deflection, and the limitation of permanent deformation have not been exceeded. This category should be specified when the door may be required to withstand repeated blasts or when entrapment of personnel is of concern and the door is a primary exit to the building.	1.0	1.2
II	The door is to be operable after the loading event, but significant permanent deformation to the door is permitted. The door must remain operable, and this category should be specified when entrapment of personnel is a concern	2–3	2.0
III	Noncatastrophic failure is permitted. The door assembly remains in the opening. No major structural failure occurs in the door panel structure, the restraining hardware system, the frame, or the frame anchorage that would prevent the door assembly from providing a barrier to blast wave propagation. However, the door will be rendered inoperable. This category should only be specified when entrapment of personnel is not a possibility.	5–10	8.0
IV	Outward rebound force and resulting hardware failure is acceptable.	Not specified	Not specified

3.5.5 Blast-Resistant Windows

Table 3.16 shows performance criteria in the form of hazard levels to building occupants for blast-loaded windows developed in the United Kingdom and used by the U.S. DoD (SAFE/SSG 1997). Table 3.17 shows similar criteria developed by the Interagency Security Committee (ISC 2004). These criteria are illustrated in Figure 3.5 and Figure 3.6. Response of windows to blast load is currently calculated with government-sponsored software that is not public domain, but is available from the U.S. government for design of U.S. government buildings on a need-to-know basis. These software programs, which include Wingard (Window Glazing Analysis Response and Design) (Applied Research Associates 2008) and HazL (Window Fragment Hazard Level Analysis) (U.S. Army Corps of Engineers 2004), model the overall window response using an equivalent SDOF system and glass flyout after window failure for a wide variety of

Table 3.16 DoD Hazard Levels for Blast-Loaded Windows

Hazard Level	Description
No break	No visible damage to the glazing or frame.
Minimal hazard	Glazing fragments inside the test structure are within a maximum distance of one meter from the window line. For laminated windows or windows with a film that is attached to the window frame, the window may be cracked and fully retained by the frame, or it may leave the frame and land outside the structure or inside the structure within 1 meter of the window opening.
Low hazard	Glazing fragments are thrown into the room for a distance of 1 to 3 meters, but do not exceed a height of 0.5 meters above the floor at the 3-meter distance. Injuries would be limited to lower body cuts, and fatalities would not be expected, although there would be some risk to persons within 1 to 2 meters of windows.
High hazard	Glazing fragments are thrown at high velocity into the occupied space and impact the vertical surface at 3 meters behind the window above a 0.5 meter height. Serious injuries, including cuts to the upper body and face from the flying fragments, would be expected. Fatalities could occur.

Table 3.17 ISC Performance Conditions and Hazard Levels for Blast-Loaded Windows

Performance Condition	Protection Level	Hazard Level	Description of Window Glazing Response
1	Safe	None	Glazing does not break. No visible damage to glazing or frame.
2	Very high	None	Glazing cracks but is retained by the frame. Dusting or very small fragments near sill or on floor acceptable.
3a	High	Very low	Glazing cracks. Fragments enter space and land on floor no further than 1 meter from the window.
3b	High	Low	Glazing cracks. Fragments enter space and land on floor no further than 3 meters from the window.
4	Medium	Medium	Glazing cracks. Fragments enter space and land on floor and impact a vertical witness panel at a distance of no more than 3 meters from the window at a height no greater than 0.6 meter above the floor.
5	Low	High	Glazing cracks and window system fails catastrophically. Fragments enter space impacting a vertical witness panel at a distance of no more than 3 meters from the window at a height greater than 0.6 meter above the floor.

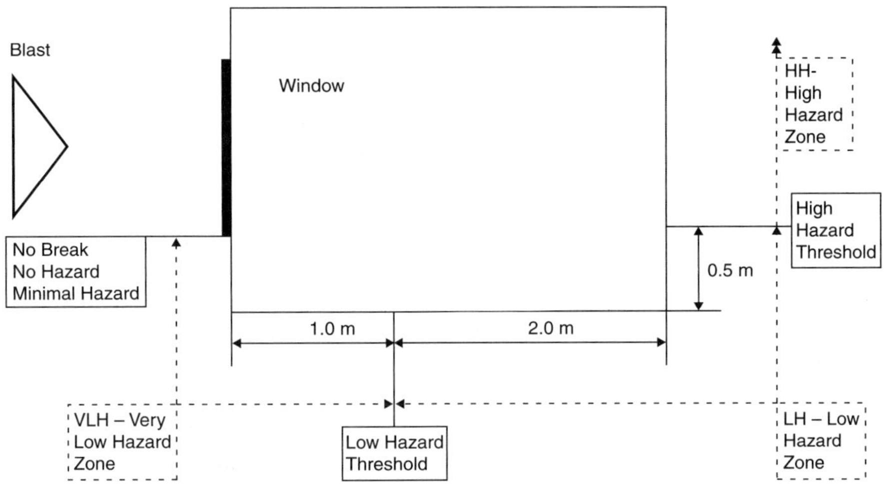

Figure 3.5 Illustration of DoD Window Hazard Levels for Blast-Loaded Windows (from ASTM 1642)

common window configurations. If the aforementioned software is not available, blast-resistant windows can be designed using an equivalent static design approach in accordance with ASTM F2248 (ASTM 2003a) and E1300 (ASTM 2007) for a medium or low protection level. This procedure is also discussed by Norville and Conrath (Norville and Conrath 2006).

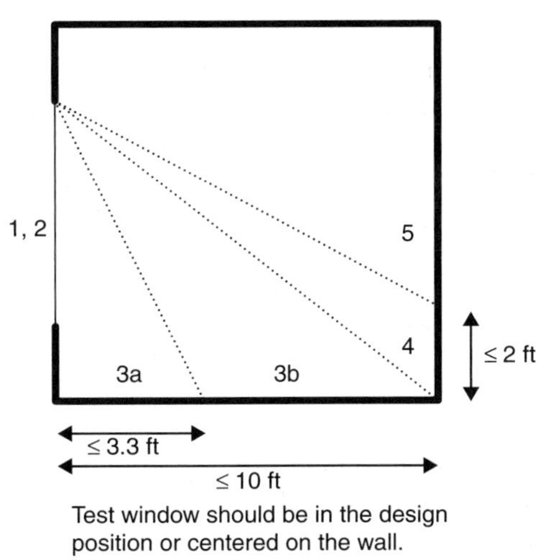

Figure 3.6 Illustration of ISC Window Hazard Levels for Blast-Loaded Windows (from ISC, 2004)

Blast-resistant windows are typically designed by vendors according to specifications from the design engineer and architect. The specifications include the design blast load history and information on the required performance criteria.

3.5.6 Response Criteria for Equivalent Static Loads

Previously discussed response criteria based on support rotations and ductility ratios apply for the typical blast design case where the component is designed using a procedure that explicitly considers the dynamic component response, such as an SDOF-based methodology or dynamic finite element analysis. Some stiff components that are not directly loaded by blast, such as connections and primary framing members in pure axial loading, have a short response time compared to the expected rise time of the dynamic reaction load transferred by supported members, and are usually designed using an equivalent static load approach. The load capacity of the stiff component or connection must exceed the equivalent static reaction load in these cases. If the dynamic analysis shows the supported member does not yield, the equivalent static reaction load may be based on the maximum elastic resistance of this member rather than the ultimate resistance. However, this approach should be used with caution, since it will be nonconservative if the actual blast load exceeds the design blast load.

The load capacity of the stiff component or connection is usually calculated using applicable strength reduction factors from LRFD (load and resistance factor design) (i.e., Ø factors), although sometimes a Ø factor of 1.0 is used at the discretion of the design engineer. Also, the load capacity can include a dynamic increase factor on the yield strength, but this is typically only 1.05 for high-strength components (i.e., A325 and A490 bolts) and should be applied with caution for cases where the capacity is controlled by a non-ductile response mode (i.e., axial buckling or anchor bolt pullout from concrete). The equivalent static reaction load is not multiplied by a load factor.

The equivalent static load approach has conservatism based on the use of a Ø factor less than 1.0 (if applicable), the use of a design load equal to the theoretical ultimate load that can be delivered by the supported component, and the fact that a static approach is used for blast design where the maximum strains are only applied for a few tenths of a second followed by load reversal. However, it is not conservative if the reaction load is applied fast enough by the supported component compared to the response time of the "stiff" supporting component (i.e. primary framing member) such that dynamic deflections occur greater than the component's failure deflection. This is not a concern for connections since they are very stiff and respond quickly compared to the application of the reaction load. Also, any component, including primary framing components, can resist the applied dynamic load regardless of the load rise time if the component has a load capacity greater than 1.2 times the peak dynamic load, and it can achieve at least a ductility ratio of 3.0 without failing, based on SDOF response charts in UFC 3-340-02 (UFC, 2008), ASCE (American Society of Civil Engineers 1997b), and Biggs (1964).

Table 3.18 Comparison of Response Criteria for Medium Damage

	DoD Criteria for Antiterrorism		ASCE Blast Design for Petrochemical Facilities[a]		UFC 3-340-02[b]	
Element Type	μ_{MAX}	θ_{MAX}	μ_{MAX}	θ_{MAX}	μ_{MAX}	θ_{MAX}
Reinforced concrete beam	—	2	—	2	—	2[c]
Reinforced concrete slab	—	2	—	2	—	1
Reinforced masonry walls	—	2	—	2		1
Prestressed concrete components	—	1[d]	—	1[d]	1	2
Hot-rolled steel beam	3	3	10	6	10	2
Cold-formed steel beams	—	3	3	3	10	2
Corrugated steel panels[e]	—	1	3	2	6	4
Open-web steel joists	—	3	2	3	4	2
Wood studs and beams	2	—	—	—	—	—
Blast door components [f]	10–20	6	5–10	8		

[a]Response criteria from draft of new edition (to be published in 2009).
[b]Design criteria for Protection Category 1. UFC 3-340-02 does not differentiate between cold-formed and hot-rolled beams.
[c]With shear reinforcement.
[d]Reinforcement index $\omega_p = (A_{ps}/bd)(f_{ps}/f'_c) < 0.15$.
[e]Panels attached to framing at both ends with some tension membrane resistance.
[f]Door not operable but remains in opening.

3.5.7 Comparisons of Published Response Criteria

The response criteria discussed in this chapter are compared in this section. Cases where different published response criteria are in agreement are indicative of more accurate correlations between the given response levels and corresponding damage or response levels, in the sense that several different organizations agree with those correlations. Cases where there is not agreement indicate that more blast testing is needed for these component types, damage levels, and response modes to more definitively establish applicable response criteria. It must also be considered that it is not always possible to have direct comparisons because of differences between the definitions of response levels by the different organizations and the stipulations associated with various response criteria.

Table 3.18 shows comparisons of response criteria for Medium/Moderate damage from the three sources discussed in this chapter. The criteria from UFC 3-340-02 are for design of Protection Category 1 (i.e., personnel protection), which can be considered approximately equivalent to a medium damage level. The comparison shows component types for which there is basic agreement of the different criteria and component types for which there is not.

Table 3.19 Comparison of Response Criteria for High or Heavy Damage

	DoD Criteria for Antiterrorism		ASCE Blast Design for Petrochemical Facilities[a]		UFC 3-340-02[b]	
Element Type	μ_{MAX}	θ_{MAX}	μ_{MAX}	θ_{MAX}	μ_{MAX}	θ_{MAX}
Reinforced concrete beam	—	5	—	5	—	8[c]
Reinforced concrete slab	—	5	—	5	—	8
Reinforced masonry walls	—	8	—	5		—
Prestressed concrete components	—	2[d]	—	2[d]	—	—
Hot-rolled steel beam	12	10	20	12	20	12
Cold-formed steel beams	—	10	12	10	20	12
Corrugated steel panels	—	4	6	4	—	—
Open-web steel joists	—	6	4	6	—	—
Wood studs and beams	3		—	—	—	—

[a]Response criteria from draft of new edition (to be published in 2009).
[b]Design criteria for Protection Category 2. UFC 3-340-02 does not differentiate between cold-formed and hot-rolled beams.
[c]With shear reinforcement.
[d]Reinforcement index $\omega_p = (A_{ps}/bd)(f_{ps}/f'_c) < 0.15$.

UFC 3-340-02 is generally more conservative than the other criteria, except for corrugated steel panels. It is difficult to get a good comparison for this case because the criteria in UFC 3-340-02 are for panels with "tension membrane capacity" without specifying any degree of tension membrane resistance, as is implied in the other two criteria. The DoD and ASCE criteria agree with each other fairly closely, except for hot-rolled steel beams and corrugated steel panels.

Table 3.19 shows comparison of response criteria for High, or Heavy, damage from the three sources discussed in Section 3.5.1, 3.5.2, and 3.5.3.. The criteria from UFC 3-340-02 are for design of Protection Category 2 (i.e., equipment protection). All of the criteria in Table 3.19 are considered to represent components that are at, or very near, complete failure where they would fall to the ground or be thrown inside the building. In this case, the response criteria in UFC 3-340-02 are not conservative compared to the criteria from the other sources.

3.6 RESPONSE CRITERIA LIMITATIONS

A limitation of response criteria is that they are only applicable for very specific response modes, since the amount of available ductility is a function of the mode that controls the component response for the specific case that is considered. Unless otherwise stated, response criteria should be assumed to apply only for components with ductile, flexural response where the component shear and connection capacities exceed maximum shear and end reaction forces, respectively, based on the maximum resistance of the component during dynamic flexural response (i.e., equivalent static reaction forces).

This means that they are not applicable for describing damage from breach (local failure through the full thickness in the component area of the highest blast loads), spall (failure of the concrete near the surface in the component area of the highest blast loads), or any other similar type of localized component response mode. Breach and spall response are most commonly a concern for reinforced concrete and masonry components subject to blast loads where the scaled standoff (i.e., charge standoff distance to the component divided by the charge weight to the 1/3 power) is less than approximately 1.0 ft/lb$^{1/3}$. Several references, such as Marchand et al. (Marchand et al. 1994), McVay (McVay 1988), and UFC 3-340-02 (UFC, 2008) have methods for predicting spall and breach in reinforced concrete slabs.

A component should be designed against blast loads so that ductile flexural response controls the response, if possible, but if this is not possible, separate response criteria that consider the specific type of response, such as response controlled by shear, brittle flexure, combined flexure and axial load, or connection failure, as applicable, should be used. Unfortunately, such criteria are not always available in published response criteria tables. Similarly, there are currently no published response criteria for many new materials that are used for blast resistant design, such as fiber-reinforced plastic (FRP), ductile polymers, and geotextile catch systems. In these situations, the response criteria must be developed on a case-by-case basis by engineers with blast design and testing experience; otherwise, very inappropriate response criteria may result. For example, available response criteria for masonry walls reinforced with conventional steel rebar have been proposed for walls reinforced with FRP, since these are both cases of reinforced masonry, but this is very nonconservative because FRP is much less ductile than reinforcing steel.

Another consideration is that the criteria in this chapter are intended primarily for response calculated with SDOF analyses. However, they may also be used for response calculated with more detailed dynamic analyses, such as multi-degree-of-freedom analysis and dynamic finite element analysis. Alternatively, design based on dynamic finite element analysis can be based on more detailed response criteria, such as maximum strain criteria, if the response criteria are validated against adequate test data or are judged to be conservative enough by engineers with significant applicable testing and blast design experience.

Finally, response criteria should also include any project-specific requirements, such as limits on deflection of a structural component from impacting building equipment or other protected items. Another case in this category is design of blast-resistant shield walls, which are placed a few feet outside an existing building wall to protect it from any blast loading. As the shield wall deflects very quickly from blast load, there is a volume decrease and corresponding pressure buildup in the annular space between the shield wall and building. The resulting pressure can cause the conventional building wall to fail if the deflection of the shield wall is not adequately controlled. This is not considered in the available response criteria.

The response criteria also do not address failure of any nonstructural items attached to blast-loaded structural components, where acceptable response of wall and roof components per the applicable response criteria may cause large enough relative accelerations between the component and attached item to rupture the connection. This is most critical for heavy overhead items. This subject has not been studied very much for blast-resistant design, but earthquake-resistant design can be used for guidance.

REFERENCES

American Society of Civil Engineers. 1997a. *Structural Design for Physical Security*. New York: American Society of Civil Engineers.

American Society of Civil Engineers, Petrochemical Committee, Task Committee on Blast Resistant Design. 1997b. *Design of Blast Resistant Buildings in Petrochemical Facilities*. New York: American Society of Civil Engineers.

American Society of Civil Engineers Structural Engineering Institute. Forthcoming. *Blast Protection of Buildings*. New York: American Society of Civil Engineers.

Applied Research Associates. 2008. Window Glazing Analysis Response and Design (Wingard) Computer Program, Version 5.5.1. Developed by Applied Research Associates for the U.S. Government Services Administration (GSA).

ASTM. 2003a. *Standard Practice for Specifying an Equivalent 3-Second Duration Design Loading for Blast Resistant Glazing Fabricated with Laminated Glass* (F2248). West Conshohocken, PA: ASTM International.

________. 2003b. *Standard Test Method for Metal Doors Used in Blast Resistant Applications (Equivalent Static Load Method)* (F2247) West Conshohocken, PA: ASTM International.

________. 2007. *Standard Practice for Determining Load Resistance of Glass in Buildings* (ASTM E1300 - 07e1). West Conshohocken, PA: ASTM International.

Biggs, J. D. 1964. *Introduction to Structural Dynamics*. New York: McGraw Hill Co.

Departments of the Army, the Navy and the Air Force (DOANAF). 1990. *Structures to Resist the Effects of Accidental Explosions* (Department of the Army Technical Manual TM 5-1300, Department of the Navy Publication NAVFAC P-397, Department of the Air Force Manual AFM 88-22). Washington, DC: Departments of the Army, the Navy and the Air Force.

Department of Defense Explosives Safety Board (DDESB). 2004. *Ammunition and Explosives Safety Standards* (DOD 6055.9). Washington, DC: U.S. Department of Defense.

Grumbach S. D., C. Naito, and R. J. Dinan. 2007. Use of precast concrete walls for blast protection of steel stud wall construction. Presented at the 78th Shock and Vibration Symposium, Philadelphia, PA, November 4–8, 2007.

Interagency Security Committee. 2004. *ISC Security Design Criteria for New Federal*

Office Buildings and Major Modernization Projects. Washington, DC: Department of Homeland Security, The Interagency Security Committee.

Marchand K. A., S. Woodson, and T. Knight. 1994. Revisiting concrete spall and breach prediction curves: Strain rate (scale effect) and impulse (pulse length and charge shape) considerations. Presented at the 26th DDESB Explosives Safety Seminar, Miami, FL, August 1994.

McVay, Mark. 1988. *Spall Damage of Concrete Structures*. Vicksburg, MS: Army Engineer Waterways Experimentation Station Structures Lab. Accession No. ADA199225 from the Defense Technical Institute Center (DTIC).

Norville H. S. and E. J. Conrath. 2006. Blast-resistant glazing design. *Journal of Architectural Engineering* 12 (3): 129–136.

Oswald C. J. 2008. Update to response criteria for blast resistant buildings in petrochemical facilities. Presented at ASCE Structures Congress 2008, Vancouver, Canada, April 24–26, 2008.

Oswald C. J. and D. T. Nebuda. 2006. Development of *Component Explosive Damage Assessment Workbook (EDAW)*. Presented at the 32nd DDESB Explosive Safety Seminar, Philadelphia, PA, August 2006.

Oswald C. J., D. T. Nebuda, D. Holgado, and M. Diaz. 2006. Shock tube testing on reinforced masonry walls. Presented at the 32nd DDESB Explosive Safety Seminar, Philadelphia, PA, August 2006.

Panagiotakos T. B. and M. N. Fardis. 2001. Deformations of reinforced concrete members at yielding and ultimate. *ACI Structural Journal* 98 (2): 135–148.

Protective Design Center. 2008a. *Component Explosive Damage Assessment Workbook (CEDAW)* (PDC TR-08-07). Omaha, NE: U.S. Army Corps of Engineers, Protective Design Center.

________. 2008b. *Single Degree of Freedom Structural Response Limits for Antiterrorism Design* (PDC TR-06-08). Omaha, NE: U.S. Army Corps of Engineers, Protective Design Center.

SAFE/SSG (Security Facilities Executive, Special Services Group). 1997. *Glazing Hazard Guide, Cubicle Standoffs, Tables and Charts*, Report No. SSG/EP/4/97. London: Special Services Group/Explosive Protection Group.

Unified Facilities Criteria Program. 2002. *Design and Analysis of Hardened Structures to Conventional Weapons Effects (FOUO)* (UFC 3-340-01). Washington, DC: U.S. Department of Defense, Unified Facilities Criteria Program.

________. 2005. *Design of Buildings to Resist Progressive Collapse* (UFC 4-023-03). Washington, DC: U.S. Department of Defense, Unified Facilities Criteria Program.

________. 2008. *Structures to Resist the Effects of Accidental Explosions* (UFC 3-340-02). Washington, DC: U.S. Department of Defense, Unified Facilities Criteria Program.

U.S. Army Corps of Engineers. 2004. Window Fragment Hazard Level Analysis (HazL), Version 1.2. Vicksburg, MS: USACE Engineer Research and Development Center.

U.S. General Services Administration. 2003. *Progressive Collapse Analysis and Design Guidelines for New Federal Office Buildings and Major Modernization Projects* (June 2003). Washington, DC: General Services Administration.

Wesevich J. W., C. J. Oswald, M. T. Edel, M. J. Lowak, and S. S. Alaoui. 2002. Compile and Enhance Blast Related CMU Wall Test Database (CMUDS). Prepared for the Office of Special Technology by Wilfred Baker Engineering, Inc., Contract No. 41756-00-C-0900, September 6, 2002.

4 Materials Performance

Andrew Whittaker and John Abruzzo

4.1 INTRODUCTION

The ASCE Standard *Blast Protection of Buildings* (American Society of Civil Engineers 2009) permits the use of reinforced concrete, masonry, structural steel, and timber in blast-resistant construction. Of these materials, only the two most common materials, namely, structural steel (Section 4.2) and reinforced concrete (Section 4.3) are discussed in this chapter. For each of these materials, constitutive models that account for strain-rate and thermal effects, procedures to account for strain-rate effects at the macro level, mechanical properties for design, and failure modes at the component level are discussed. Only those grades of steel, concrete and rebar materials, and components used in the construction of typical buildings are discussed. Steel plates, light-gage and cold-formed framing are not structural components in typical buildings, and the interested reader is directed to UFC 3-340-02 (U.S. Department of Defense 2008) for relevant information.

Section 4.4 discusses the use of strength-reduction factors for the computation of strength of reinforced concrete and steel components and provides the rationale for the use of $\phi = 1$in the ASCE Standard.

A list of key references and resource documents is provided at the end of this chapter.

4.2 STRUCTURAL STEEL

4.2.1 Stress-Strain Relationships

Generic uniaxial stress-strain relationships are available for structural steels in textbooks (e.g., Bruneau et al. 1997, Brockenbrough and Merritt 2006) and on the World Wide Web. Sample stress-strain relationships are presented in Figure 4.1 (from Brockenbrough and Merritt 2006). Such relationships are established by ASTM-standard tests and can be described using yield stress (σ_y), tensile (or ultimate) strength (σ_u), yield strain (ε_y), strain at the onset of material hardening, strain at tensile strength, and strain at rupture. ASTM specifications for structural steels (e.g., A36, A500, A572, A913, and A992) specify minimum values for yield and tensile strengths and strain at rupture. Limits can be set on

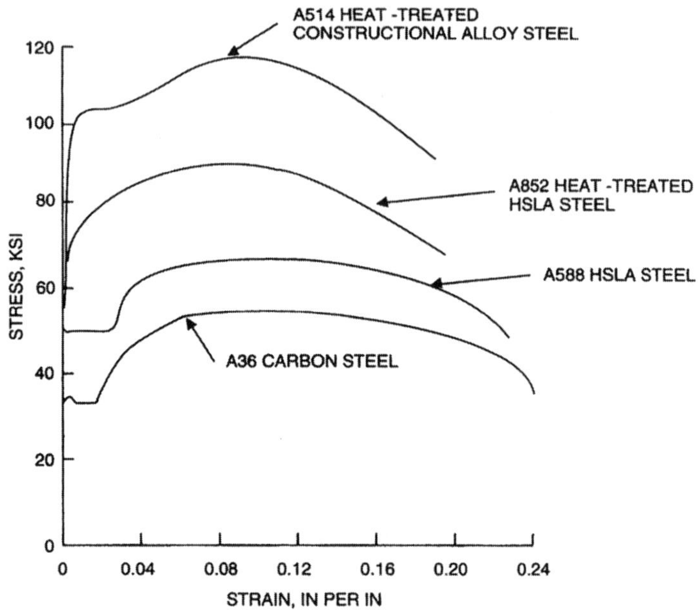

Figure 4.1 Stress-Strain Relationships for Selected Structural Steels at Room Temperature and Low Strain Rate (Brockenbrough and Merritt 2006, with permission)

maximum yield strength and maximum ratios of yield-to-tensile-strength can be specified for selected ASTM-designated steel. Tests to determine stress-strain relationships, $\sigma = f(\varepsilon)$, are generally conducted at low speeds and produce results at which strain-rate effects are considered negligible. Data from such relationships, after modification for strain-rate effects, are routinely used to compute mechanical properties for design.

Rapid or high strain-rate loading affects the mechanical properties of structural steels. High strain-rate data are reported in the literature (e.g., Campbell and Ferguson 1970, Brockenbrough and Merritt 2006, Chang and Lee 1987, Klepaczko 1994, Lee and Liu 2006). The typical effects of increased strain rate on the response of structural steels are an increase in yield stress; an increase in ultimate strength, albeit smaller than for yield stress; and a reduction in the elongation at rupture. The elastic modulus is not substantially affected by strain rate. Sample data for selected steels at three temperatures, from Brockenbrough and Merritt (2006), are presented in Figure 4.2.

4.2.2 Constitutive Models for Structural Steel

Finite element analysis of steel components and structures for blast effects generally involves the use of either solid or shell elements to construct a beam (column) cross section and span or a beam-column element. If solid (shell) elements

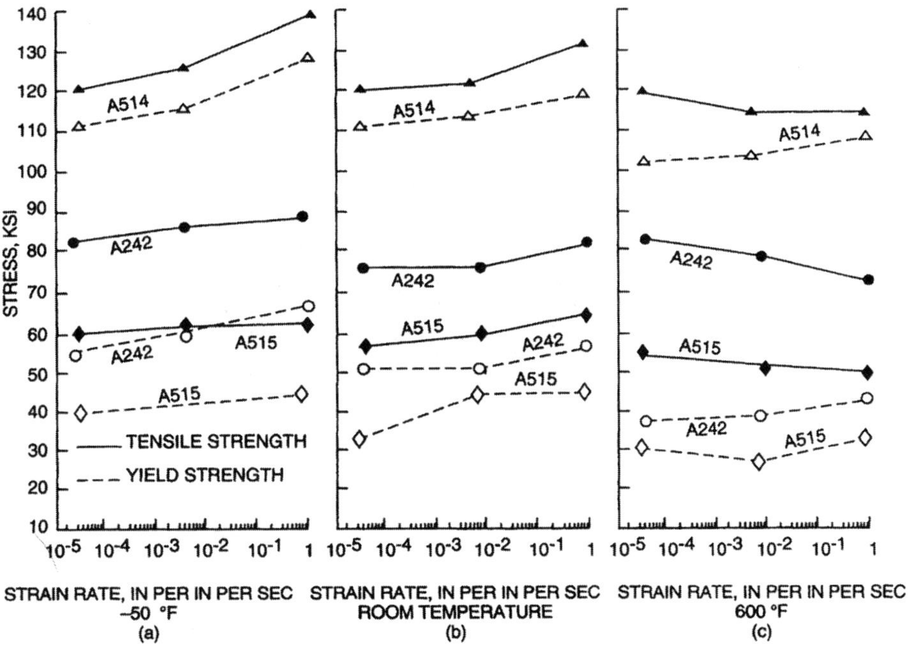

Figure 4.2 Effects of Strain Rate on Yield Stress and Tensile Strength at −50°F, Room Temperature, and +600°F (Brockenbrough and Merritt 2006, with permission)

are used to discretize a cross section, as shown in Figure 4.3, an appropriate constitutive model must be selected for the structural steel. If beam-column elements are used to describe a structural component, the effects of strain rate and temperature must be accommodated indirectly, as described in Section 4.2.3.

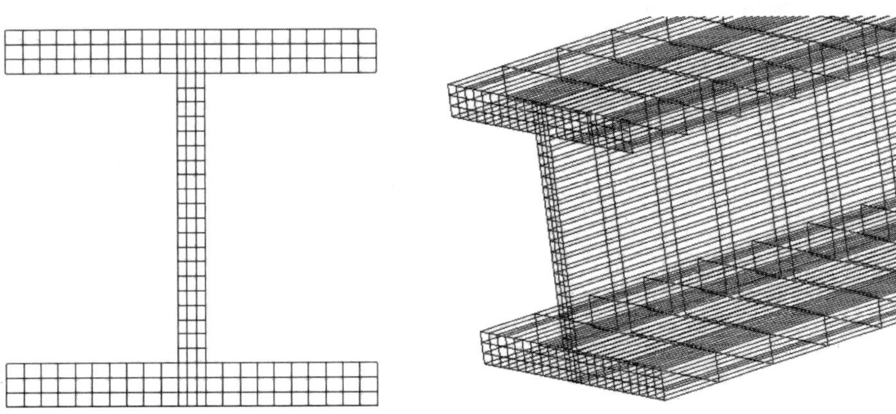

Figure 4.3 Finite Element Model of a W-Shape Steel Cross Section

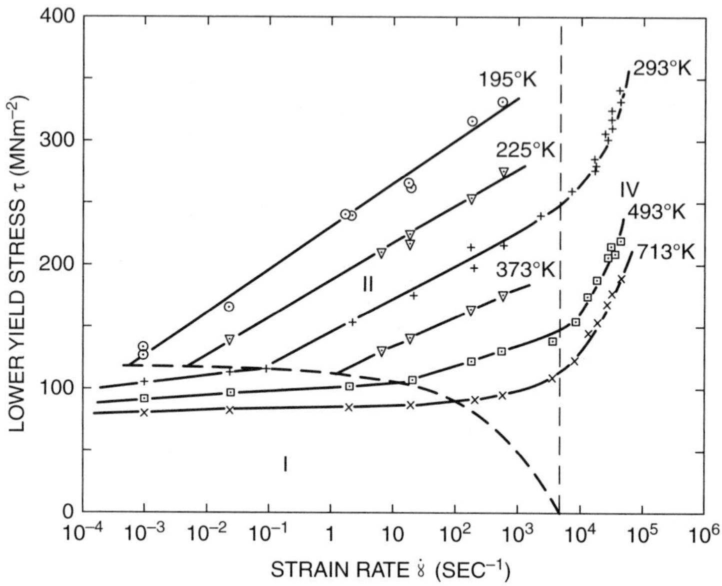

Figure 4.4 Effect of Strain Rate on Mild Steel (Campbell and Ferguson 1970, with permission)

Most of the finite element codes used for commercial blast analysis of structural components and systems include a family of constitutive models for metals that relate stress (σ) and strain (ε), strain rate ($\dot{\varepsilon}$), and temperature (T), namely,

$$\sigma = f(\varepsilon,\ \dot{\varepsilon},\ T) \tag{4.1}$$

Some of these models are empirical, with coefficients established by curve-fitting to experimental data. Empirical models seek to reproduce test data collected at alternate strain rates and temperatures, such as the dataset of Figure 4.4 for mild steel reported by Campbell and Ferguson (1970). Others are physically based. Meyer (1992) and Meyers (1994) identify many of the constitutive models in use today.

One of the most widely used empirical models is that of Johnson and Cook (1983) who proposed that the basic relationship between stress and strain at low strain rate

$$\sigma = \sigma_0 + A\varepsilon^a \tag{4.2}$$

be modified by independent terms related to strain rate and temperature, namely,

$$\sigma = (\sigma_0 + A\varepsilon^a)(1 + B\mathrm{In}\frac{\dot{\varepsilon}}{\dot{\varepsilon}_0})(1 - [T^*]^b) \tag{4.3}$$

where σ_0 is the yield stress, a is a hardening coefficient, A and B are experimentally determined factors, $\dot{\varepsilon}_0$ is a reference strain rate, T^* is a normalized temperature calculated per Equation 4.4, and b is a coefficient determined by curve-fitting to data.

$$T^* = \frac{T - T_r}{T_m - T_r} \tag{4.4}$$

In Equation 4.4, T is the temperature at which the stress is calculated, T_r is a reference temperature at which the yield stress is measured, and T_m is the melting point. The five parameters (σ_0, A, B, a, b) in the model of Equation 4.3 are determined experimentally. Johnson and Cook (1983), Nicholas and Rajendran (1990) and Meyers (1994) report values of these parameters for a number of materials, but not structural grades of steel used in buildings, bridges, and infrastructure. The Johnson-Cook model treats the effects of strain rate and temperature independently, which may be invalid for many metals at very high strain rates (Zukas 2004, Nicholas and Rajendran 1990).

Other constitutive models are physically based and provide a relationship between stress and strain that accounts for the effects of lattice structure, grain size, strain rate, and temperature. Two models that are implemented in LS-DYNA (Livermore Software Technology Corporation 2003) are Zerilli-Armstrong (Zerilli and Armstrong 1987, Lee and Liu 2006) and Mechanical Threshold Stress, both of which are described by Meyers (1994) in some detail.

The use of either empirical or physically-based constitutive models in finite element analysis enables the user to track the effects of strain rate and temperature over the depth and length of a cross section at each time step in the analysis.

4.2.3 Component Level Strain Rate and Temperature Effects

Most applications of the Standard will involve the use of simplified nonlinear models of structural components and indirect treatment of strain-rate effects. Thermal effects are generally ignored because (a) most computer codes used to compute reflected pressure histories on components do not return temperature histories, and (b) the time frame for the conduction of heat through the depth of a steel component is often much longer than the duration of the imposed overpressure.

The most widely used nonlinear component model has a single-degree-of-freedom (SDOF). The properties of the SDOF system are computed using procedures developed by Biggs (1964) and his co-workers at the Massachusetts Institute of Technology in the 1950s. These procedures can accommodate pinned and fixed boundary conditions and typically assume a uniform distribution of applied pressure, which is suitable for far-field detonations only. The response of the nonlinear SDOF system is a function of the mass per unit length, flexural stiffness, span, boundary conditions, history of reflected pressure, and component resistance function, which is a function of strain rate.

Consider the cross section of Figure 4.3 and assume a linear distribution of strain across the depth of the section. For W-shape cross sections subjected to major-axis bending, axial strain and strain rate will be greatest at the exterior edges of the flanges and zero at the neutral axis—strain rate will vary over the depth of the cross section. For W-shape sections subjected to axial compression or tension, axial strain and strain rate will be uniform across the depth of the section.

UFC 3-340-02 (U.S. Department of Defense 2008) provides guidance on the inclusion of strain-rate effects in the computation of component axial and flexural strength in the form of stress-increase factors. Information is provided for ASTM A36 (carbon structural steel), A588 (high-strength, low-alloy structural steel up to 345 MPa minimum yield point), and A514 (high-yield-strength quenched and tempered alloy steel plate) steels, for flexure (bending) and axial force (tension or compression) and low-pressure (far-range detonation, incident overpressure less than 100 psi, lower strain rate) and high-pressure (near-range detonation, incident overpressure much greater than 100 psi, higher strain rate) loadings.

The UFC presents a simple equation for strain rate, which is assumed to be constant from zero strain to the yield strain. The equation is reproduced in Equation 4.5, below. The assumption of constant strain rate to yield is not unreasonable given the coarseness of the analysis that generally accompanies the use of the stress-increase factors.

$$\dot{\varepsilon} = \frac{f_{ds}}{Et_e} \tag{4.5}$$

where $\dot{\varepsilon}$ is the average strain rate in the elastic range of response, E is the elastic modulus for steel (29,000 ksi, 200 GPa), t_e is the time to yield in the cross section, and f_{ds} is the dynamic design stress, which is a function of minimum specified yield stress, a stress-increase factor, and a factor to relate expected strength to minimum specified strength that varies as a function of steel grade. The dynamic design stress is discussed further in Section 4.2.4

Stress-increase factors, which are termed Dynamic Increase Factors (DIF) in the UFC and denoted by the symbol c, are used to increase the yield stress and tensile strength of structural steels. Table 4.1, which is adapted from UFC 3-340-02, presents values for c. The values of c that are applied to yield stress are larger for flexure than axial force because blast-induced axial forces are likely applied indirectly through beam and girder reactions for which the rise time will be longer and the load effects staggered in time. If a component is subjected to combined axial force and bending moment (e.g., a column subjected to bending due to direct air-blast loading on the column flange and axial force due to indirect loading through air-blast-induced beam/girder reactions), one value of c should be used for component strength calculations and deformation calculations, unless the component yield strength can be updated step-by-step in the analysis. The chosen value of c should reflect the dominant contributor to the combined stress

Table 4.1 Dynamic Increase Factors for Structural Steel (Adapted from DoD 2008)

ASTM Grade	Specified Minimum Yield Stress[1] (ksi)	DIF, c, for Yield Stress: Flexure, Low Pressure[2]	Flexure, High Pressure[2]	Axial, Low Pressure[3]	Axial, High Pressure[3]	DIF, c, for Tensile Strength
A36	36	1.29	1.36	1.19	1.24	1.10
A588[4]	50	1.19	1.24	1.12	1.15	1.05
A514	100	1.09	1.12	1.05	1.05	1.00

[1]Typical value for standard W-shape sections.
[2]Assumed strain rate of 0.10 s^{-1} and 0.30 s^{-1} for low and high pressure, respectively.
[3]Assumed strain rate of 0.02 s^{-1} and 0.05 s^{-1} for low and high pressure, respectively.
[4]The values presented in the shaded cells are estimates only.

ratio in the beam-column. The values of c for ultimate or tensile strength are assumed to be independent of strain rate and are close to unity.

The dynamic increase factors presented in the last column of Table 4.1 are close to unity and could be set equal to unity with no loss of robustness in the resultant component design. Components subjected to blast loadings that sustain strains associated with the tensile strength of the material likely do so near the stage of maximum displacement, where the velocity and thus strain rate are close to or equal to zero.

It is worthwhile to compare the strain-rate multipliers in UFC 3-340-02 of Table 4.1 with data and equations provided by others. Moncarz and Krawinkler (1981) provide an equation for the yield strength of ASTM A36 steel as a function of strain rate in the range of 0.0002 to 0.1 s^{-1}, namely:

$$\sigma_y = \sigma_0(0.973 + 0.45\dot{\varepsilon}^{0.33}) \tag{4.6}$$

where σ_0 is the yield stress at a (low) strain rate of 0.0002, and all other terms have been defined previously. Using this equation, the ratio of yield stress at $\dot{\varepsilon} = 0.0002$ to $\dot{\varepsilon} = 0.1\ s^{-1}$ is 1.2. For the data of Campbell and Ferguson of Figure 4.4, the ratio of yield stress at $\dot{\varepsilon} = 0.0002$ to $\dot{\varepsilon} = 0.1\ s^{-1}$ is also approximately 1.2 (at a temperature of 273°K or 20°C).

4.2.4 Mechanical Properties for Design

Moment-curvature relationships for steel cross sections have been developed using the data from uniaxial tension tests of steel coupons, but such relationships generally ignore the presence of residual stresses in the cross section that result from the fabrication of the rolled section (Bruneau et al. 1997). Simplified bilinear moment-curvature relationships are generally sufficient for blast analysis computations, as described below.

The flexural, axial, and shear strengths of steel components are typically defined in terms of cross section properties (e.g., area, web area, elastic moduli, plastic moduli) and either yield stress or tensile strength. For blast-resistant design in accordance with the Standard, these stresses can be increased to a dynamic design stress by factors to account for strain-rate effects (see Section 4.2.2) and material strengths in excess of minimum specified stresses or strengths.

UFC 3-340-02 presents a widely accepted procedure for computing so-called dynamic design stresses. The focus of the UFC presentation is flexure. Alternate computations are made for ductility ratios of less than or equal to 10 and ductility ratios greater than 10. Therein, ductility is defined as the ratio of maximum deflection to yield deflection. Although large lateral deflections may be acceptable to a designer, careful consideration must be given to geometry of the cross section at these deflections, because the cross section may no longer be prismatic due to either element deformation due to direct air-blast loading (see Figure 4.5) or cross section instability caused by flange (web) local buckling or lateral-torsional buckling. If the component cross section is distorted, the design equations presented herein, in UFC 3-340-02 and the AISC *Steel Construction Manual* (American Institute of Steel Construction 2006), cannot be used for component checking.

For ductility ratios of 10 and less, the dynamic design stress, f_{ds}, can be computed per Equation 4.7 as the product of the minimum specified yield stress, f_y, a ratio of the expected yield stress to the minimum specified yield stress, R_y, and a Dynamic Increase Factor, c, namely,

$$f_{ds} = f_{dy} = f_y R_y c \tag{4.7}$$

where R_y can be taken as either (a) 1.1 for f_y of 50 ksi (345 MPa) or less and 1.0 otherwise, or (b) per Table I-4-1 of the AISC *Seismic Provisions for Steel Buildings* (American Institute of Steel Construction 2005) for new structural steel, c is given in Table 4.1 or established on the basis of test data, and f_{dy} is the dynamic yield stress. For evaluation of older structural members constructed with A36 steel, R_y should be set equal to 1.1 unless test data indicate that a greater value is appropriate.

For ductility ratios of greater than 10, the UFC acknowledges the increase in steel stress due to strain hardening per Equation 4.8. The relationship between f_{ds} and ductility ratio must be considered approximate because displacements and peak strains do not scale linearly. Importantly, the technical basis for the increase in stress beyond the dynamic yield stress is not provided in the UFC.

$$f_{ds} = f_{dy} + \frac{f_{du} - f_{dy}}{4} \tag{4.8}$$

where f_{dy} is the dynamic yield stress per Equation 4.7, and f_{du} is the dynamic ultimate stress. In the UFC, the dynamic ultimate stress is taken as the product

Figure 4.5 Steel Cross Section Distorted Due to Air-Blast Loading (Courtesy of American Institute of Steel Construction)

of the specified minimum tensile strength and c per Table 4.1. Alternately, the dynamic design stress at high deformation can also account for the expected increase in tensile strength above the specified minimum value, namely,

$$f_{ds} = f_{dy} + \frac{cR_t f_y - f_{dy}}{\alpha} \tag{4.9}$$

where values of R_y are provided in Table I-4-1 of *Seismic Provisions for Steel Buildings* (American Institute of Steel Construction 2005), and α is specified by the user, but can be taken as 4.

4.2.5 Failure Modes of Structural Components

Introduction This section of the handbook presents introductory information on typical failure modes (flexural response, shear response, combined flexural and axial response, instability) of structural shapes, with emphasis on W-shapes loaded in the plane of the web. The focus is on design of new structural components rather than assessment of existing components.

For assessment of existing structural components whose geometries and/or materials do not comply with modern standards such as AISC (American

Institute of Steel Construction 2005, 2006), the designer is encouraged to use first-principles mechanics to determine whether performance is acceptable. Knowledge of the component deformation demands may enable a relaxation of the element compactness and component bracing rules required by *Seismic Provisions for Structural Steel Buildings* (AISC 2005) and adopted by-and-large below.

Flexure Procedures for the computation of flexural strength of prismatic cross sections as a function of unbraced length of the compression flange and element compactness are well established and reported in textbooks and the literature. For the simplest case, where the unbraced length is zero, the cross section is compact, and the strain rate is low, the flexural strength is computed as the product of the plastic section modulus and the minimum specified yield stress.

For component checking involving blast load effects, the flexural strength can be calculated as the product of the plastic section modulus and the dynamic design stress per Section 4.2.3, provided the cross section is seismically compact per Table I-8-1 of AISC (2005) and laterally braced per Section 9.8 of AISC (2005). Lateral bracing of both flanges is generally required to address rebound.

Shear The capacity of a steel beam or column in shear can be computed as the product of the shear area and the dynamic yield stress in shear. For a W-shape loaded in the plane of its web, the shear area shall not exceed the web area. The dynamic yield stress in shear, f_{dv}, may be computed as

$$f_{dv} = 0.60 f_{dy} \tag{4.10}$$

where f_{dy} is calculated per Equation 4.7. At the connection of such a W-shape beam to the flange of a W-shape column, the distributions of stress and strain will not follow beam theory, and a significant percentage of the shear force in the beam will be transmitted to the column via the beam flanges (e.g., Kim et al. 2002).

Flexure and Axial Load For checking components subjected to blast load effects that produce flexure and axial load, the procedures of Section 9 of AISC (2005) and AISC (2006) should be followed to compute coexisting axial and flexural strengths.

Splices in components subjected to both flexure and axial load should develop the capacity of the cross section computed assuming the dynamic design stress.

Flexure-axial-shear interaction need not be considered for a W-shape section loaded in the plane of its web if the blast-induced shear force is less than shear capacity of the cross section calculated using Equation 4.10 and the ratio of the plastic section modulus of the web alone to that of the W-shape is small.

Element and Component Instability Element instability (e.g., flange and web local buckling, web crippling and tearing) and/or component instability (e.g., lateral-torsional buckling) can prevent a cross section from attaining its plastic capacity.

Design and detailing of new structural components to resist blast load effects should follow the AISC *Seismic Provisions for Structural Steel Buildings* (AISC 2005) unless the component deformations are small. For component ductility demands of 3 and greater, cross sections should be seismically compact per Table I-8-1 of the AISC *Seismic Provisions*. For component ductility demands of less than 3, cross sections should be compact per Table B4.1of the AISC *Steel Construction Manual* (AISC 2006).

Crippling or tearing of webs in W-shape cross sections may result from the application of concentrated loads by other framing members. Design against crippling or tearing should conform with Section J10 of AISC (2006), with no allowance for increase in yield stress above the expected value due to strain rate because the rise time of the loading will likely be much longer than the duration of the air-blast loading.

Lateral bracing of W-shape flanges to delay lateral-torsional buckling should generally follow the rules of AISC (2005) for seismic design. Alternate limits on unbraced length that vary as a function of the component deformation are provided in UFC 3-340-02 and can be used in lieu of those presented in Section 9.8 of AISC (2005).

4.3 REINFORCED CONCRETE

4.3.1 Stress-Strain Relationships for Concrete

Many textbooks present stress-strain relationships for concrete in uniaxial compression. Sample relationships are presented in Figure 4.6 from Wight and MacGregor (2008) for concrete with compressive strength (f_c') ranging between 4500 psi (31 MPa) and 17500 psi (120 MPa). As with structural steel, the stress-strain relationships are established using an ASTM-standard test, which is conducted at a low stress rate of the order of 30 psi per second and causes failure in 90 to 180 seconds. An increase in compressive strength is accompanied by increase in the modulus of elasticity (secant modulus in the figure to a stress of approximately$0.5f_c'$), an increase in the strain at maximum stress, an increase in the slope of the softening branch of the stress-strain relationship, and a decrease in the maximum concrete strain. The modulus of elasticity in units of psi for normal-weight concrete with a density of 145 lb/ft^3 is often taken per ACI 318 (American Concrete Institute 2008) as

$$E_c = 57000\sqrt{f_c'} \tag{4.11}$$

where f_c' is in units of psi. Values of compressive strength are used to compute mechanical properties for design after modification for strain-rate effects, as described in Section 4.3.5 below.

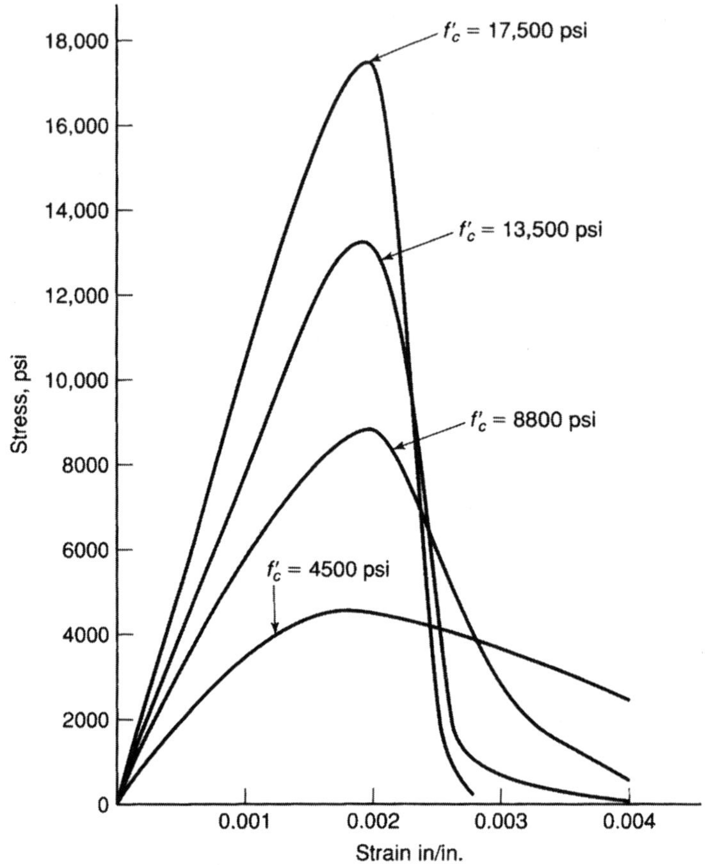

Figure 4.6 Stress-Strain Relationships for Concrete in Uniaxial Compression at Low Speed (From Wight and MacGregor 2008, with permission)

Wight and MacGregor present analytical approximations to the uniaxial stress-strain relationship for concrete, which are not repeated here, and a stress-strain relationship for normal-weight concrete in tension, which is reproduced in Figure 4.7. The tensile strength of concrete, f_t', ranges between 8 and 15 percent of the compressive strength, with the value dependant on the type of test used for the measurement (Wight and MacGregor 2008).

The stress-strain relationship for plain concrete is a function of strain rate. Bischoff and Perry (1991), Ross et al. (1995), Malvar and Ross (1998), Hentz et al. (2004), Schuler et al. (2006), and Hao and Zhou (2007), among many others, have reported on the effect of strain rate on either compressive or tensile strength. Sample results are presented in Figure 4.8 (Bischoff and Perry 1991) for compressive strength and Figure 4.9 (Schuler et al. 2006) for tensile strength. The vertical axis in each figure is the ratio of dynamic stress to static stress.

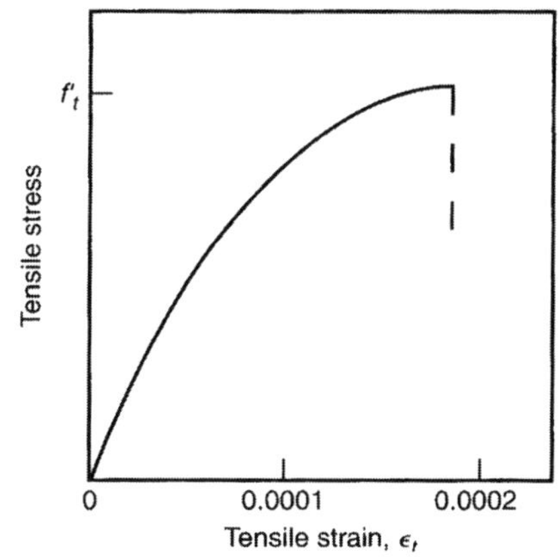

Figure 4.7 Generic Stress-Strain Relationship for Concrete in Tension (From Wight and MacGregor 2008, with permission)

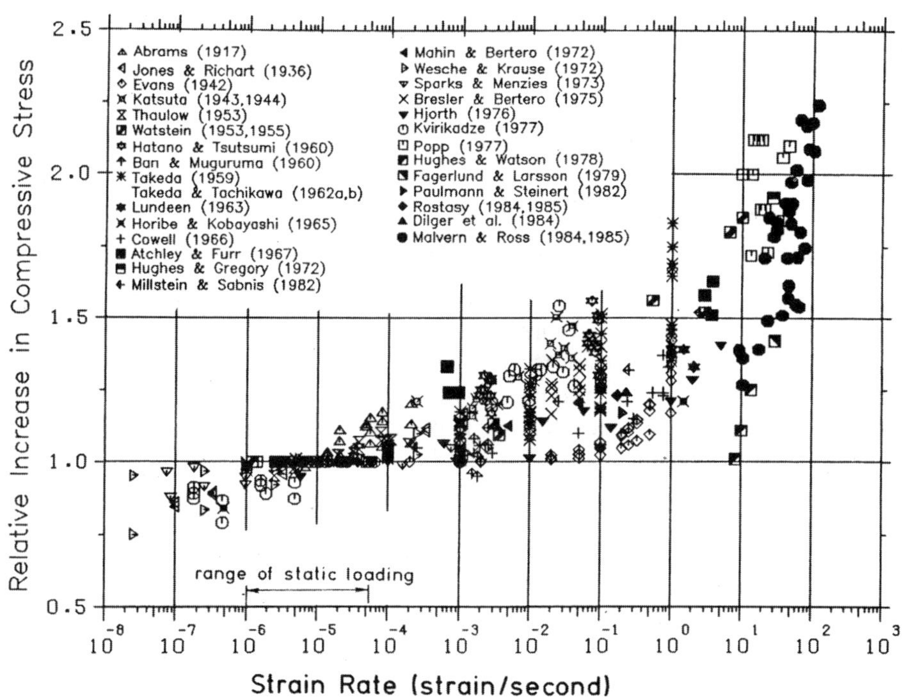

Figure 4.8 Effect of Strain Rate on Compressive Strength (From Bischoff and Perry 1991, with permission)

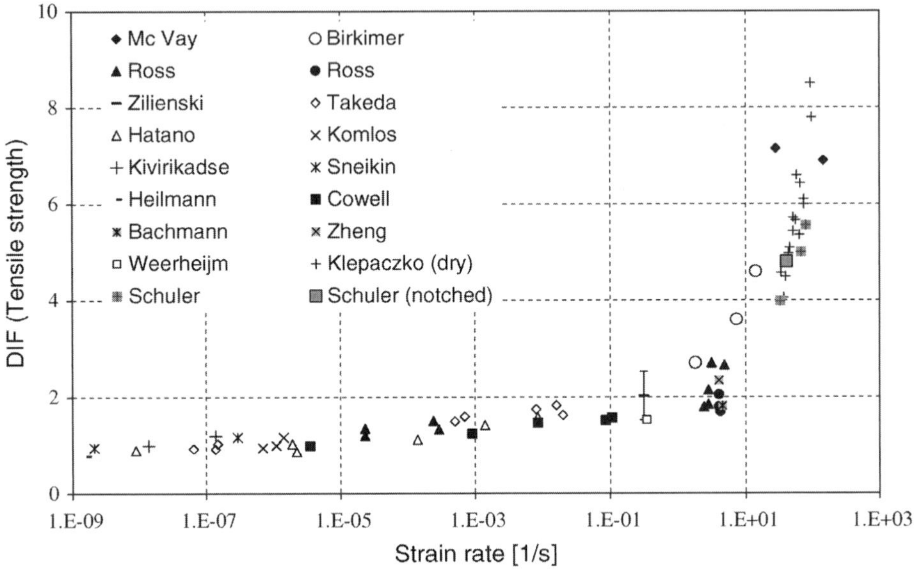

Figure 4.9 Effect of Strain Rate on Tensile Strength of Concrete (From Schuler et al. 2006, with permission)

Appreciable increases in both compressive and tensile strength are evident at strain rates greater than 0.1 s^{-1}.

4.3.2 Stress-strain Relationships for Reinforcement

Typical stress-strain relationships for steel reinforcement (or rebar) are presented in standards and textbooks. Figure 4.10 presents a typical stress-strain relationship for Grade 60 rebar from Malvar (1998). Grade 60 reinforcement is produced in accordance with a number of ASTM standards, including ASTM A615 and ASTM A704. ASTM A706 rebar is often used for seismic and blast applications where weldability, ductility, and bendability are important (Wight and MacGregor 2008).

As with structural steel, the stress-strain relationship for steel reinforcement is a function of strain rate. Sample data from Malvar (1998) showing the effect of strain rate on yield stress and tensile strength are presented in Figure 4.11 and 4.12. The greatest increases in yield stress and tensile strength are observed for the lower grades of rebar, and the percentage increases in yield stress are greater than those in tensile strength.

4.3.3 Constitutive Modeling of Concrete and Rebar

Introduction Finite element models of reinforced concrete components and structures for air-blast analysis will generally use either solid elements to mesh a

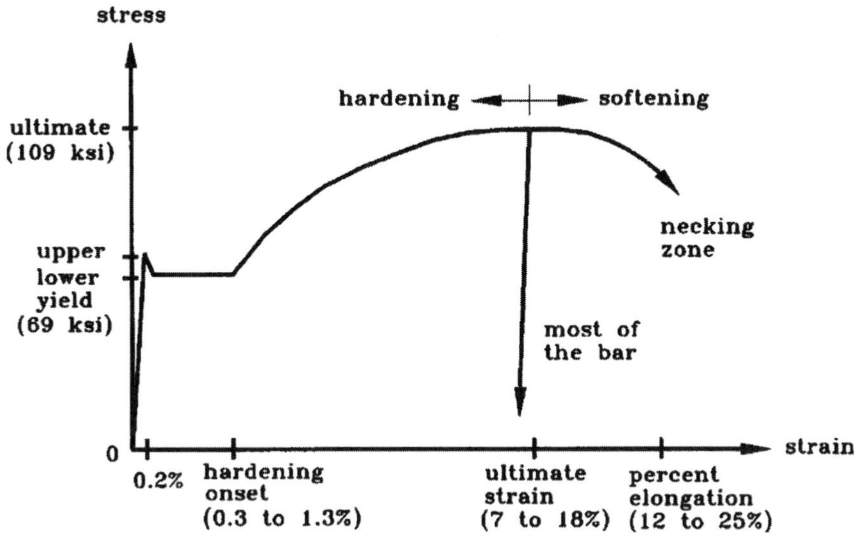

Figure 4.10 Typical Stress-Strain Relationship for Grade 60 Steel Reinforcement (From Malvar 1998, with permission)

cross section along a span or a beam-column element. If solid elements are used to discretize a cross section, appropriate constitutive models must be selected for the concrete and reinforcement, and algorithms and elements introduced to model bond between the concrete and reinforcement. The following subsections of Section 4.3.3 introduce constitutive models for plain concrete and reinforcement. If beam-column elements are used to model and analyze a reinforced

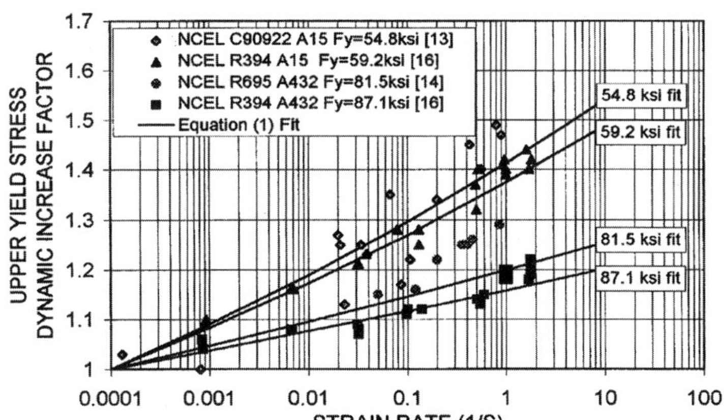

Figure 4.11 Effect of Strain Rate on Yield Stress of Steel Reinforcement (From Malvar 1998, with permission)

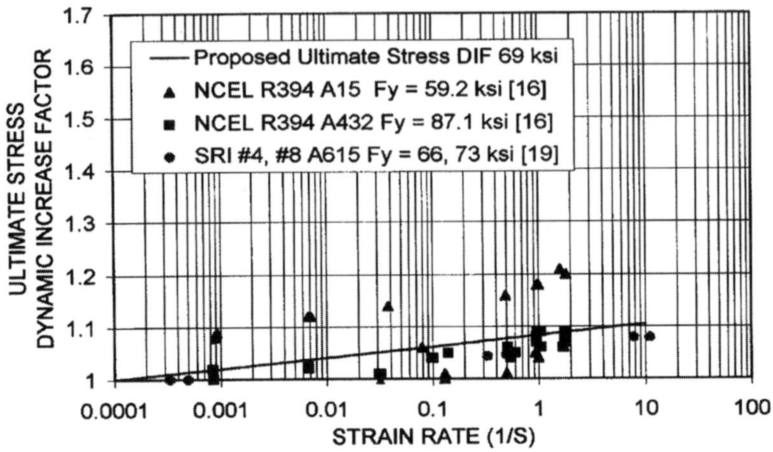

Figure 4.12 Effect of Strain Rate on Tensile (Ultimate) Strength of Steel Reinforcement (From Malvar 1998, with permission)

concrete subjected to blast loading, the effects of strain rate (and temperature if necessary) must be accommodated indirectly, as described in Section 4.3.4.

Plain Concrete Many material models for plain concrete at low strain rate have been developed. Hao and Zhou (2007) note that these concrete material models are often modified for high strain-rate loadings by multiplying important low strain-rate properties such as compressive and tensile strength by a dynamic increase factor that is strain-rate-dependent. The interested reader is directed to the literature, including Century Dynamics (2005), LSTC (Livermore Software Technology Corporation 2003), Hao and Zhou (2007), and Tu and Lu (2009), for information on these models, their implementation in hydrocodes and finite element codes, and detailed bibliographies. Only increases in compressive and tensile strength due to strain-rate effects are summarized below.

Recommendations for the compressive strength and/or tensile strength of concrete as a function of strain rate can be found in Soroushian et al. (1986), CEB (Comité Euro-International du Béton 1993), Malvar and Ross (1998), and Hao and Zhou (2007), among others.

For compressive strength, CEB (1993) writes that the ratio of dynamic to static strength ($C(\dot{\varepsilon})$), is

$$\begin{aligned} C(\dot{\varepsilon}) = \frac{f_{cd}}{f_{cs}} &= \left(\frac{\dot{\varepsilon}}{\dot{\varepsilon}_s}\right)^{1.026\alpha} \qquad \text{for } \dot{\varepsilon} \leq 30\ \text{s}^{-1} \\ &= \gamma \left(\frac{\dot{\varepsilon}}{\dot{\varepsilon}_s}\right)^{0.33} \qquad \text{for } \dot{\varepsilon} > 30\ \text{s}^{-1} \end{aligned} \tag{4.12}$$

where f_{cd} is the dynamic compressive strength at strain rate $\dot{\varepsilon}$ in the range of $30 \times 10^{-6}\text{s}^{-1}$ to $300\,\text{s}^{-1}$, f_{cs} is the static compressive strength at a reference strain rate $\dot{\varepsilon}_s$ of $30 \times 10^{-6}\text{s}^{-1}$, $\log \gamma = 6.156\alpha - 2$, $\alpha = 1/(5 + 9f_{cs}/f_{co})$, and $f_{co} = 1450$ psi (=10 MPa). Malvar and Crawford (1998) noted that the CEB model for compressive strength properly fitted the data available at the time of their writing.

Gebbeken and Rupert (2000) proposed the multiplier of Equation 4.13 for strength increases in compression (C) and tension (T),

$$C(\eta) = T(\eta) = \left\{(1 + \tanh[0.4(\log \eta - 2)]) \times \left(\frac{F_m}{W_y} - 1\right)\right\} W_y \tag{4.13}$$

where η is a strain rate normalized by a reference value of 1 s^{-1}, log is the common logarithm, F_m is a variable that limits the increase at very high strain rates, and W_y is a variable that controls the shape of the function. Values of the variables are different in tension and compression (Gebbeken 2009).

Johnson and Holmquist (1992) presented the multiplier of Equation 4.14 for strain-rate-related strength increases in tension and compression,

$$C(\eta) = T(\eta) = 1 + c\text{Ln}(\eta) \tag{4.14}$$

where c is a constant equal to 0.007 (Hao and Zhou 2007), Ln is the natural logarithm, and all terms have been defined previously. An identical equation was adopted by Riedel et al. (1999) for the RHT model and implemented in AUTODYN (Century Dynamics 2005).

For tensile strength, CEB (Comité Euro-International du Béton 1990) writes that the ratio of dynamic to static strength ($T(\dot{\varepsilon})$), is

$$\begin{aligned} T(\dot{\varepsilon}) = \frac{f_{td}}{f_{ts}} &= \left(\frac{\dot{\varepsilon}}{\dot{\varepsilon}_s}\right)^{1.016\delta} && \text{for } \dot{\varepsilon} \leq 30\ \text{s}^{-1} \\ &= \beta\left(\frac{\dot{\varepsilon}}{\dot{\varepsilon}_s}\right)^{0.33} && \text{for } \dot{\varepsilon} > 30\ \text{s}^{-1} \end{aligned} \tag{4.15}$$

where f_{td} is the dynamic tensile strength at strain rate $\dot{\varepsilon}$ in the range of $3 \times 10^{-6}\text{s}^{-1}$ to $300\,\text{s}^{-1}$, f_{ts} is the static tensile strength at a reference strain rate $\dot{\varepsilon}_s$ of $3 \times 10^{-6}\ \text{s}^{-1}$, $\log \beta = 7.112\delta - 2.33$, $\delta = 1/(10 + 6f_{cs}/f_{co})$, and $f_{co} = 10\,\text{MPa}$ (=1450 psi). Malvar and Ross (1998) modified the CEB equations as follows,

$$\begin{aligned} T(\dot{\varepsilon}) = \frac{f_{td}}{f_{ts}} &= \left(\frac{\dot{\varepsilon}}{\dot{\varepsilon}_s}\right)^{\delta} && \text{for } \dot{\varepsilon} \leq 1\ \text{s}^{-1} \\ &= \beta\left(\frac{\dot{\varepsilon}}{\dot{\varepsilon}_s}\right)^{0.33} && \text{for } \dot{\varepsilon} > 1\ \text{s}^{-1} \end{aligned} \tag{4.16}$$

where f_{td} is the dynamic tensile strength at strain rate $\dot{\varepsilon}$ in the range of 10^{-6}s^{-1} to $160\ \text{s}^{-1}$, f_{ts} is the static tensile strength at a reference strain rate $\dot{\varepsilon}_s$of 10^{-6}s^{-1}, $\log \beta = 6\delta - 2$, $\delta = 1/(1 + 8 f_{cs}/f_{co})$, and $f_{co} = 10$ MPa (=1450 psi). Hao and Zhou (2007) proposed an alternate formulation based on the data of Figure 4.9, namely,

$$\begin{aligned} T(\dot{\varepsilon}) &= 1.0 && \text{for } \dot{\varepsilon} \leq 1\ \text{s}^{-1} \\ &= 2.06 + 0.26 \log(\dot{\varepsilon}) && \text{for } 10^{-4}\ \text{s}^{-1} \leq \dot{\varepsilon} \leq 1\ \text{s}^{-1} \\ &= 2.06 + 2.0 \log(\dot{\varepsilon}) && \text{for } 1\ \text{s}^{-1} < \dot{\varepsilon} \leq 10^{3}\ \text{s}^{-1} \end{aligned} \tag{4.17}$$

where all terms are defined previously.

Rebar Materials models for rebar are generally identical to those for structural steels as described in Section 4.2.2: Zhou et al. (2008) used the Johnson and Cook (1983) material model to characterize the high-strain-rate response of rebar in slabs constructed of conventional concrete and steel-fiber-reinforced concrete.

4.3.4 Component Level Strain-Rate Effects

Similar to steel structures, most reinforced concrete components will be analyzed and designed for air-blast effects using nonlinear single-degree-of-freedom (SDOF) models and an indirect treatment of strain-rate effects, as introduced in Section 4.2.3.

UFC 3-340-02 (U.S. Department of Defense 2008) provides guidance on the inclusion of strain-rate effects for the computation of component axial, flexural, and shear strength. The tensile strength of concrete is ignored for axial and flexural strength calculations. Strain-rate effects are addressed using stress-increase or dynamic increase factors for material strength. Information is provided for concrete compressive and tensile strength and steel reinforcement yield stress and tensile (ultimate) strength in alternate formats, including functions of strain-rate and either far-design (low pressure) or close-in design (high-pressure) loadings.

As with calculations for structural steel, the UFC presents simple calculations for average strain rate for concrete and rebar, which is assumed to be constant. The calculations are based on an estimate of the time to yield of reinforcement and the strain in the materials at yield of the cross section. For rebar, the average strain rate is computed as

$$\dot{\varepsilon}_s = \frac{f_{dy}}{E t_e} \tag{4.18}$$

where $\dot{\varepsilon}_s$ is the average strain rate in the elastic range of response, E is the elastic modulus for steel rebar (29,000 ksi, 200 GPa), t_e is the time to yield of the cross section (generally computed using response charts), and f_{dy} is the dynamic yield stress, which is a function of minimum specified yield stress, a stress (dynamic) increase factor, and a factor to relate expected strength to minimum specified strength that varies as a function of rebar grade. The variable f_{ds} is the rebar equivalent of the dynamic design stress for structural steel that was introduced in Section 4.2.4. For concrete, the average strain rate, $\dot{\varepsilon}_c$, is computed as

$$\dot{\varepsilon}_c = \frac{\varepsilon_{c0}}{t_e} \tag{4.19}$$

where ε_{c0} is the strain at maximum concrete stress, which can be taken as 0.002 (see Figure 4.6).

Calculations of strain rate made using Equations 4.18 and 4.19 should be used with care, in the knowledge that they are at best approximate. Consider a singly reinforced concrete beam for which the neutral axis depth at yield of the cross section is approximately 0.30 times the effective depth of the tension rebar, and the yield strain in the rebar after accounting for strength in excess of the nominal yield stress and rate effects is 0.0025: the maximum concrete strain at yield of the tension rebar is approximately 0.0011, and the concrete strain at the approximate location of the resultant compressive force is approximately 0.0007, and substantially smaller than the 0.002 recommended by the UFC. Alternate values of strain rate for rebar and concrete would be calculated for doubly reinforced cross sections (with differing values of strain rate in the tension and compression rebar), columns subjected to blast loads producing flexure and compression or tension, walls, and slabs.

Table 4.2, which is adapted from UFC 3-340-02, presents Dynamic Increase Factors (DIF) for rebar (yield stress and tensile strength) and concrete in compression for flexure (bending), axial compression, shear (diagonal tension failure), shear (direct tension), and bond. The UFC writes that the values given in the table for flexure assume a strain rate of 0.10 s^{-1} and 0.30 s^{-1} for both rebar and concrete in the low-pressure (far-field) and high-pressure (near-field) ranges, respectively; for reinforced concrete members in compression (i.e., columns), the

Table 4.2 Dynamic Increase Factors for Reinforced Concrete (Adapted from DoD 2008)

	Low Pressure			High Pressure		
	Rebar			Rebar		
Action	Yield	Tensile	Concrete	Yield	Tensile	Concrete
Flexure	1.17	1.05	1.19	1.23	1.05	1.25
Compression	1.10	1.00	1.12	1.13	1.00	1.16
Shear—DT	1.00	1.00	1.00	1.10	1.00	1.00
Shear—Direct	1.10	1.00	1.10	1.10	1.00	1.10
Bond	1.17	1.05	1.00	1.23	1.05	1.00

corresponding strain rates are 0.02 s^{-1} and 0.05 s^{-1}, respectively. These default strain rates are identical to those assumed for structural steel and reported in the footnotes to Table 4.1.

As with components of structural steel, the UFC recommends greater increases for rate effects for flexure than compression, noting that "slabs, beams and girders filter the dynamic effects of the blast load.... Dynamic load reaching columns... [has]... a long rise time of load... [which] results in lower strain rate." Further, the UFC writes that the values for shear and bond are smaller than those for bending or compression for the purpose of "the need to prevent brittle shear and bond failure and to account for uncertainties in the design process for shear and bond."

It is not clear whether the use of three significant figures in Table 4.2 is warranted. Given the range of strain rate associated with the categories of low and high pressure, the use of two significant figures is likely appropriate. For lower levels of protection, reinforced concrete components are permitted to sustain maximum deformations associated with the tensile strength of the rebar. However, component velocity and thus strain rate at the stage of maximum deformation are zero, and material dynamic increase factors should be set to unity.

If a reinforced concrete component is subjected to combined axial force and bending moment (e.g., a column subjected to bending due to direct air-blast loading on the column flange and axial force due to indirect loading through air-blast-induced beam/girder reactions), one value of the dynamic increase factor should be used for component strength calculations and deformation calculations, unless the component yield surface can be updated step-by-step in the analysis. The chosen value of the factor should reflect the dominant contributor to the combined strength ratio in the beam-column.

4.3.5 Mechanical Properties for Design

The flexural, axial, and shear strength of reinforced concrete components are typically computed using cross section properties and rebar stresses and concrete strengths. For blast-resistant design in accordance with the Standard, these stresses and strengths can be increased to a dynamic design stress by factors to account for strain-rate effects (see Section 4.3.3) and material strengths in excess of minimum specified values.

UFC 3-340-02 presents procedures for computing dynamic design stresses and strengths. As for structural steel, the stresses and strengths are tied to deformations but support rotation computed using SDOF analysis is used instead of ductility ratio (see Section 4.2.4). Table 4.3 presents the UFC recommendations for dynamic design stresses and strengths for the purpose of design, organized by action: flexure, shear (diagonal tension and direct), and compression. For flexure and support rotations, θ_m, of less than 2 degrees, compression rebar can be ignored for the calculation of flexural strength; for support rotations in excess of 2 degrees, the concrete is assumed to have crushed, and the flexural strength is provided by a couple between the tension and compression rebar. An increase

Table 4.3 Dynamic Design Stresses for Reinforced Concrete (Adapted from DoD 2008)

Action	Rebar Type	Support Rotation, Degrees	Dynamic Design Stress: Rebar, f_{ds}	Dynamic Design Stress: Concrete, f'_{dc}
Flexure	Tension and Compression	$0 < \theta_m \leq 2$	f_{dy} [1]	f'_{dc}
		$2 < \theta_m \leq 6$	$f_{dy} + (f_{du} - f_{dy})/4$	[2]
		$6 < \theta_m \leq 12$	$(f_{du} + f_{dy})/2$	[2]
Shear—DT[3]	Stirrups	$0 < \theta_m \leq 2$	f_{dy}	f'_{dc}
		$2 < \theta_m \leq 6$	f_{dy}	f'_{dc}
		$6 < \theta_m \leq 12$	f_{dy}	f'_{dc}
Shear—DT	Lacing	$0 < \theta_m \leq 2$	f_{dy}	f'_{dc}
		$2 < \theta_m \leq 6$	$f_{dy} + (f_{du} - f_{dy})/4$	f'_{dc}
		$6 < \theta_m \leq 12$	$(f_{du} + f_{dy})/2$	f'_{dc}
Shear—Direct	Diagonal	$0 < \theta_m \leq 2$	f_{dy}	f'_{dc}
		$2 < \theta_m \leq 6$	$(f_{du} + f_{dy})/2$	[4]
		$6 < \theta_m \leq 12$	f_{dy}	[4]
Compression	All	[5]	f_{dy}	f'_{dc}

[1]For tension rebar only.
[2]Concrete assumed crushed and moment resistance provided by tension-compression rebar couple.
[3]Diagonal tension.
[4]Concrete considered ineffective and shear resisted by diagonal rebar only.
[5]Component capacity not a function of support rotation.

in support rotation corresponds to an increase in rebar strain (and stress), as indicated in the table. The equation used to compute dynamic design stress for support rotations $2 < \theta_m \leq 6$ degrees is identical to Equation 4.8 for structural steel at large deformations.

Dynamic design stresses (strength) for rebar are computed in a similar manner to that described in Section 4.2.4 for structural steel. Equation 4.7 is used to compute the dynamic yield stress. For ASTM A615 Grade 60 and A706 rebar, the UFC recommends R_y equal to 1.1, which results in an expected yield stress of 66 ksi (455 MPa). Again, as with structural steel, the dynamic ultimate stress for rebar is taken as the product of the specified minimum tensile strength and a dynamic increase factor. For concrete, the dynamic compressive strength is taken as the product of the specified (static) compressive strength and a dynamic increase factor; no allowance is made for concrete strength in excess of the minimum specified value.

The dynamic increase factors for steel rebar and concrete can be established using the values provided in Table 4.2, using the data presented in Sections 4.3.1 and 4.3.2, or using charts presented in the UFC, two of which are reproduced below for concrete compressive strength (Figure 4.13) and rebar yield stress (Figure 4.14). The curves provided in these figures express dynamic increase

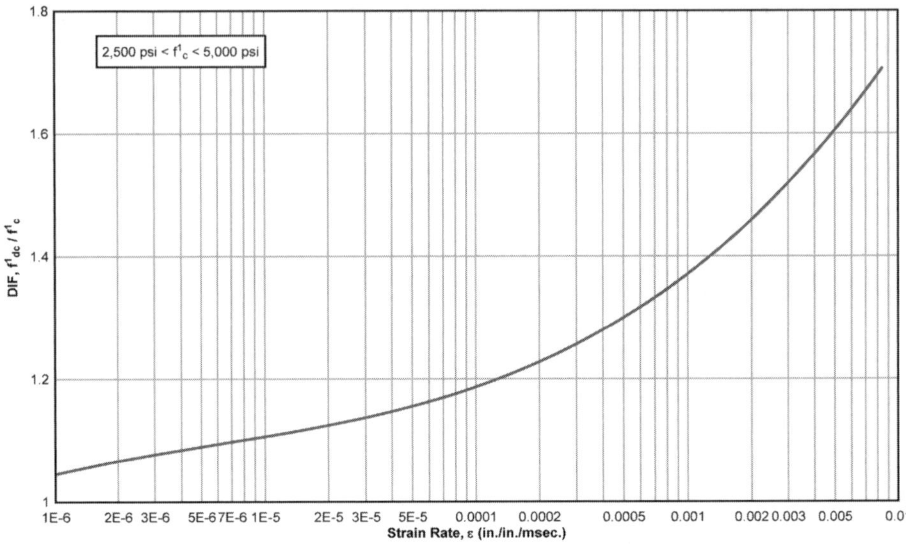

Figure 4.13 Effect of strain rate on compressive strength of concrete with f_c' in the range of 2500 to 5000 psi (From DoD 2008)

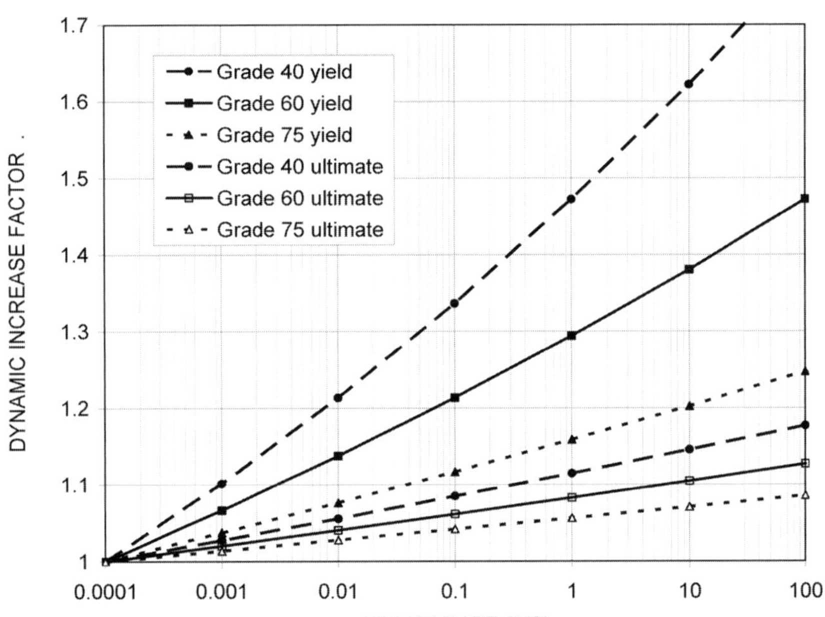

Figure 4.14 Effect of strain rate on the yield stress and tensile strength of rebar (From DoD 2008)

factors as a function of strain rate. Figure 4.13 represents a reasonable lower bound on the data of Bischoff and Perry and the CEB recommendations of Equation 4.12 for a concrete strength of 50 MPa (7250 psi) and strain rates of 10 s^{-1} and less. Note that the strain rate in this figure is presented in units of msec. The curves of Figure 4.14 for the yield and tensile strength of rebar are in close agreement with the recommendations of Malvar (1998).

4.3.6 Component-Level Failure Modes

Introduction This section of the handbook presents introductory information on typical failure modes (flexural response, shear response, combined flexural and axial response) of reinforced concrete members. The focus is on design of new structural components rather than assessment of existing components. The reader is referred to the literature for information on external reinforcement of concrete elements for blast resistance using steel plate and carbon fiber polymeric materials (e.g., Crawford et al. 2001, Buchan and Chen 2007).

Exactly as with structural steel, the designer is encouraged to use first-principles mechanics to determine whether performance is acceptable. The failure mechanisms considered here assume that the core of the reinforced concrete cross section is intact. Spalling and scabbing may occur, and the depth of reinforced sections should be restricted to core dimensions in such cases. Reinforced concrete components are susceptible to brisance, or shattering, for near-field detonations. The likelihood of shattering should be checked, and the reader is directed to the literature for appropriate procedures.

The following subsections address the common failure modes of reinforced concrete components under blast loading. Much additional information on failure modes can be found in the literature, including UFC 3-340-02 (U.S. Department of Defense 2008), Krauthammer's *Modern Protective Structures* (Krauthammer 2008), Mays and Smith's *Blast Effects on Buildings* (Mays and Smith 1995), and Smith and Hetherington's *Blast and Ballistic Loading of Structures* (Smith and Hetherington 1994).

Flexure Procedures for the computation of flexural strength of prismatic cross sections are well established. Flexural failure can occur in either brittle or ductile modes, depending upon the volume of steel reinforcement in the flexural element. Economical design of reinforced concrete components to resist blast loads generally requires ductile detailing to accommodate the large expected deformations while maintaining near-peak strength. Confinement in the form of seismic hoops or lacing may be required.

For component checking involving blast effects, the flexural strength can be computed using the dynamic design stresses of Table 4.3 and compression-tension force couples. If the support rotations are less than 2 degrees, flexural strength is computed using traditional procedures that assume that the tension rebar yields prior to the concrete in the compression block crushing. For support rotations in the range of 2 to 6 degrees, the cover concrete is assumed to be lost, equal amounts of compression and tension rebar should be provided, and the flexural strength is computed using a force-couple involving the compression and tension rebar. For support rotations in excess of 6 degrees and less than 12 degrees, equal amounts of compression and tension rebar must be provided, the flexural strength is computed as a force-couple between the compression and tension rebar, and the rebar layers must be adequately tied. Shear reinforcement in the form of ties (stirrups) or lacing is generally required to sustain support rotations greater than 2 degrees. UFC 3-340-02 provides guidance on the design of such reinforcement.

For robustness, the minimum longitudinal reinforcement in slabs and beams must meet the requirements of ACI 318 for load combinations not involving blast effects. UFC 3-340-02 also requires a minimum amount of tension and compression flexural rebar to be provided in slabs, but not less than that required to resist the effects of direct air-blast loading and rebound. These limits for slabs tend to be smaller than those of Section 10.5 of ACI 318, although a minimum amount of compression rebar must be provided in all cases.

Shear Four modes of shear failure in reinforced concrete components must be addressed: (1) diagonal tension, (2) diagonal compression, (3) direct shear, and (4) punching shear. Critical cross sections are at a distance (a) equal to the effective depth of the cross section from the support for diagonal tension and compression checks for loadings that produce compression in the support and at the face of the support otherwise, and (b) at the face of the support for direct shear. Punching shear is checked for slabs around isolated supports (e.g., columns) and applied loads. Concrete and rebar strengths for component checking can be increased for strain-rate effects per Tables 4.2 and 4.3.

Design against failure by diagonal compression or tension follows the procedure set forth in Chapter 11 of ACI 318. Diagonal compression failure is avoided in beams (slabs) equipped with stirrups by limiting the nominal shear stress at the critical section to $10\sqrt{f'_{dc}}$, where f'_{dc} is the concrete dynamic design stress per Tables 4.2 and 4.3. Diagonal tension failure is avoided by ensuring that the sum of the shear resistance provided by the concrete and the stirrups exceeds applied shear force. ACI 318 Sections 11.2.1.1, 11.2.1.2, and 11.2.2.3.can be used to compute the shear resistance provided by concrete under no axial load, axial compression, and axial tension, respectively. ACI 318 Sections 11.4.7.2 and 11.4.7.4 present equations for establishing the required area of shear reinforcement perpendicular to the axis of the member (i.e., stirrups) and inclined to the axis of the member (i.e., lacing bars), respectively. The effective depth of the section used in these calculations of concrete and rebar shear resistance is a function of the expected support rotation and the likelihood of loss of cover concrete. Section 4-18.2 of UFC 3-3340-02 provides appropriate guidance for the calculation of effective depth.

UFC 3-340-02 requires that shear reinforcement be provided in slabs that are required to develop large deflections (associated with support rotations greater than 2 degrees). Shear reinforcement must always be provided to resist shear stresses in excess of the shear capacity of the concrete. Section 4.18.4 of the UFC provides appropriate guidance for minimum shear reinforcement in slabs.

Direct shear failure has been observed in blast testing of reinforced concrete slabs. Failure occurs along a vertical plane at load or stiffness discontinuities such as the intersection of a slab and its supporting beam or wall. Krauthammer (2008) provides a discussion of direct shear failure and presents a model to predict the relationship between imposed shear force and deformation (slip across a vertical plane). Section 4.19 of UFC 3-340-02 presents design equations for direct shear capacity that includes contributions from concrete and diagonal (inclined) bars. (Stirrups that are placed perpendicular to the plane of a slab provide no resistance to direct shear because the failure plane is vertical.) Concrete is assumed to provide resistance to direct shear per Equation 4.20 if the support rotations are less than 2 degrees (i.e., there is little to no damage to the cover concrete) and the slab is not subjected to net tension:

$$V_c = 0.16 f'_{dc} bd \tag{4.20}$$

where b is the slab width and d is the effective depth of the cross section. For support rotations greater than 2 degrees, the direct shear must be resisted by diagonal (inclined) reinforcing bars only, where the required rebar area per width b is

$$A_d = \frac{(V_u - V_c)}{f_{ds} \sin \alpha} \tag{4.21}$$

where A_d is the area of inclined rebar per width b, α is the angle between the inclined rebar and the plane of the longitudinal rebar, V_u is the direct shear force per width b, V_c is the concrete resistance width b, and f_{ds} is the dynamic design stress for the inclined rebar per Table 4.3. The UFC sets no upper bound on the applied direct shear stress, likely because the direct shear force is resisted by inclined bars only for shear stress in excess of $0.16 f'_{dc}$. Further, the UFC equations do not recognize the benefit of axial compression on the cross section, which will be present in some components (e.g., columns) and not in others (e.g., slabs). Upper bounds on direct shear stress are identified in the literature with sample values of $0.35 f'_c$ (Ross and Krawinkler 1985) for static loads and $0.6 f'_{dc}$ (Crawford et al. 2001) for blast loadings. Normal compressive force could be included in the formulation using an approach similar to the shear-friction design method of Section 11.6 of ACI 318 or as proposed by Crawford et al. (2001).

Punching shear must be checked for slabs supported on isolated supports such as columns. The procedures of Chapter 11 of ACI 318 should be used to design components to avoid punching shear failure. Section 4.20.1 of UFC 3-340-02 provides appropriate guidance for the calculation of effective depth for the punching shear check.

Flexure and Axial Load Reinforced concrete components subjected to flexure and axial load can fail due to either excessive concrete strain (a compression-dominated component) or rebar rupture (flexure-dominated component). The latter can be treated as a flexural element with consideration of axial load. The procedures of Chapter 10 of ACI 318 should be used to proportion axial-load-dominated components. Appropriate consideration should be given to strain-rate effects per Section 4.3.4 to compute material stresses for component strength calculations.

Reinforced concrete components such as slabs may be restrained against movement in their plane by adjacent framing. Such restraint can give rise to membrane effects. Krauthammer (2008) notes that compressive membrane forces at relatively low lateral displacements can enhance the ultimate capacity of a slab by increasing the flexural strength of the slab at the critical sections. At large lateral displacements, associated with the loss of core concrete, resistance to blast loadings can be provided by the layers of rebar acting as a cable net. For such a net to form, the rebar must be appropriately developed into the supports, and the adjacent framing must be capable of developing the associated reactions.

Scabbing Scabbing is the loss or separation of cover concrete from the core concrete associated with large deformations in the component.

Spalling Spalling is similar to scabbing but occurs as a result of the reflection of a compressive shock wave from the rear face of a concrete component. If the resulting tensile stresses exceed the tensile strength of the concrete, concrete will spall. The likelihood of spalling can be computed using codes such as CONWEP (Hyde 1992).

4.4 STRENGTH-REDUCTION FACTORS FOR STEEL AND REINFORCED CONCRETE

Strength-reduction factors, denoted ϕ in most materials standards, are used to pre-multiply *nominal* component strength to compute *design* strength. Design strengths are generally compared with factored load effects to determine the adequacy of a design. The values assigned to strength-reduction factors range between 0.6 and 0.9, with smaller values assigned to component actions that are brittle (e.g., shear) and larger values to component actions that are ductile (e.g., flexure in the absence of axial force).

There is no industry-wide consensus on the values of ϕ for blast-resistant design. Traditional practice has set ϕ equal to 1.0, and this decision can be rationalized on the bases that (a) the weapon effect (air-blast) is idealized and likely conservative, especially for near-field detonations, (b) equivalent SDOF models are used for load-effect calculations, and (c) strain-rate effects are estimated conservatively.

Blast analysis calculations performed using a finite element code use material constitutive models to assess stress and strain at the cross section level, deformations at the component level, and displacements at the global level. Best estimate (mean) material properties are used as input to these constitutive models with no reduction to design values, that is, $\phi = 1$.

REFERENCES

American Concrete Institute (ACI). 2008. *Building Code Requirements for Structural Concrete* (ACI-318-08) *and Commentary* (ACI 318R-08). Farmington Hills, MI: American Concrete Institute.

American Institute of Steel Construction (AISC). 2005. *Seismic Provisions for Structural Steel Buildings* (ANSI/AISC-341s1-05). Chicago, IL: American Institute of Steel Construction.

______. 2006. *Steel Construction Manual.* 13th Ed. Chicago, IL: American Institute of Steel Construction.

______. 2009. *Blast Protection of Buildings* (ASCE Standard). Reston, VA: American Society of Civil Engineers.

Biggs, J. 1964. *Introduction to Structural Dynamics.* New York: McGraw-Hill.

Bischoff, P. H. and S. H. Perry. 1991. Compressive behavior of concrete at high strain rates. *Materials and Structures* 24 (6): 425–450.

Brockenbrough, R. L. and F. S. Merritt. 2006. *Structural Steel Designers Handbook.* New York: McGraw-Hill.

Bruneau, M., C. M. Uang, and A. S. Whittaker. 1997. *Ductile Design of Steel Structures.* New York: McGraw-Hill.

Buchan, P. A. and J. F. Chen. 2007. Blast resistance of FRP composites and polymer strengthened concrete and masonry structures—A state-of-the-art review. *Composites Part B: Engineering* 38 (5–6): 509–522.

Campbell, J. D. and W. G. Ferguson. 1970. The temperature and strain rate dependence of the shear strength of mild steel. *Philosophical Magazine* 21 (169): 63–82.

Century Dynamics. 2005. *AUTODYN Users Manual Version 6.1.* Horsham, West Sussex, UK: Century Dynamics.

Chang, K. C. and G. C. Lee. 1987. Strain rate effect on structural steel under cyclic loading. *Journal of Engineering Mechanics* 113(9): 1292–1301.

Comité Euro-International du Béton (CEB). 1993. *CEB-FIP Model Code 1990.* Trowbridge, Wiltshire, UK: Redwood Books.

Crawford, J. E., L. J. Malvar, K. B. Morrill, and J. M. Ferritto. 2001. Composite retrofits to increase the blast resistance of reinforced concrete buildings (P-01-13). Proceedings of the 10th International Symposium on Interaction of the Effects of Munitions with Structures 2001. San Diego, CA.

Gebbeken, N. and M. Ruppert. 2000. A new material model for concrete in high dynamic hydrocode simulations. *Archive of Applied Mechanics* 70 (7): 2509–2524.

Gebbeken, N. 2009. Personal communication.

Hao, H. and X. Q. Zhou. 2007. Concrete material model for high-rate dynamic analysis. Proceedings of the 7th International Conference on Shock and Impact Loads on Structures, Beijing, October 2007.

Hentz, S. F., V. Donze, and L. Daudeville. 2004. Discrete element modelling of concrete submitted to dynamic loading at high strain rates. *Computers and Structures* 82 (29 30): 2509–2524.

Hyde, D. W. 1992. *User's Guide for Microcomputer Program CONWEP, Application of TM5-855-1: Fundamental of Protective Design for Conventional Weapons*. Instruction Report SL-88-1. Vicksburg, MS: Structures Laboratory, U.S. Army Waterways Experiment Station.

Johnson, G. R. and W. H. Cook 1983. A constitutive model and data for metals subjected to large strains, high strain rates and high temperatures" Proceedings of the 7th International Symposium on Ballistics 1983. The Hague, Netherlands.

Johnson, G. R. and T. J. Holmquist. 1992. A computational constitutive model for brittle materials subjected to large strains, high strain rates and high pressures. In *Shock-Wave and High-Strain-Rate Phenomena in Materials*, 1075–1081, edited by M. A. Meyers, L. E. Murr, and K. P. Staudhammer. New York: Dekker.

Kim, T., A. S. Whittaker, A. J. S. Gilani, V. V. Bertero, and S. M. Takhirov. 2002. Cover-plate and flange-plate steel moment-resisting connections. *Journal of Structural Engineering* 128 (4): 483–491.

Klepaczko, J. R. 1994. An experimental technique for shear testing at high and very high strain rates. The case of a mild steel. *International Journal of Impact Engineering* 15 (1): 25–39.

Krauthammer, T. 2008. *Modern Protective Structures*. Boca Raton, FL: CRC Press.

Lee, W.-S. and C.-Y. Liu. 2004. The effects of temperature and strain rate on the dynamic flow behavior of different steels. *Materials Science and Engineering A* 426 (1–2): 101–113.

Livermore Software Technology Corporation (LSTC). 2003. *LS-DYNA User's Manual*. Livermore, CA: Livermore Software Technology Corporation.

Malvar, L. J. 1998. Review of static and dynamic properties of steel reinforcing bars. *ACI Materials Journal* 95 (5): 609–616.

Malvar, L. J. and J. E. Crawford. 1998. Dynamic increase factors for concrete. Proceedings of the 28th Department of Defense Explosives Safety Seminar 1998. Orlando, FL.

Malvar, L. J. and C. A. Ross. 1998. Review of strain rate effects for concrete in tension. *ACI Materials Journal* 95 (6): 735–739.

Mays, G. C. and P. D. Smith. 1995. *Blast Effects on Buildings*. London: Telford.

Meyer, L. W. 1992. Constitutive equations at high strain rates. In *Shock-Wave and High-Strain-Rate Phenomena in Materials*, 49–68, edited by M. A. Meyers, L. E. Murr, and K. P. Staudhammer. New York: Dekker.

Meyers, M. A. 1994. *Dynamic Behavior of Materials*. New York: Wiley.

Moncarz, P. and H. Krawinkler. 1981. Theory and application of experimental model analysis in earthquake engineering (Report No. 50). Palo Alto, CA: John A. Blume Earthquake Research Center, Stanford University.

Nicholas, T. and A. M. Rajendran. 1990. Material characterization at high strain rates. Chapter 3 in *High Velocity Impact Dynamics*, edited by J. Zukas. New York: Wiley-Interscience.

Reidel, W., K. Thoma, S. Hiermaier, and E. Schmolinske. 1999. Penetrating of reinforced concrete by BETA-B-500 numerical analysis using a new macroscopic concrete model for hydrocodes. Proceedings of the 9th International Symposium on the Interaction of the Effects of Munitions with Structures 1992, 315–322, Berlin.

Ross, C. A., J. W. Tedesco, and S. T. Kuennen. 1995. Effects of strain rate on concrete strength. *ACI Materials Journal* 92 (1): 37–47.

Ross, T. J. and H. Krawinkler. 1985. Impulsive direct shear failure in RC slabs. *Journal of Structural Engineering* 111 (8): 1661–1676.

Schuler, H., C. Mayrhofer, and K. Thoma. 2006. Spall experiments for the measurement of the tensile strength and fracture energy of concrete at high strain rates. *International Journal of Impact Engineering* 32 (10): 1635–1650.

Smith, P. D. and J. G. Hetherington. 1994. *Blast and Ballistic Loading of Structures*. Oxford: Butterworth-Heinemann.

Soroushian, P., K. B. Choi, and A. Alhamad. 1986. Dynamic constitutive behavior of concrete. *ACI Journal* 83 (2): 251–259.

Tu, Z. and Y. Lu. 2009. Evaluation of typical concrete material models used in hydrocodes for high dynamic response simulations. *International Journal of Impact Engineering* 36 (1): 132–146.

U.S. Department of Defense (DoD). 2008. *Structures to Resist the Effects of Accidental Explosions* (United Facilities Criteria UFC 3-340-02). Washington, D.C.: U.S. Department of Defense.

Wight, J. K. and J. G. MacGregor. 2008. *Reinforced Concrete: Mechanics and Design*. 5th Ed. Upper Saddle River, New Jersey: Prentice Hall.

Zerilli, F. J. and R. W. Armstrong. 1987. Dislocation-mechanics-based constitutive relations for material dynamics calculations. *Journal of Applied Physics* 61 (5): 1816–1825.

Zhou, X. Q., V. A Kuznetsov, H. Hao, and J. Waschl. 2008. Numerical prediction of concrete slab response to blast loading. *International Journal of Impact Engineering* 35 (10): 1186–1200.

Zukas, J. A. 2004. *Introduction to Hydrocodes*. Oxford: Elsevier.

5 Performance Verification

Curt Betts

5.1 INTRODUCTION

The purpose of this chapter is to familiarize designers who are interested in blast design with verification of performance for blast and weapons effects and for stopping vehicles using barriers. The purpose is not to provide guidance on how to do any of that testing. It is merely to make designers understand when verification and validation should be provided, to make them effective consumers of performance verification services, and to ensure they understand how to validate claims made either by other designers or by vendors. This chapter will address verification by both testing and analysis. It will also address the topic of peer review.

5.2 PERFORMANCE VERIFICATION

Before the subject of performance verification can be addressed, it is important to point out that verification has a specific meaning in common use, and it is different from validation. Both involve confirmation, but by convention they confirm different things in different ways. While their differences in meaning have been the subject of controversy for decades, the following convention will be used in this chapter.

Verification refers to the practice of ensuring that a model works for a given set of conditions. For example, in blast-resistant designs, members may be assumed to be loaded uniformly within a particular scaled range, and they may be assumed to respond to those loadings in flexure. If a member is closer than that scaled range, models based on uniform loadings or flexural responses are not applicable for the member response. Those models cannot be verified for that application. Verification can also apply to ensuring that a given configuration falls within the assumptions for meeting a particular performance condition. For example, for tensile membrane action to be applicable, certain detailing needs to be verified. Yet another way of applying verification is by checking that all design requirements have been addressed, that calculations are correct, and that the design actually reflects what the design analysis says it should.

Validation refers to the practice of ensuring that models are technically correct by comparing them to other validated models or test results. For example, there are many validated single-degree-of-freedom models for use in blast-resistant design. Any new model should be validated against previously validated models by running identical member configurations and loadings to ensure that both codes give similar answers. Alternatively, code results can be compared against test data to ensure that the models give answers that are consistent with the test data. Of course, that assumes that the test data are themselves validated or verified, which will lead to the next section of this chapter.

5.3 TESTING

This section will address testing of both building components and vehicle barriers. In both cases there are validated test procedures that can be used to evaluate performance. Those test procedures are commonly based on some sort of standard test, such as one included in a national standard.

5.3.1 Vehicle Barrier Testing

Vehicle barriers are discussed in Chapter 12, where they are described as anti-ram structures. Due to the complexities of the structural systems associated with such barriers, the soil–structure interaction, and the vehicle–structure interaction, designing vehicle barriers solely through analytical means is not sufficient for validation. Analytical means can be used to design the barriers, but proper validation requires those models to be validated by vehicle testing.

Any validation testing of vehicle barriers must be based on a standard test method to be repeatable, to avoid bias, and to ensure that common performance measures are reported for all vehicle barriers. One such test standard is ASTM F2656, *Vehicle Crash Testing of Perimeter Barriers*. It, like any good standard test, includes specific requirements on how the tests must be set up, how the tests are run, and how performance is measured. Unlike some standard test methods, there is a degree of subjectivity to vehicle barrier tests, which is controlled by the test director who has to decide such things as where the barrier is impacted and then to determine the validity of the test. Issues such as those are why such standards commonly require testing laboratories to be accredited. Figure 5.1 shows the end result of a vehicle barrier test.

ASTM F2656, as an example of a test standard, includes details on the vehicles themselves, including gross vehicle weight, age of the vehicles, vehicle soundness, how ballast needs to be assembled and attached, and in the case of one of the vehicles, the wheel base. The test standard further establishes requirements for the information to be provided to the test director on the barrier and information on how the barrier needs to be installed. The standard further establishes detailed criteria for measuring the performance of the barriers, including the information that has to be collected and evaluated by the test director. That

Figure 5.1 Vehicle Barrier Test

information includes vehicle acceleration, velocity, penetration, and vehicle and barrier deformation. It also includes the apparatus necessary to measure those quantities.

While test standards like ASTM F2656 provide for validation of designs, there is also an element of verification associated with the barriers. Verification comes into play when a barrier is specified for installation after testing. Verification at that point includes reviewing the barrier drawings to ensure that the barrier that is being installed is the barrier that was tested, that it is installed in a manner that is consistent with the tested configuration, and that its deployment will result in its being employed in a manner that is consistent with the way it was tested.

5.3.2 Building Components

Testing of building components similarly requires standard testing procedures to ensure that tests are run consistently and with standard, comparable performance measures. Tests can be run at full scale or sub-scale, and using either actual explosives or blast simulators.

Full-scale testing is the most accurate method to test building components. It involves building a component to be tested or an entire building to the actual scale with the threat explosive, and at the applicable standoff distance. Through

Figure 5.2 Full–Scale Building Testing

such tests the full-scale behavior of the building or building component can be measured or observed to validate compliance with a performance standard.

One common means to set up such a test is to build a building or a test frame in which the articles to be tested are installed, and to place the threat explosive at the standoff distance at which the response is to be observed. In the case of building a whole building, its response and the response of its components are directly and unambiguously observed. Figure 5.2 shows a full-scale building that was tested.

In the case of a test frame, the results may need to be evaluated carefully. Where such a test frame is narrow in comparison to how it will be installed in a real building, a phenomenon referred to as "clearing" may result in inaccuracies in the observed component response. Clearing may be an issue when reflected pressure effects are being measured. With small or narrow targets, the reflected pressure may be created on the front face of the target, but it may not persist because the finite boundaries of the target allow the blast wave to propagate around its edges such that the duration of the load in the target is reduced (the clearing time), resulting in a reduced load on the component compared to what it would see were it in a much wider or infinite surface. Clearing is not always an issue, however. For components such as windows that commonly respond very quickly to blast waves, the clearing time may not be an issue. Because of the potential uncertainties in the component response, results from tests where clearing is an issue need to be evaluated carefully. Figure 5.3 shows test frames.

Figure 5.3 Test Frames in Arena Configuration

Full-scale testing is often done in what is referred to as an arena test, as shown in Figure 5.3. In such a test, a number of test objects or test cells or frames are placed around a test explosive in roughly a circular pattern. Different test specimens may be at different standoff distances from the explosive, allowing multiple results from the same test.

Sub-scale testing involves reducing the scale of dimensions associated with the test, including the test specimen, the distance from the explosive to the test specimen, and the explosive itself. In those tests, dynamic similitude must be maintained, which for explosives is done according to what is called Hopkinson scaling. In Hopkinson scaling, quantities are scaled in proportion to the cube root of the explosive weight. By scaling, similar blast phenomena can be observed with smaller explosives at closer distances. For example, refer to Table 5.1 below, which shows the pressures and impulses for two different explosive weights at two different standoff distances.

Note that the scaled range (standoff distance / the cube root of the explosive weight) is the same for both tests and that the pressures are the same. That shows that a test can be run for one explosive weight and can then be scaled upward or downward to reflect the effects of different explosive weights at different distances. This is a common practice in the blast community. Similarly, other dimensions such as wall heights can be scaled, allowing a wide range of options in scaled modeling and testing.

It is important to note, however, that not all quantities scale alike. Note that the impulses associated with the two equivalent pressures are not the same. This illustrates why scaled testing has to be evaluated and verified carefully. While such a pair of tests as those in Table 5.1 may be valid for components that are

Table 5.1 Hopkinson Scaled Pressures and Impulses

Quantity	Test #1	Test #2
Weight (W)	1000 lbs TNT	220 lbs TNT
Range (standoff distance)	100 ft	60.4 ft
Scaled range (Range / $W^{1/3}$)	10	10
$W^{1/3}$	10 $lb^{1/3}$	6.04 $lb^{1/3}$
Pressure	9.56 psi	9.56 psi
Impulse	80.9 psi-msec	48.7 psi-msec

most responsive to peak pressures, they may not be valid for components that are more responsive to impulse.

In either full-scale or sub-scale testing, the phenomena that are to be observed need to be measured carefully and accurately. Most importantly, it is imperative that the measurements that best reflect the response to be observed are measured. In all cases the basic geometry of the test needs to be recorded accurately, which includes the dimensions and construction of the test subject, how it installed into the test cell or target building, the construction of the test cell or building, the size and composition of the explosive, and the distance and angle of incidence from the explosive to the target. In addition, atmospheric conditions such as temperature, humidity, altitude, and wind speed and direction must be recorded.

In measuring test results, there are many measurements that a test report should include, depending on the nature of the test. Very importantly, the blast wave must be fully characterized using pressure gages and similar measuring devices. Then the test specimen response parameters must be measured. Those measurements depend on the response to be observed. If the member is intended to remain intact with either elastic or inelastic deformation, the deflection of the member must be measured. That will allow for calculation of support rotations or ductility ratios (see Chapter 3). Those are commonly measured using strain gages or displacement transducers.

If the response is expected to include failure of the test specimen with some expected hazard level of the resulting fragments, additional measurements would be required. Those would include measuring the sizes, weights, and translations of the fragments, and their accelerations. Measuring devices for those include accelerometers and high-speed digital cameras. Figure 5.4 shows a setup for both high-speed digital camera recording and deflection. The camera cannot be seen, but the stripes painted in the wall allow for accurate evaluation of the velocities of the fragments.

Another option for measuring blast response of components is to use a blast simulator. The most common blast simulators are tubes into which test specimens are mounted at one end. The blast simulating "driver" is at the end of the tube opposite the test specimen, and it commonly drives air into the test specimen at high velocities that simulate a blast wave. The measurements of the responses of the test specimens are similar to those in open-air tests. Tests in blast simulators can usually be done much less expensively and in much quicker progression than open-air blast tests, and they can be operated in areas with far fewer restrictions than those in which open-air blasts tests are run. It is important to recognize, however, that the blast simulator loadings should be validated against open-air tests to ensure that the loadings are representative of actual conditions. Figure 5.5 shows a shock tube blast simulator with a wall specimen installed into it

There are few test standards for blast testing, but two commonly used standards are ASTM test standards for doors and windows. ASTM F1642, *Glazing and Glazing Systems Subject to Airblast Loadings*, is described in Chapter 10. It describes test setup, test execution, and performance standards for windows subjected to blast loads or simulated blast loads. ASTM F2247, *Standard Test*

Figure 5.4 Digital Photography and Deflection Measurement

Figure 5.5 Shock Tube Blast Simulator

Method for Metal Doors Used in Blast Resistant Applications, provides similar standards for doors.

It is very important that test results undergo rigid verification, which includes both validation against other test results and evaluating test reports. The latter is particularly important. Somebody who understands test reports should carefully review all of the conditions under which the tests were run, how the measurements were made, and how the test specimen responded. He or she must ensure that the test was run in a valid manner and that the results are meaningful. In addition to evaluating the conduct and setup of the test, he or she must also verify that the results are valid based on issues such as accuracy of the loading, whether scaling was taken into effect correctly, and whether clearing was an issue or was accounted for. This is less difficult for tests that are conducted in accordance with a particular standard, but those tests must also be evaluated carefully. In many ways explosive testing is more of an art than a science.

5.4 ANALYSIS

There are many ways to analyze the response of building elements to blast shocks, ranging from approximations to highly detailed analyses such as finite element analysis, and for each of those there are multiple computer codes for performing that analysis. There are also hand methods and charts. Analysis is the least expensive means for evaluating performance, but because of the nonlinearity of such analyses and the uncertainties in the loadings, dynamic material properties, and dynamic structural response, analyses require stringent verification and validation.

Verification includes evaluating the applicability of specific methods to individual situations and ensuring that the models are applied correctly. Depending on the assumptions upon which models are based, they may have limited applicability. Optimally, the applicability should be verified before they are applied, but if not, their applicability must be verified in evaluating the results of the analyses.

While the applicability of a model to a given situation needs to be verified, it is at least as important that models be validated. Models can be validated by comparing their results to the results of other models that have been validated for identical situations. For example, a computer code to be validated can be used to analyze a beam, a wall, a window, or any other building component of a specific geometry and configuration, and those results can be compared to those derived by using a previously validated code. This is particularly the case with finite element analyses, where there are so many variables that have to be considered in the models that results can appear reasonable, but turn out to be fallacious.

Results can also be compared to test results, which can serve as benchmarks. In either case, validation gives the user both confidence in the model and confidence that its application is reasonable for that situation. This is particularly useful for finite element models.

In addition to analyzing buildings and building components to blast loads, there are models for analyzing vehicle barriers. Because they model such complex situations as vehicle–structure interaction, structural response, and soil–structure interaction, those models tend to be very complicated. They often include finite element analysis models. Because they're so complicated, it is especially important that they be validated. Usually that validation is done by either comparing them to existing test results or performing testing after the analysis to validate the analysis results. The most common use of analysis models for vehicle barriers, therefore, is to design the barriers and validate the results through testing.

5.5 PEER REVIEW

Because analysis and testing for blast loads involve such complexity, it is very important that test results and analyses be validated through peer review. Peer review is particularly important for structures with unique structural features or structures that have high importance or high economic value.

In addition, because the number of experienced blast consultants is still limited, peer review is particularly important where analysis is done by less experienced designers.

Verification and validation through peer review can take multiple forms. As a minimum, it should include detailed review of drawings to ensure that details are consistent with the assumptions in the design analyses and that the drawingsare properly detailed, to allow the expected performance to be achieved. Verification should also evaluate the applicability of the models used and the accuracy with which they were applied. It should also include checks of at least major elements of the design calculations.

Peer review should be done by people who have significant experience in applying the models used for the applications on which they were applied.

REFERENCES

ASTM. 2003. *Standard Test Method for Metal Doors Used in Blast Resistant Applications* (F2247). West Conshohocken, PA: ASTM International.

________. 2004. *Glazing and Glazing Systems Subject to Airblast Loadings* (F1642). West Conshohocken, PA: ASTM International.

________. 2007. *Vehicle Crash Testing of Perimeter Barriers* (F2656). West Conshohocken, PA: ASTM International.

II Blast Phenomena and Loadings

6 Blast Phenomena

Paul F. Mlakar and Darrell Barker

6.1 INTRODUCTION

Blast is a pressure disturbance caused by the sudden release of energy. People often think of blasts in terms of explosions such as the detonation of an explosive charge. However, there are many other blast sources that have the potential to cause damage. For example, chemicals may undergo a rapid decomposition under certain conditions. These events are often referred to as runaway reactions. Flammable materials mixed with air can form vapor clouds that when ignited can cause very large blasts. Blasts are not always caused by combustion; they can also result from any rapid release of energy that creates a blast wave, such as a bursting pressure vessel from which compressed air expands, or a rapid phase transition of a liquid to a gas.

The loads resulting from a blast are created by the rapid expansion of the energetic material, creating a pressure disturbance or blast wave radiating away from the explosion source, as shown in Figure 6.1. Blast pressure is more properly overpressure, because it is relative to ambient conditions, rather than an absolute pressure. Shock waves are high-pressure blast waves that travel through air (or another medium) at a velocity faster than the speed of sound. Shock waves are characterized by an instantaneous increase in pressure followed by a rapid decay. Pressure waves are lower amplitude and travel below the speed of sound. Pressure waves are characterized by a more gradual increase in pressure than a shock wave, with a decay of pressure much slower than a shock wave. In most cases, shock waves have a greater potential for damage and injury than pressure waves.

As a blast wave travels away from the source, the pressure amplitude decreases, and the duration of the blast load increases. Overexpansion at the center of the blast creates a vacuum in the source region and a reversal of gas motion. This negative pressure region expands outward, causing a negative pressure (below ambient), which trails the positive phase. The negative phase pressure is generally lower in magnitude (absolute value) but longer in duration than the positive phase. Generally speaking, positive phase blast loads are more consequential than negative phase loads, the latter of which is often ignored.

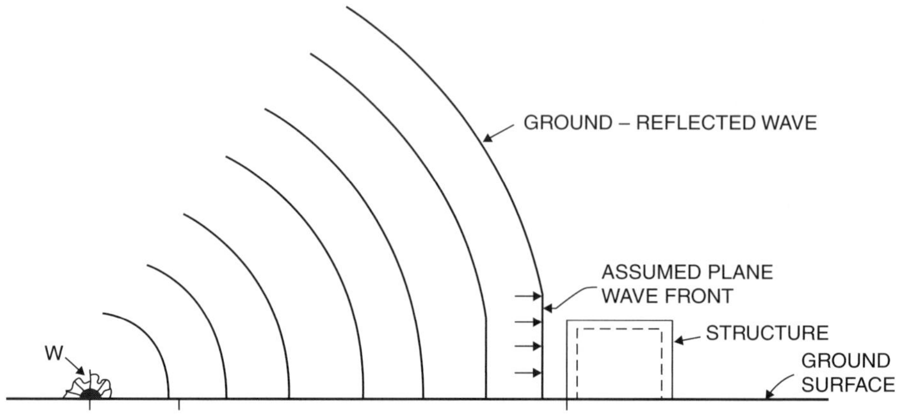

Figure 6.1 Propagating Blast Wave

Expansion of the blast wave causes air particles to move outward during the positive phase and inward during the negative phase. The flow of air particles creates a pressure analogous to that caused by wind. The pressure produced by this flow is referred to as the dynamic pressure. This pressure is lower in magnitude than the shock or pressure wave and imparts a drag load similar to wind loads on objects in its path.

As the shock or pressure wave strikes a wall or other object, a reflection occurs, increasing the applied pressure on the surface. This reflected pressure is considerably higher than the incident or free-field pressure wave. At the free edges of a reflecting surface, the discontinuity between the forward traveling incident blast wave and rearward traveling reflected blast wave creates a rarefaction, or pressure relief wave. Rarefactions travel inward from the outer edges across the face of the reflecting surface. The rarefaction waves relieve the positive reflected pressure down to the free-field or side-on pressure, plus drag pressure. The peak reflected pressure is not affected, only the duration The time required for the rarefaction waves to completely relieve the reflected pressure is termed the clearing time. This clearing time varies across the surface. It should be noted that a rarefaction wave does not instantaneously clear reflected pressure; rather, the relief is somewhat gradual and takes longer than the time required for the leading edge of the rarefaction to travel to the center of the reflected surface. If the clearing time exceeds the positive phase blast wave duration, clearing does not affect the positive phase loads.

6.2 SOURCES OF BLASTS

Blasts involving chemical reactions can be classified by their reaction rates into two primary groups: deflagrations and detonations. A *deflagration* is an

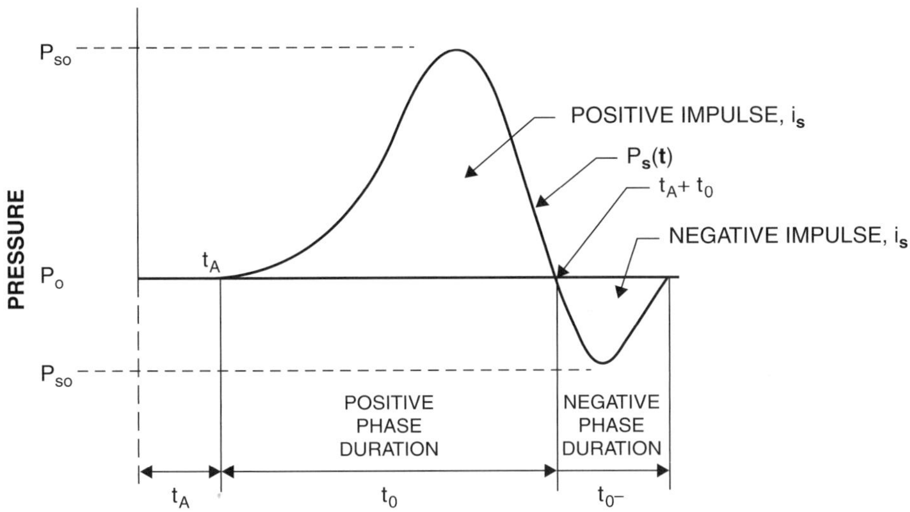

Figure 6.2 Pressure Wave from Deflagration

oxidation reaction that propagates at a rate less than the speed of sound in the unreacted material. The corresponding blast wave is often termed a *pressure* wave and has a finite rise time, as illustrated in Figure 6.2. A "fast" deflgration can create a more sudden rise in pressure. By contrast, in a *detonation*, the reaction front propagates supersonically, usually many times faster than the speed of sound. This blast wave is termed a *shock* wave and has an instantaneous rise in pressure, as seen in Figure 6.3. Since pressure is closely related to reaction rate, detonation pressures are usually many times higher than deflagration pressures.

A blast involving an explosive is an exothermic chemical reaction, usually involving an oxidizer and a fuel. Explosion reactions can be produced by a wide array of materials, some familiar, such as trinitrotoluene (TNT), while others are less well known. The reaction rate is dependent on the chemical and physical properties of the energetic material, reactant proportions and homogeneity, geometry of the material, characteristics of the "container" in which the material resides, method and energy of initiation, and other initial conditions. A deflagration may be initiated by a "soft" ignition source such as friction, spark, or open flame. Under certain conditions, a deflagration can transition into a detonation. Alternatively, detonations can be directly initiated in an explosive material that exceeds certain minimum geometry constraints if it is impinged upon by a shock source of sufficient strength.

Deflagrations and detonations may involve oxidizers and fuels that are oxygen-deficient, or materials that may produce flammable gases as a product of reaction. In either case, the unreacted or flammable products may mix with air and result in secondary burning. The secondary burning does not contribute

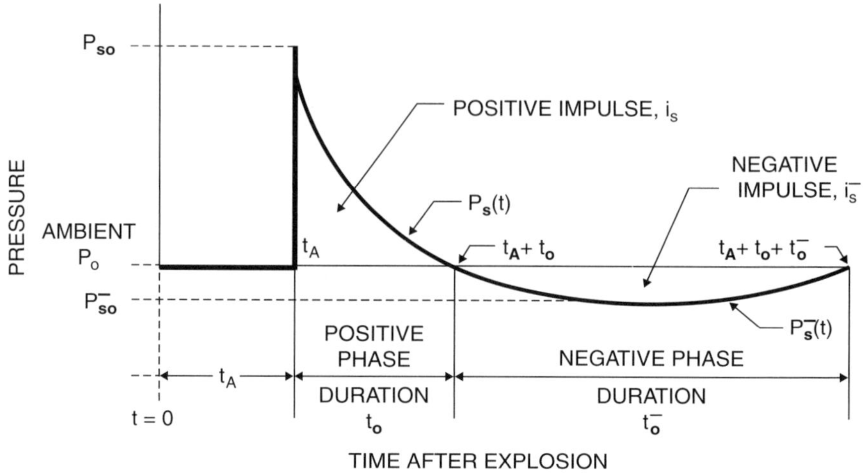

Figure 6.3 Shock Wave from Detonation

significantly to the blast pressures for external explosions but can be a major consideration for predicting internal explosion blast pressures.

Industrial explosive, propellant, and pyrotechnic manufacturing provides a wide array of energetic material configurations and materials. These products may include gas generators for airbags, packaged general-purpose explosive charges, shaped high-explosive charges for cutting, or blasting agents used in mining. Some materials used in these manufacturing processes may not be well characterized in terms of potential explosive potential or reactivity hazards. Therefore the structural analyst should take care to understand the full range of blast loads that may be produced, and to design and apply appropriate factors of safety to account for uncertainties.

Terrorist threats principally involve solid materials such as plastic explosives or improvised explosives such as ammonium nitrate and fuel oil (ANFO). These materials can readily produce detonations, but the blast strength may be low due to inefficient configuration or the presence of contaminants. Flammables can potentially be used, but they are more difficult to transport and successively initiate.

For blasts originating from a chemical reaction, the material involved and its state, the proportions of fuel and oxidizer (if applicable), container strength and configuration, and the method and energy of initiation are among the parameters that determine blast characteristics. Explosive materials can be broadly categorized based on their state. Solid materials typically produce high-pressure, short-duration loads, while combustible gases of comparable energy produce lower-pressure, longer-duration loads. Energy output and standoff distance in the configuration of interest are key to accurately determining blast loads acting on a structure. Blast sources are described by group in the following paragraphs.

High explosives are materials that are intended to produce detonation events with supersonic reaction fronts. The reactions proceeding through the explosive are self-sustaining if the charge is of sufficient diameter and properly initiated. Reaction rate, or detonation velocity, varies with material type and is a key factor in detonation pressure for a material. Detonation velocities typically range from 3000 to 30,000 ft/s (1–9 km/s). High explosives can be classified by their sensitivity to initiation into primary, secondary, and tertiary explosives. Primary explosives are very sensitive to initiation—for example, lead azide used in detonators. Secondary high explosives include TNT, composition C4, and dynamite. Tertiary explosives are much less sensitive to initiation and require a powerful booster charge to initiate them. The blasting agent ANFO is an example of a tertiary explosive.

For convenience in predicting blast pressures, the energy release of high explosives is commonly measured as a value relative to that of TNT. This "TNT equivalence" is used to determine a TNT charge weight capable of producing the same explosion energy, blast pressure, or blast impulse as the explosive of interest. The TNT equivalence is different for energy, peak pressure, and impulse, and separate TNT equivalent energy values are reported for many materials. Equivalencies for a number of these materials appear in Table 6.1.

High-explosive detonations may be accidental or intentional events. Accidents involving high explosives can occur in processing, handling, and transportation. Intentional detonations can include explosives testing, military weapons, demolition, specialized cutting or explosive forming, and terrorist acts. In the case of intentional detonations, structures, such as test structures, may be required to withstand multiple events.

Confined explosive charges will create gas pressures in the structure in addition to shock waves. If the structure is not adequately vented, the effects of these gas pressure loads may exceed those of the shock loads.

High explosives come in many forms determined by their chemical and physical properties and intended use. Most explosives have additives to aid in stabilizing the material chemically or physically. Powdered explosives may be combined with a plastic binder to form plastic-bonded or PBX explosive. Extrudable explosives have a viscous form due to the addition of rubber resins to the explosive material. This addition allows the explosive to be molded or extruded to produce a particular shape.

Certain liquids, such as nitromethane, can also detonate. Nitromethane is relatively insensitive and must be initiated by a strong ignition source such as a high-explosive booster. It does have safety advantages over other materials and may be handled as a simple flammable liquid in many cases. The shock front produced by nitromethane is well formed and produces blast loads with approximately 100% TNT equivalence.

Propellants and pyrotechnics, also known as low explosives, are energetic materials that do not typically detonate and are used to produce gas, smoke, flash, or sound. Both solid and liquid rocket propellants, gun propellants, and black powder are examples of low explosives. Pyrotechnics include many different

Table 6.1 TNT Equivalence of High Explosives

Explosive	Density Mg/m^3	Equivalent Moss for Pressure	Equivalent Mass for Impulse	Pressure Range MPa
Amatol (50/50)	1.59	0.97	0.87	NA[1]
Ammonia Dynamite (50 percent Strength)	NA[1]	0.90	0.90[2]	NA[1]
Ammonia Dynamite (20 percent Strength)	NA[1]	0.70	0.70[2]	NA[1]
ANFO (94/6 Ammonium Nitrate/Fuel Oil)	NA[1]	0.87	0.87[2]	0.03 to 6.90
AFX-644	1.75	0.73[2]	0.73[2]	NA[1]
AFX-920	1.59	1.01[2]	1.01[2]	NA[1]
AFX-931	1.61	1.04[2]	1.04[2]	NA[1]
Composition A-3	1.65	1.09	1.07	0.03 to 0.35
Composition B	1.65	1.11	0.98	0.03 to 0.35
		1.20	1.30	0.69 to 6.90
Composition C-3	1.60	1.05	1.09	NA[1]
Composition C-4	1.59	1.20	1.19	0.07 to 1.38
		1.37	1.19	1.38 to 20.70
Cycloid (75/25 RDX/TNT)	1.71	1.11	1.26	NA[1]
(70/30)	1.73	1.14	1.09	0.03 to 0.35
(60/40)	1.74	1.04	1.16	NA[1]
DATE	1.80	0.87	0.96	NA[1]
Explosive D	1.72	0.85[2]	0.81	0.01 to 0.30
Gelatin Dynamite (50 percent Strength)	NA[1]	0.80	0.80[2]	NA[1]
Gelatin Dynamite (20 percent Strength)	NA[1]	0.70	0.70[2]	NA[1]
H-6	1.76	1.38	1.15	0.03 to 0.70
HBX-1	1.76	1.17	1.16	0.03 to 0.14
HBX-3	1.85	1.14	0.97	0.03 to 0.17
HMX	NA[1]	1.25	1.25[2]	NA[1]
L.X-14	NA[1]	1.80	1.80[2]	NA[1]
MINOL II	1.82	1.20	1.11	0.02 to 0.14
Nitrocellulose	1. 65 to 1. 70	0.50	0.50[2]	NA[1]
Nitroglycerine Dynamic (50 percent Strength)	NA[1]	0.90	0.90[2]	NA[1]
Nitroglycerine (NQ)	1.72	1.00	1.00[2]	NA[1]
Nitromethane	NA[1]	1.00	1.00[2]	NA[1]

(continued)

Table 6.1 *(Continued)*

Explosive	Density Mg/m³	Equivalent Moss for Pressure	Equivalent Mass for Impulse	Pressure Range MPa
Octol (75/25 HMX/TNT)	1.81	1.02	1.06	NA[1]
(70/30)	1.14	1.09	1.09[2]	0.01 to 0.30
PBX-9010	1.80	1.29	1.29[2]	0.03 to 0.21
PBX-9404	1.81	1.13	1.13[2]	0.03 to 0.69
		1.70	1.70	0.69 to 6.90
PBX-9502	1.89	1.00	1.00	NA[1]
PBXC-129	1.71	1.10	1.10[2]	NA[1]
PBXN-4	1.71	0.83	0.85	NA[1]
PBXN-107	1.64	1.05[2]	1.05[2]	NA[1]
PBXN-109	1.67	1.05[2]	1.05[2]	NA[1]
PBXW-9	NA[1]	1.30	1.30[2]	NA[1]
PBXW-125	1.80	1.02[2]	1.02[2]	NA[1]
Pentolite. (Cast)	1.64	1.42	1.00	0.03 to 0.69
	1.68	1.38	1.14	0.03 to 4.14
	NA[1]	1.50	1.00	0.69 to 6.90
PETN	1.77	1.27	1.27[2]	0.03 to 0.69
Picrotol (52/48 Ex D/TNT)	1.63	0.90	0.93	0.03 to 4.10
RDX	NA[1]	1.10	1.10[2]	NA[1]
RDX/Wax (98/2)	1.92	1.16	1.16[2]	NA[1]
RDX/AL/Wax (74/21/5)	NA[1]	1.30	1.30[2]	NA[1]
TATB	NA[1]	1.00	1.00[2]	NA[1]
Tetryl	1.73	1.07	1.07[2]	0. 02 to 0.14
Tetrytol (75/25 Tetryl/TNT)	1.59	1.06	1.06[2]	NA[1]
TNETB	1.69	1.13	0.96	0.03 to 0.69
TNETB/AL (90/10)	1.75	1.23	1.11	0.03 to 0.69
(78/22)	1.18	1.32	1.32[2]	NA[1]
(65/35)	123	1.38	1.38[2]	NA[1]
TNT	1.63	1.00	1.00	Standard
Torpex	1.85	1.23	1.28	0.01 to 0.30
Tritonal (80/20 TNT/AL)	1.72	1.07	0.96	0.03 to 0.69

[1]NA - Data not available.
[2]Value is estimated.

materials such as fireworks, smoke grenades, flares, and the gas generator material in automobile air bags. The relatively low reaction rates of these materials result in low blast pressure output because they typically don't detonate. Fire hazards and gas generation concerns dominate the safety analysis in most cases, but detonations or deflagrations cannot be ruled out in all cases.

Figure 6.4 Vapor Cloud Explosion in 9/11 Pentagon Crash

Predicting blast loads for these materials can be challenging. There is much less data available for propellants and pyrotechnics than for high explosives, with respect to blast loads. Burn rates, gas generation, and energy output rates are available in handbooks for common propellants (Cooper and Kurowsky 1996) These values can be used to perform a burn simulation and determine the gas pressure load in a confined area. In some cases, TNT equivalencies for these materials have been used. However, this is problematic because peak pressure for propellants and pyrotechnics is much lower than for TNT, but the impulse of propellents can be higher in a confined space than for the TNT equivalent. Care must be taken when equivalencies are based on small-scale tests, as large quantities may react in a more energetic manner.

Vapor cloud explosions involve the release of a flammable material which, when mixed with air and given the proper conditions, forms an ignitable material, potentially leading to an explosion. The vapor cloud explosion following the 9/11 Pentagon crash is shown in Figure 6.4. Only fuel within flammability limits participates in the explosion, which is a minor percentage of the total quantity of fuel released (Baker et al. 1991). Much of the fuel vapor is too lean or too rich to combust and does not produce blast overpressure. Blast output of a vapor cloud explosion is highly variable and depends on the fuel, size of the flammable vapor cloud, presence of congestion (obstacles such as structural members, pipes, and vessels) in the flammable region, and presence of confining surfaces. Release of a flammable material into the open air, absent congestion or confinement, typically results in a fireball that does not produce damaging overpressures.

Confinement and congestion in the flammable cloud create turbulence and accelerate the combustion, thereby increasing the flame speed. Peak overpressure is a strong function of flame speed; thus, a key element in vapor cloud explosion prediction is maximum flame speed. Vapor cloud detonations are rare and only known to occur with high-reactivity materials, strong ignition sources, and/or severe cases of confinement and congestion. The vast majority of vapor cloud explosions are deflagrations. A particularly noteworthy characteristic of deflagrations is that high flame speeds only occur in confined/congested regions of the flammable cloud. Outside of congested areas, flame speed decreases. As a result, deflagration blast loads are produced in confined/congested areas, rather than at the point of release or dispersion. In the rare case of vapor cloud detonations, the blast loads are produced by the detonatable material and are typically considered to be centered in the dispersed cloud and not at the point of release.

Overpressures produced by vapor cloud explosions are substantially lower than those produced by high explosives. For example, a vapor cloud detonation may reach 16–18 atmospheres overpressure, compared with 100,000 atmospheres or greater for most high explosives. However, the energy of vapor cloud explosions can be very large, as evidenced by some of the devastating accidents which have occurred. In Flixborough, U.K., a 30,000-kg release of cyclohexane caused an explosion that resulted in 28 fatalities, multiple injuries, and destruction of more than 150 structures. The large energy in hydrocarbons produces a commensurately large impulse. Blast predictions made using a TNT equivalent approach tend to be inaccurate due to the disparity between the pressure and impulse compared with high explosives. As a result, TNT equivalence analyses of vapor cloud explosions are not recommended. A more accurate method is to perform a dispersion analysis of the flammable release to determine fuel in the flammability limits, select appropriate speed or reaction severity, then use nondimensional scaled energy curves to determine the pressure and impulse at various standoff distances. Computational fluid dynamics (CFD) analysis is another method of determining blast pressures for vapor releases.

Vessels designed to contain fluids under pressure have the potential to create hazardous overpressures and fragments if the vessel fails while under pressure. A *pressure vessel burst* may occur at normal working pressure due to a mechanical integrity problem, weakening, or other external factors (such as impact); or at elevated pressure due to excessive pressurization from an external source, an uncontrolled internal chemical reaction, heating vessel contents with insufficient pressure relief, or other means. This type of accident results in loss of contents and may lead to a follow-on explosion when the contents subsequently mix with air, if the contents are flammable or pyrophoric.

In the far field, the blast load from a vessel that fails at high pressures (several thousand psi [1000 psi = 6895 kPa] or higher) is similar to that caused by high explosives. In the near field, overpressures are less than those produced by high explosives, and the use of TNT equivalent prediction procedures will overestimate pressure. Key parameters for estimating blast effects from bursting vessels are the volume of the material in the vessel, the failure pressure, the composition of the contents, and the temperature of the contents.

Fragments represent an additional hazard from bursting vessels. Failure at working pressure or overpressurization at a relatively slow rate usually results in a small number of fairly large fragments. In some cases, one or both ends of a cylindrical pressure vessel may be propelled long distances by a rocketing mechanism. Large fragments may be thrown as a result of the expansion of the vessel contents. Containment of these fragments, which can be quite challenging, may be required to provide adequate safety. Fragment hazard prediction methods are discussed further in Chapter 8 of this handbook.

A boiling liquid expanding vapor explosion (BLEVE) occurs when a vessel containing a liquid stored above its boiling point for atmospheric pressure fails and a rapid depressurization of the liquid occurs. Upon depressurization, a portion of the superheated liquid rapidly boils, and the expanding vapors contribute to the blast wave. Compressed gases in the vessel prior to failure also contribute to the blast wave as in a bursting pressure vessel. A common BLEVE scenario is a pressure vessel engulfed in an extended fire, causing weakening of the vessel shell to the point of failure. A BLEVE may involve a flammable liquid, and mixing of the flammable material with air and subsequent ignition may cause a fireball. A detailed discussion of BLEVE events and of simplified blast prediction methods is provided in CCPS's *Guidelines for Evaluating the Characteristics of Vapor Cloud Explosions, Flash Fires, and BLEVES* (Center for Chemical Process Safety 1994).

6.3 CHARACTERISTICS OF BLAST WAVES

6.3.1 Key Parameters

An idealized pressure-time history indicating the key parameters of a blast load is shown in Figure 6.3. Figure 6.5 shows a comparison between free-field, or side-on, and reflected pressure-time histories. The parameters shown in these figures are defined as:

P_o = Ambient pressure

P_{so} = Peak positive side-on overpressure

P_{so}^- = Peak negative side-on overpressure

$P_s(t)$ = Time varying positive overpressure

$P_s^-(t)$ = Time varying negative overpressure

P_r = Peak reflected overpressure

I_s = Positive-phase-specific impulse, the integration of the positive phase pressure-time history

i_s^- = Negative-phase-specific impulse, the integration of the negative phase pressure-time history

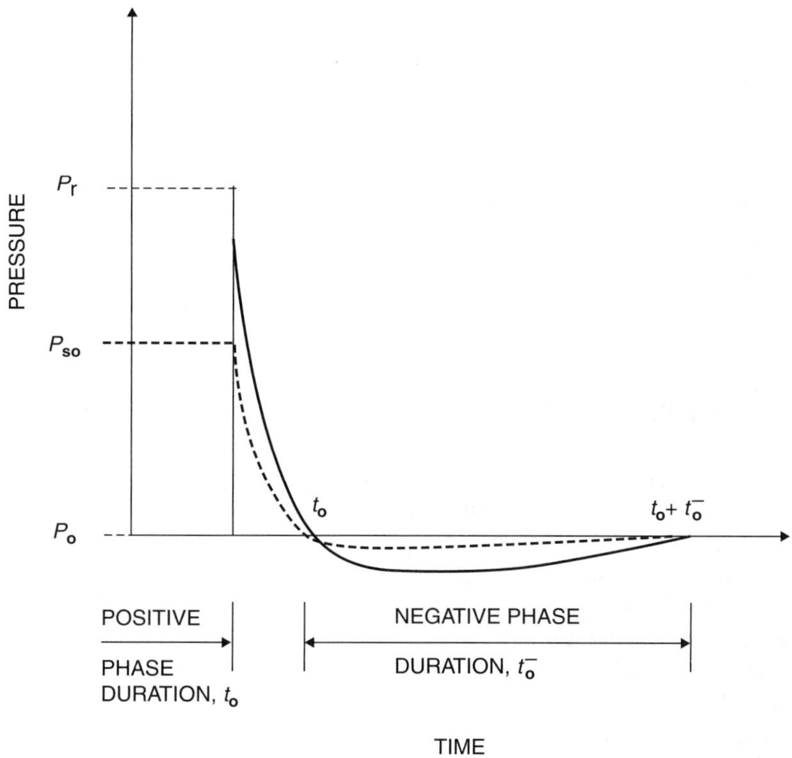

Figure 6.5 Comparison of Free Field and Reflected Blast Loads

t_a = Time of arrival

t_o = Positive phase duration

t_o^- = Negative phase duration

The term "overpressure" refers to a gauge pressure, or the blast pressure relative to ambient pressure.

Free-field loads are those produced by blast waves sweeping over surfaces unimpeded by any objects in their path. This load is also referred to as side-on when the blast wave sweeps over a wall or other object parallel to its direction of travel. Side-on pressure terms are indicated by a "so" subscript in the figures in this chapter, as well as in most references.

6.3.2 Scaling

Blast pressures, load duration, impulse, shock wave velocity, arrival times, and other blast parameters are frequently presented in scaled form. Research has shown that scaling laws can be applied to explosions, allowing data from one explosion test to be applied to a geometrically similar (larger or smaller) case. As a result, scaling has tremendous utility in blast predictions, allowing compilations of explosion test and numerical modeling data to be used to predict loads for any combination of explosion energy and standoff distance within the range of data. The most common form of scaling is called "cube root scaling" owing to the fact that blast parameters are scaled by the cube root of the explosion energy. Both dimensional and dimensionless scaling are used, and care must be taken with all unit parameters to ensure the scaling is correctly applied for the blast curves being used.

6.4 PREDICTION OF BLAST PARAMETERS

This section describes the principal methods for predicting blast parameters, with some discussion of practical applications and limitations. A more detailed treatment is given for high explosives and vapor cloud explosions because these tend to be the most common design threats for protective construction. Definition of the blast hazard is discussed in Chapter 2, and calculation of the loads imposed on the structure in Chapter 7.

In many cases, blast load prediction is a multi-step process. Often an initial estimate of blast loads will be made using simplified techniques and a conservative estimate of explosion energy. It may become obvious at this point that the predicted loads are so low they are not a contributing factor in structural design, or will not cause undue cost or complexity in the design. Thus, the initial loads are judged to be adequate for analysis or design. If loads are high enough to drive the design, it may be appropriate to perform a more detailed blast load prediction which uses less conservative estimates of energy and more detailed prediction techniques.

The basic simplified approach involves predicting free-field loads using empirical or semi-empirical methods. Blast curves that relate explosion energy, standoff distance, and blast wave parameters are typically used in this initial load-prediction step. These curves are readily available for high explosives and vapor cloud explosions.

6.4.1 High Explosives

A significant amount of data has been produced that quantifies the relationships among charge weight, standoff distance, and blast parameters for both positive and negative phases. A Unified Facilities Criteria publication, *Structures to Resist the Effects of Accidental Explosions*, produced by the Department of Defense, contains the relationships in the form of scaled blast curves (U.S. Department of Defense 2008). This document is publicly available, and an electronic version also exists to aid in the use of the numerous tables and figures.

The empirical blast parameter curves provided in this document plot air-blast parameters versus scaled distance for both air-burst and surface-burst configurations. A charge detonated on the ground will produce higher blast loads than an air burst due to the reflection of the shock by the ground surface. For unyielding surfaces with the charge located on the surface, the effect is a doubling of the charge weight, since the energy of the blast directed to the ground is fully reflected. For soft soils or charges located above the surface, the reflection factor is less than two, since some of the energy is absorbed by the substrate. Some guidance for selecting the reflection factor is given in the document, but it is conservative to assume a fully reflecting surface.

The most commonly used approach to present blast wave relationships for high explosives is the Hopkinson-Cranz, or cube root, scaling method. Figure 6.6 is a simplified version of TNT blast curves that provides the parameters for charges located at ground level. The cube root term results from geometric scaling laws in which charge diameter varies in proportion to all distances, and thus the charge weight is proportional to the cube of the charge diameter. To use these empirical curves, one computes the scaled distance by dividing the standoff distance from the charge to the point of interest by the cube root of the charge weight. For explosives, this takes the form of

$$Z = R/W^{1/3} \text{ where :} \tag{6.1}$$

Z = scaled distance (ft/lb$^{1/3}$)
R = standoff distance (ft)
W = explosives weight (lb)

The curves provide pressures, P, which are the same at a given scaled distance. The curves also provide scaled times, $t/W^{1/3}$, and scaled impulses, $I/W^{1/3}$. The actual times and impulses are then found by multiplying the scaled values by $W^{1/3}$.

In accordance with the common practice for blast design, the variables in Equation 6.1 and Figure 6.6 have physical dimensions. However the procedure is rigorously founded on dimensionless modeling. It is also based on atmospheric conditions at sea level. While this suffices for most practical situations, the effects of differing ambient conditions can be treated more comprehensively through Sachs or energy scaling (Baker 1973).

As an example of cube root scaling, let us consider the Oklahoma City Bombing. This event has been reported to have been equivalent to the detonation of 4,000 lbs of TNT at essentially the ground surface. If a location of interest is 100 ft away, the scaled distance is

$$Z = 100/(4000)^{1/3} = 6.30\,\text{ft/lb}^{1/3}$$

From Figure 6.6 we have an incident pressure of $P_o = 24.9$ psi and a reflected pressure of $P_r = 79.5$ psi at this scaled range. This figure also provides the scaled positive phase duration $t_o/W^{1/3} = 1.77$ msec/lb$^{1/3}$. From this, the total positive

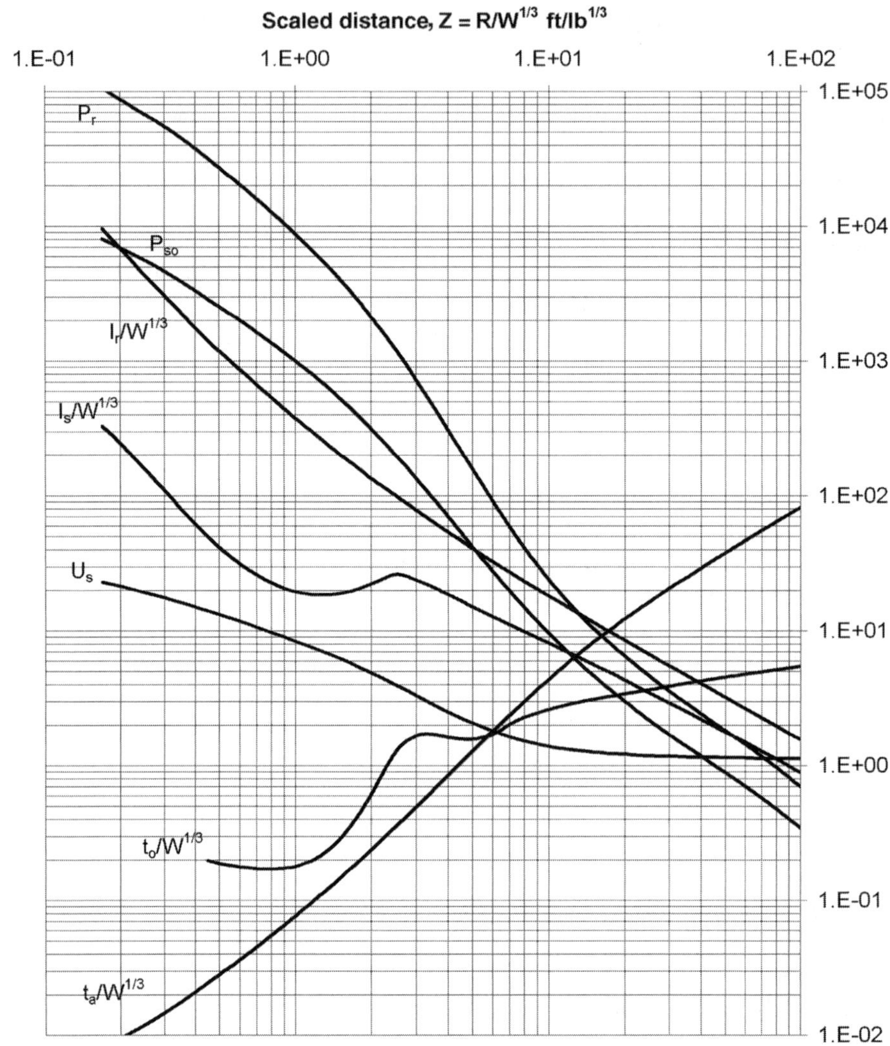

Figure 6.6 Blast Parameters for TNT Surface Bursts

phase duration is

$$t_o = 1.77 \times (4000)^{1/3} = 28.1\,\text{msec}$$

The figure similarly indicates the scaled positive-phase-specific impulse to be $I_s = 12.17$ psi $\times$ msec/lb$^{1/3}$ and the total impulse is

$$I_s = 12.17 \times (4000)^{1/3} = 193.2\,\text{psi} \times \text{msec}$$

Some structures, such as explosives or propellant processing bays, mail rooms, and loading docks, often require blast design for containment of internal explosion effects. Containment designs must consider that shock waves emanate from the charge, strike wall and roof components, and reflect to impact other surfaces. Each surface is subjected to multiple shock waves.

Internal explosions also produce gas and heat, which cause a pressure increase within the containment. The peak gas pressure, P_g, developed is a function of the charge weight/free volume ratio in the containment. This pressure buildup is relatively slow compared to the load duration associated with shock waves. For simplicity, the peak gas or quasi-static pressure is assumed to rise instantaneously to the peak gas pressure but is not additive to shock pressures. A curve of peak gas pressuresversus weight/volume is shown in Figure 6.7.

For example, let us consider the detonation of $W = 200$ lbs of TNT within a structure having a free volume of $V_f = 50{,}000$ ft^3. This loading density is $W/V_f = 200/50{,}000 = 4 \times 10^{-3}$ lb/ft^3. From Figure 6.7 the peak gas pressure is $P_g = 47.2$ psi.

The duration of the gas pressures is a function of the vent area available and the rate at which temperature decays in the containment area. Vent area can change over time, depending on how fast the vent cover moves away from the enclosure. The vent velocity is a function of the applied shock impulse and the frangibility and mass of the vent cover. The duration is determined by first computing the peak gas pressure from Figure 6.7. Then the gas impulse, I_g, is estimated from a set of graphs as a function of the scaled vent area, scaled mass of the vent cover, loading density, and scaled shock impulse (U.S. Department of Defense 2008). An equivalent triangular gas duration is then

$$t_g = 2I_g/P_g$$

Analytical methods have been developed to predict blast loads. These methods fall into two groups: semi-empirical and hydrocode. The semi-empirical approach utilizes a physics-based model to compute selected blast parameters with coefficients that are "tuned" to match test data. These models are limited to configurations and charge weight ratios for which data are available, but offer the advantage of quick run times compared with more detailed techniques. Semi-empirical codes may offer limited abilities to model shock diffraction, shielding, and reflection. Semi-empirical methods have been developed primarily by defense-related agencies and are restricted to distribution to the government and its contractors.

Hydrodynamic codes, or hydrocodes, utilize a grid of computational cells to track detonation propagation through an explosive charge and shock wave propagation through a medium. Using principles of material behaviors, fundamental thermodynamics, and fluid dynamics, hydrocodes predict pressure, density, and other key parameters. This type of analysis is much more complex than the use of empirical relationships. Some of the available tools have user interfaces that greatly simplify the analysis, especially for standard materials, but considerable effort and expertise are typically required to conduct a competent analysis.

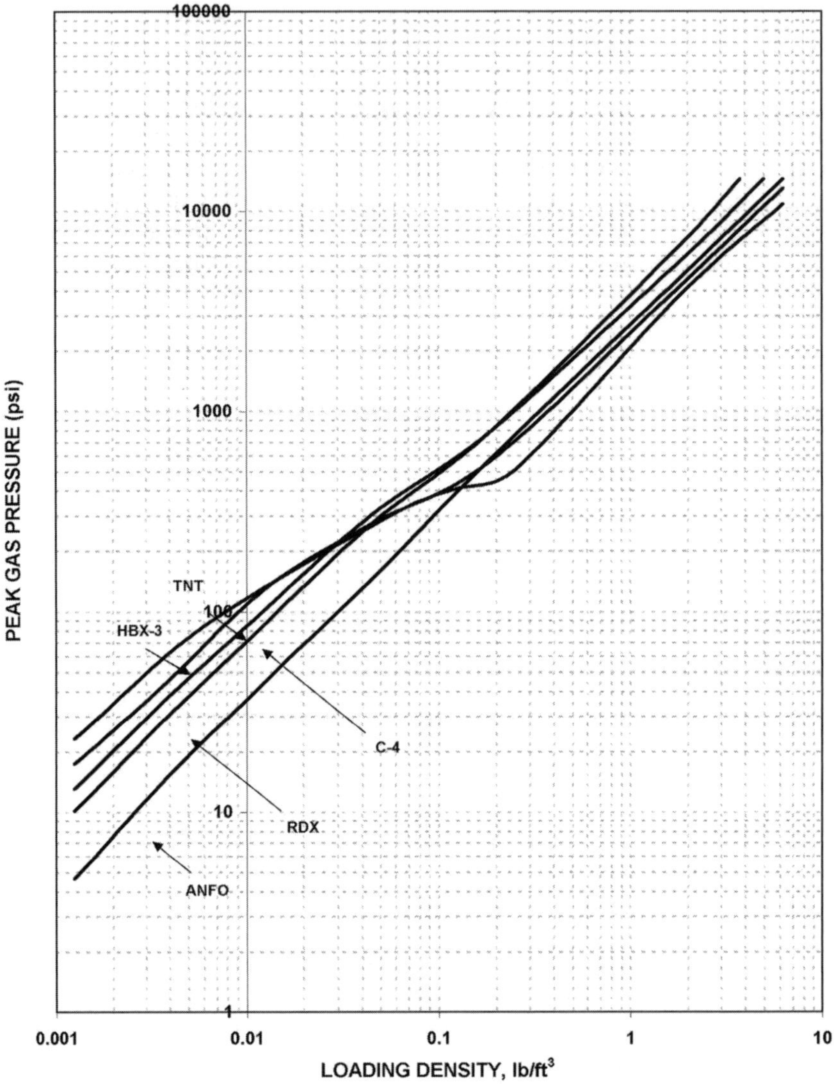

Figure 6.7 Peak Gas Pressure for Internal Detonations

The output from hydrocodes will be in the form of pressure-time series for selected locations in the model. Application of these loads to the structural model can be tedious. If the structural analysis involves a time series calculation, the time varying loads may be transferred into the structural analysis, but the number of data points to be transferred may be excessive. Normally, a simplification is made to the applied load history to greatly reduce the number of load points to a series of lines approximating the "real" curve. This reduced multipoint curve is more complex than the linear decay curve used for the empirical relationship analysis. A coupled analysis may be performed in the hydrocode

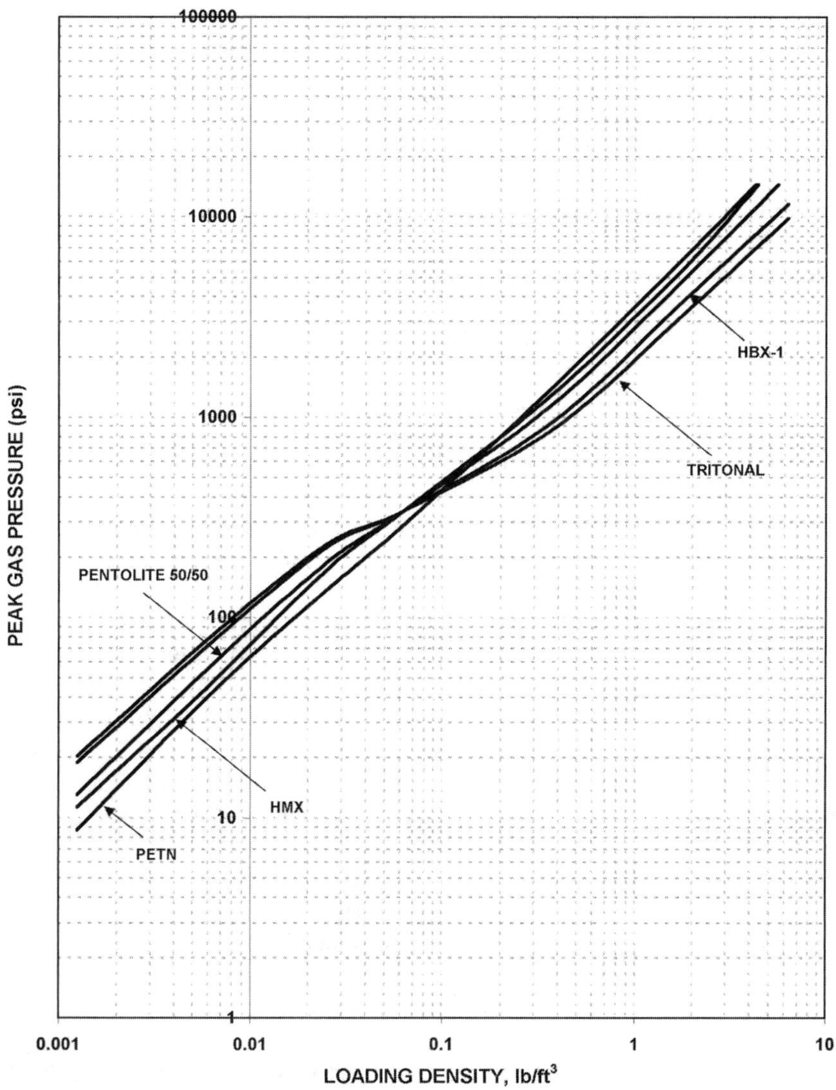

Figure 6.7 (*Continued*)

which models blast propagation and structural response. This approach offers additional capability for detailed response analysis but with significant increase in complexity compared with simulation of blast loads alone.

6.4.2 Bursting Pressure Vessels

Bursting pressure vessel blast predictions are most commonly conducted in one of two ways: simplified analytical methods or computation fluid dynamics (CFD) models. In either case, it is necessary to select the conditions under which the vessel would be expected to fail—overpressure/overfilling, runaway reaction, external heating, mechanical or impact load, corrosion, or other mechanisms. This selection is important because it determines the composition, pressure, temperature, phase, and energy of the vessel contents, all of which can have a significant effect on the predicted blast loads. In any case, as long as the vessel pressure is above two atmospheres at failure and the pressure is released abruptly, the resultant blast will be a shock wave (i.e., suddenly applied).

The most recent simplified method was published by Tang (Tang et al. 1996) and consists of a series of nondimensionalized curves to predict the side-on pressure and impulse at a particular standoff as a function of pressure ratio (defined as burst pressure divided by ambient pressure), ratio of specific heats of the vessel contents (a property of the gaseous contents of the vessel), expansion energy potential of the vessel contents, and the temperature ratio (the ratio of the vessel internal temperature at failure to the ambient temperature). The vessel energy term in this analysis should be corrected for ground reflection, liquid fraction in the vessel, and energy consumed by fragment production. In addition, the side-on pressures need to be modified to account for reflection and clearing effects, as described in the next chapter.

If more detail is necessary, a CFD model can be applied. These consider the thermodynamics and gas properties to compute the production and propagation of blast waves produced by a bursting vessel. It is important to note that there are complex interactions between the failing vessel and the fluid escaping the vessel that cannot be adequately modeled by any publicly available current program. In addition, the failure of the vessel is not wholly deterministic, as cracks can propagate along micro-cracks and grain boundaries in a manner that cannot usually be predicted. As a result, these numerical analyses will rely heavily on the experience and expertise of the analyst. The advantage of using a hydrocode is that it can directly produce load-time distributions on arbitrary surfaces and account for shielding and focusing effects produced by adjacent structures.

6.4.3 Vapor Cloud Explosions

Prediction of blast loads for vapor cloud explosions can be more complex than loads for high explosive detonations or bursting pressure vessels. The first step in determining blast load from vapor cloud explosions is to develop the release scenario for the flammable material. This selection process is typically done with input from process engineers and plant operations. Dispersion modeling is performed to model the release scenario and determine the size, shape, and concentration of the flammable vapor cloud. This dispersion is overlaid onto areas of confinement and congestion to define the flammable mass that may participate in the explosion. In some cases, cloud dispersion is not evaluated, and the confined/congested volume is conservatively assumed to be filled with a stoichiometric mix of flammable gas. This approach may be justified if it is evident that credible release scenarios can fill the volume. Blast loads may be determined using simplified methods consisting of graphs (blast curves) of pressure and impulse, or of duration versus scaled standoff, or through more complex modeling using CFD.

In the blast curve method, the scaled standoff is computed by using distance from the center of the explosion to the point of interest and the energy content of the confined/congested flammable mass. Scaled pressure and impulse values are read from blast charts containing flame speed curves. These curves range from low flame speeds that produce almost negligible blast pressure to the supersonic flame speed of a detonation producing the highest pressure. Determination of maximum flame speed requires knowledge of the amount of confinement and congestion in the flammable portion of the vapor cloud, and the fuel reactivity.

The two most commonly used methods are the Baker-Strehlow-Tang (BST) (Tang and Baker 1999) and TNO Multi-energy Method (MEM) (Van den Berg 1985). Figure 6.8 shows the overpressure, and Figure 6.9 shows the impulse as a function of scaled distance and flame speed for BST. To use the curves, compute the scaled distance $R_{\text{bar}} = R(p_o/E)^{1/3}$ in which R = standoff distance, p_o = ambient pressure, and E = heat of detonation. Then obtain the scaled pressure $P_{\text{bar}} = (P/p_o)$ from Figure 6.8 and unscale to determine the pressure P. Next, determine the scaled impulse $i_{\text{bar}} = i\, a_o/(E\, p_o)^{1/3}$ in which i = impulse and a_o = ambient sound velocity. The impulse, i, follows by unscaling. The equivalent triangular duration is then $t_o = 2i/P$.

The process is similar for MEM, using Figures 6.10 and 6.11 for pressure and duration respectively. Impulse is $i = Pt_0/2$.

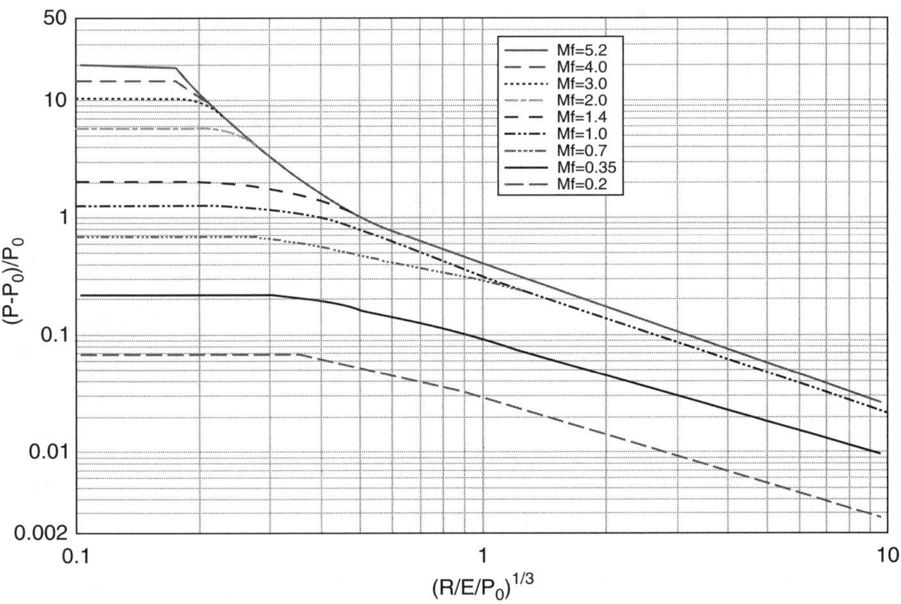

Figure 6.8 Peak Pressure for Vapor Cloud Explosions—BST

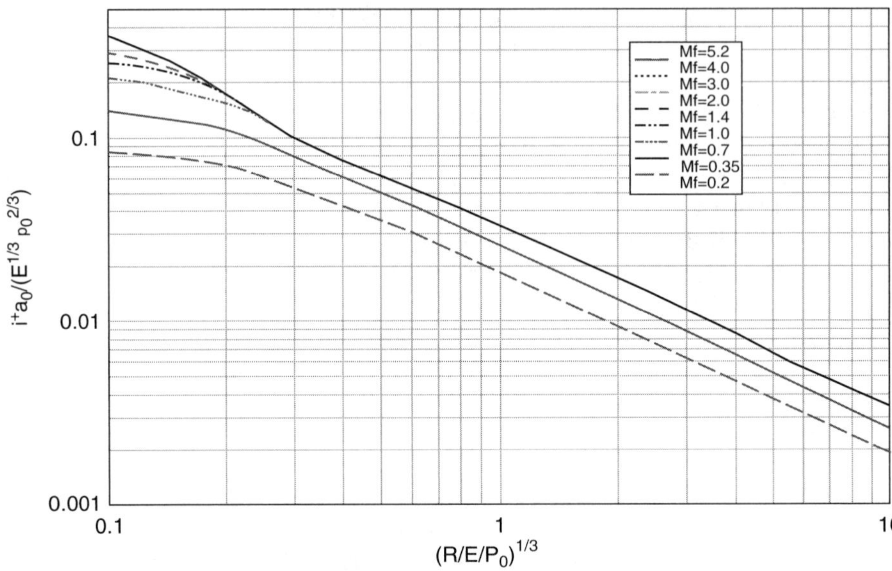

Figure 6.9 Impulse for Vapor Cloud Explosions—BST

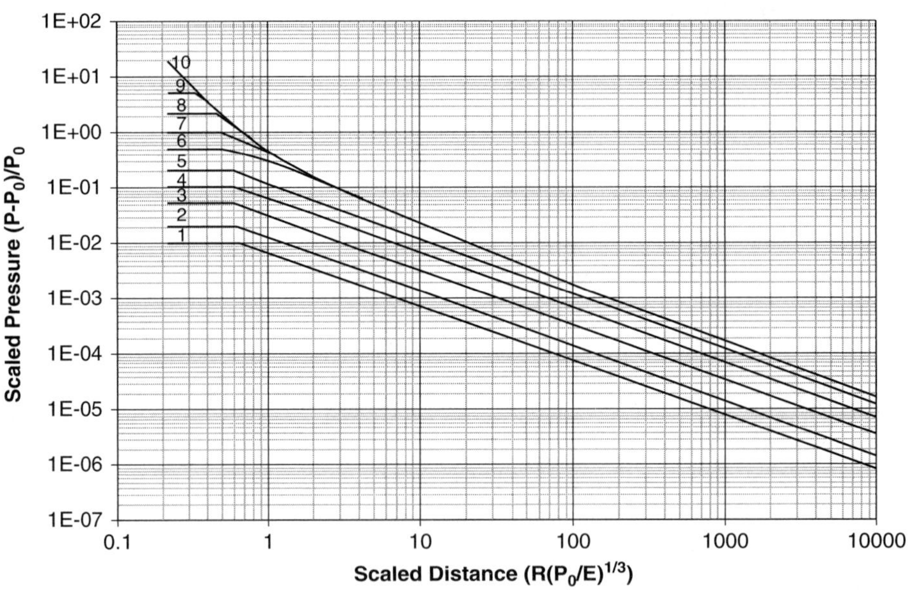

Figure 6.10 Peak Pressure for Vapor Cloud Explosions—TNO MEM

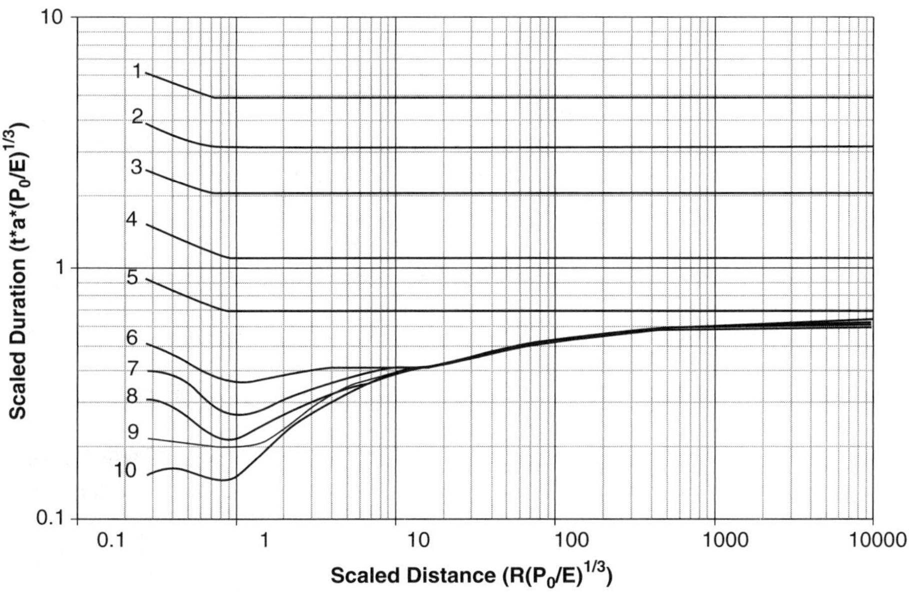

Figure 6.11 Duration for Vapor Cloud Explosions—TNO MEM

A different approach is to develop a CFD model of the process area. A grid of computational cells is constructed, and the controlling combustion chemistry and gas kinetics equations are solved to track the progression of the reaction and resulting pressures. The end result is a pressure-time series at selected points. It is important for the structural analyst to know if the pressure-time plots are free-field or applied loads. An advantage of this approach is that it allows the analyst to include the effects of blockage and impingement of objects, which can substantially influence the dispersion and explosion effects. This capability is not available in the simplistic methods using scaled energy curves.

6.5 SUMMARY

Blast phenomena are characterized by a rapid release of energy resulting in a pressure wave that propagates outward from the source. The source can be a high explosive, propellant, or other gases that are released suddenly. Shock waves are characterized by an instantaneous rise in pressure from ambient. Pressure waves, are produced by deflagrations resulting from combustion of low explosives and flammable gases. The characteristics of a blast waves can be predicted with relatively simple empirical methods or more complex computational fluid dynamics. Once free-field or side-on blast pressure and impulse are defined, the applied loads on structures can be determined, as described in the next chapter.

REFERENCES

Baker W. E. 1973. *Explosions in Air*. Austin: University of Texas Press.

Baker, Q.A. and W. E. Baker. 1991. Pros and cons of TNT equivalence for industrial explosion accidents. Presented at AIChE International Conference on Workshop on Modeling and Mitigating the Consequences of Accidental Releases of Hazardous Materials, New Orleans, LA, May 21–24, 1991.

Center for Chemical Process Safety. 1994. *Guidelines for Evaluating the Characteristics of Vapor Cloud Explosions, Flash Fires, and BLEVES*. New York: American Institute of Chemical Engineers.

Cooper, W. P. and R. S. Kurowski. 1996. *Introduction to the Technology of Explosives*. New York: Wiley-VCH.

Tang M. J. and Q. A. Baker. 1999. A new set of blast curves from vapor cloud explosion. *Process Safety Progress* 18 (4): 235–240.

Tang, M. J., C. Y. Cao, and Q. A. Baker. 1996. Calculation of blast effects from bursting vessels." Presented at the 30th Annual AIChE Loss Prevention Symposium, New Orleans, LA, February 1996.

U.S. Department of Defense. 2008. *Structures to Resist the Effects of Accidental Explosions* (UFC 3-340-02). Washington, DC: U.S. Department of Defense.

Van Den Berg, A. C. 1985. The multi-energy method—A framework for vapor cloud blast prediction." *Journal of Hazardous Materials*, 12 (1): 1–10.

7 Blast Loading

Paul F. Mlakar and William Bounds

7.1 INTRODUCTION

In the previous chapter, blast phenomena and the characteristics of blast waves were discussed. In this chapter, methods for determining the dynamic blast loading on structural elements from these characteristics are discussed. Emphasis is placed on simple approximate procedures that apply widely in practical blast-resistant design. More comprehensive methods are also discussed.

7.2 EMPIRICAL METHOD

Although research into the effects of explosions dates back to 1870, most development to determine the blast loading on buildings and other similar structures was started in the 1950s and 1960s by the U.S. military. Several publicly available military manuals were distributed during this period (TM 5-856 and TM 5-1300) that presented empirically-derived charts and equations. Several papers and publications published during that period (Newmark 1956, Biggs 1964, ASCE 1985) also provided information for design.

The empirical method consists of published equations, graphs, tables, and figures that allow one to determine the principal loading of a blast wave on a building or a similar structure. Software has also been developed to automate calculations based on this same source information. The basic advantage of empirical methods is speed and simplicity. Applicable factors and coefficients can be looked up in tables, blast loading can be calculated manually, and a building blast load can be determined in a matter of minutes. Empirical methods are also well suited for situations where the accuracy of the blast source location and magnitude is uncertain. More comprehensive methods, such as computational fluid dynamics (CFD), require specialized software, operator training, and, potentially, weeks of data input and verification. The disadvantage of empirical methods is the lack of flexibility in application. Most data are based on plain rectangular target structures located in open terrain. Explosions are assumed to be either an air blast or surface blast. The blast wave is typically based on high explosives producing a shock wave, as opposed to a deflagration that produces a pressure wave. Empirical methods remain popular for the preliminary evaluation of complex design situations and for the design situations where the location and magnitude of the blast source are not well understood. The analyst must make a judgment to pursue an efficient design calculation, and to balance the accuracy of input information and the desired accuracy of results.

The most extensive and widely referenced publication for empirical design is UFC 3-340-02 (formerly TM5-1300) (U.S. Department of Defense 2008). This design manual addresses accidental explosions related to munitions manufacturing, handling, and storage. Nevertheless, many of the procedures are applicable to buildings and structures designed for other blast scenarios. Basic information from UFC 3-340-02 is provided in this document for external loading. UFC 3-340-02 also provides load determination information on other configurations, such as partially open cubicles and interior explosions. Information on empirical methods is available from a number of other sources such as Mays and Smith's *Blast Effects on Buildings* (Mays and Smith 1995), Biggs's *Introduction to Structural Dynamics* (Biggs 1964), and ASCE's *Structural Design for Physical Security* and *Design of Blast Resistant Buildings in Petrochemical Facilities* (ASCE Physical Security 1999, ASCE Petrochemical 1997).

The empirical blast load example calculations provided in this chapter assume a blast wave interaction with a rectangular structure of finite size, such that the structure is blast-loaded on all sides. Furthermore, due to the structure depth, or building dimension parallel to the direction of the blast wave, there will be a significant net lateral force applied to the structure. Mays and Smith describe this as one of three blast loading situations to keep in mind. The second situation is a blast wave interacting with a relatively small structure, such as a vehicle, that is effectively engulfed with blast pressure acting on all sides of the structure at once. The third situation is for a blast wave acting on a relatively large structure, such as a large office building, where the magnitude of the blast wave varies significantly across the surfaces of the structure. Some surfaces of these structures may see little if any external blast loading.

Empirical methods typically rely on a straight-line equivalent of the actual blast pressure-time relationship, as well as the use of a straight-line equivalent loading on the structure of interest. The approximation is fairly close, and the resulting calculations are much simpler.

Details and examples illustrating implementation of empirical methods are included in following sections of this chapter. Key input parameters for the determination of building loads, as presented in Chapter 6, are: the peak side-on overpressure, P_{so}, the positive phase duration, t_o, and the shock front velocity, U_s.

One other parameter necessary for the determination of building blast loads is the dynamic wind pressure, q_o. Dynamic wind is the movement of air particles resulting from a shock wave. The effect is additive to the blast overpressure and is a function of the free-field blast overpressure and the obstruction's shape. Open-frame structures and small buildings where a blast wave will produce quick envelopment are most sensitive to dynamic wind. Values of q_o can be determined

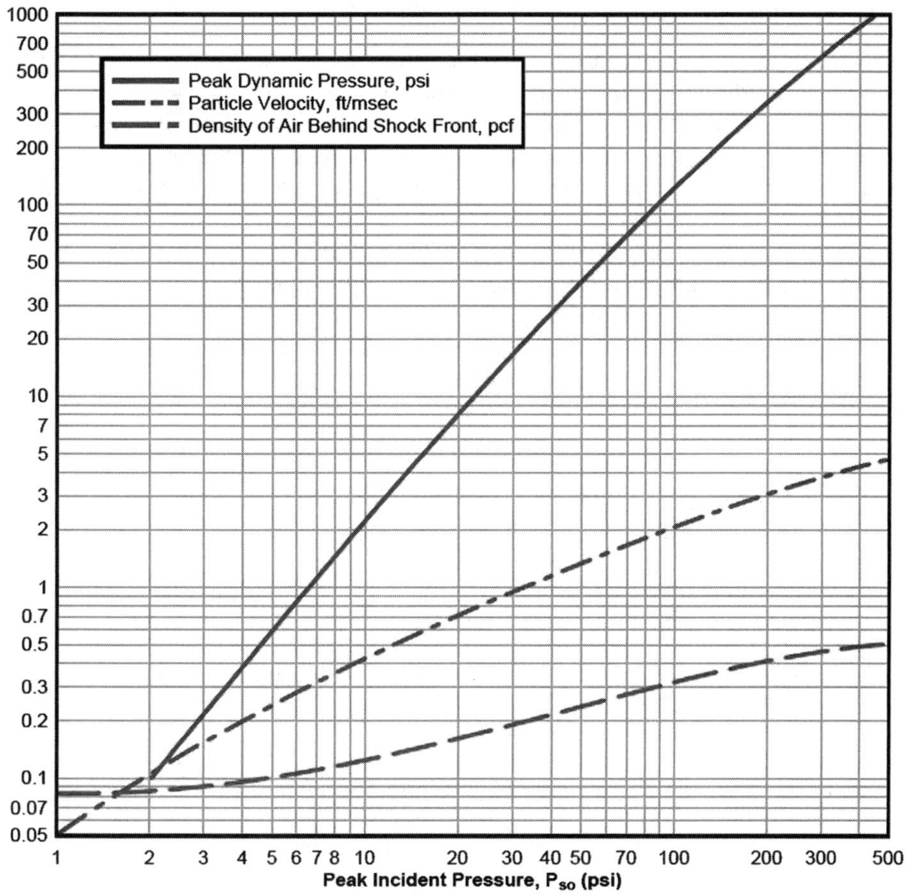

Figure 7.1 Peak Incident Pressure versus Peak Dynamic Pressure, Density of Air Behind the Shock Front, and Particle Velocity (UFC 3-340-02)

from Figure 7.1. Alternatively, in the low overpressure range, and at sea level atmospheric pressure, the following equation from Newmark can be used.

$$q_o = 0.022\,(P_{so})^2 \tag{7.1}$$

The pressure exerted on a structural element is the dynamic wind pressure multiplied by a drag coefficient. The drag coefficient, C_d, is a function of the shape and orientation of the obstructing element. Newmark lists approximate values of C_d for open-frame structural elements as 2 for structural shapes, 1.25 for box shapes, and 0.8 for cylinders. Values of C_d for enclosed rectangular buildings are provided in the following sections.

7.2.1 Empirical Method—Basic Blast Wave Example

The example calculation presented in the following sections is based on the blast wave and building parameters described below. The given blast wave will be applied normal to the long side of a rectangular flat-roof building. It is further determined that the distance to the explosion and the length of the building are such that the overpressure and duration do not change significantly over the length of the building.

Given charge weight, $W = 5$ tons (TNT) $= 10{,}000$ lb:
The charge weight, W, should include any applicable safety factor
Given distance from source to building, $R = 215$ ft
The blast wave parameters, computed using Figure 6.6 as follows:
To use Figure 6.6, compute the scaled distance:

$$Z = R/W^{1/3} = (215\text{ ft})/(10{,}000\text{ lb})^{1/3} = 10.0$$

Using $Z = 10$, use Figure 6.6 to determine the following parameters:
Peak side-on overpressure, $P_{so} = 10$ psi
Shock front velocity, $U_s = 1.4$ ft/ms
Scaled positive phase duration, $t_o/W^{1/3} = 2.6\text{ ms/lb}^{1/3}$
Compute the positive phase duration:

$$t_o = (t_o/W^{1/3})W^{1/3} = (2.6\text{ ms/lb}^{1/3})(10{,}000\text{ psi})^{1/3} = 56\text{ ms}$$

Compute the positive phase wave length:

$$L_w = (U_s)t_o = (1.4\text{ ft/ms})(56\text{ ms}) = 78\text{ ft}$$

With $P_{so} = 10$ psi, use Figure 7.1 to determine the peak dynamic wind pressure, $q_o = 2.1$ psi

Alternatively, use Equation 7.1 to calculate the peak dynamic wind pressure,

$$q_o = 0.022\,(P_{so})^2 = 0.022\,(10\text{ psi})^2 = 2.2\text{ psi}$$

Loads are to be computed on an aboveground rectangular building with the following dimensions,

Building height, $B_H = 20$ ft

Building width, $B_W = 150$ ft

Building depth, $B_D = 50$ ft

7.3 FRONT WALL LOADS

The wall facing the explosion source is subjected to a reflection effect as the blast wave impacts the facing wall and reflects back towards the blast source. The reflection effect amplifies the blast pressure on the front or facing side of the building. Because the overpressure at the top and side edges of the front wall is

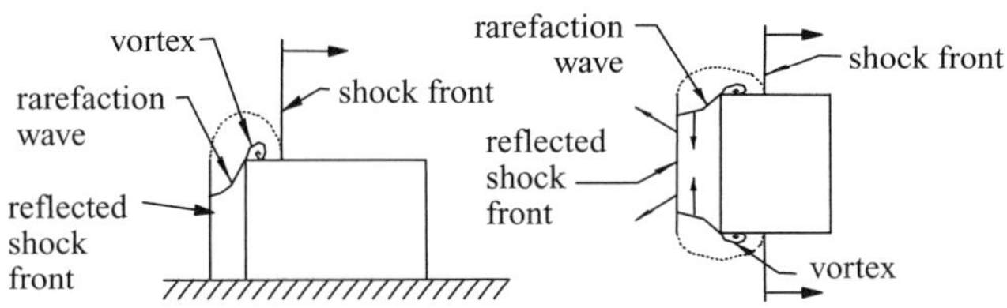

Figure 7.2 Blast Wave at Front Wall (TNO Green Book)

less than the reflected overpressure (Figure 7.2), a decay in the reflection effect takes place that starts at the edges and works inward. The effect is completely removed after what is called the clearing time, t_c. The clearing time is a function of the height and width of the front wall.

The peak reflected overpressure, P_r, can be determined from Figure 7.3 using the free-field peak overpressure. Alternatively, for free-field peak overpressures

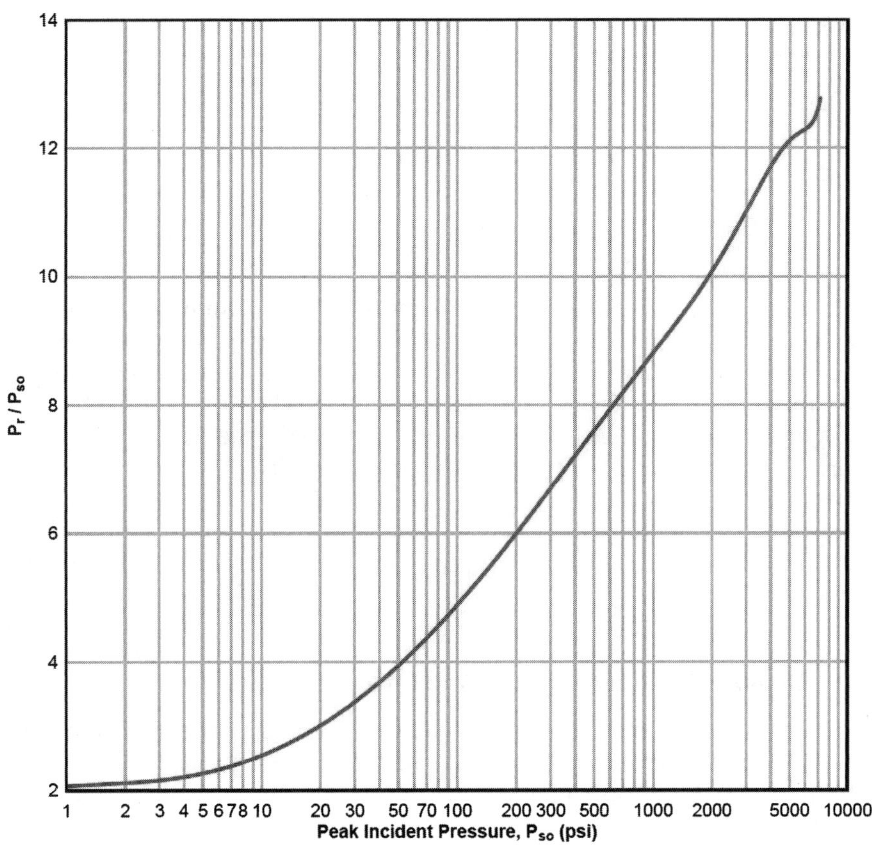

Figure 7.3 Peak Incident Pressure versus the Ratio of Normal Reflected Pressure/ Incident Pressure for a Free-Air Burst (UFC 3-340-02)

less than 40 psi and for sea level atmospheric pressures, the peak reflected overpressure can be determined using the following equation form Newmark:

$$P_r = [2 + 0.05\,(P_{so})]P_{so} \tag{7.2}$$

The clearing time can be calculated using the following equation from UFC 3-340-02.

$$t_c = 4S/[1 + S/G]C_r \tag{7.3}$$

In the preceding equation, S is the lesser of building height or building width. G is the greater of building height or width, and C_r is the velocity of sound and can be determined using Figure 7.4.

For the calculation of front wall dynamic wind pressure, a drag coefficient, C_d, equal to 1.0 is used with the q_o value determined from Figure 7.1.

To compute the remainder of the pressure-time curve, the following equations are used for stagnation pressure, P_s, impulse, I_s, and the effective duration, t_e, based on a simplified straight-line approximation. These values are illustrated in Figure 7.5.

$$P_s = P_{so} + C_d\,(q_o) \tag{7.4}$$

$$I_s = 0.5\,(P_r - P_s)t_c + 0.5P_s t_o \tag{7.5}$$

$$t_e = 2I_s/P_r \tag{7.6}$$

The blast wave angle of incidence affects the blast pressure load on the front wall. This angle is taken as 0° for a blast wave traveling perpendicular into the plane of the front wall where the full reflected overpressure is applied, and taken as 90° for a blast wave traveling parallel to a surface, where the free-field, or side-on overpressure is applied. For intermediate values of the angle of incidence, Figure 7.6 provides coefficients to calculate the applied pressure for use in the following equation.

$$P_{r\alpha} = (C_{r\alpha})(P_{so}) \tag{7.7}$$

7.3.1 Empirical Method—Front Wall Loading Example

The front wall is assumed to span vertically from foundation to roof. The calculation will be for a typical wall segment one foot wide.

With $P_{so} = 10$ psi, use Figure 7.3 to determine the reflected overpressure ratio, $P_r/P_{so} = 2.5$.

$$P_r = (P_r/P_{so})P_{so} = (2.5)(10 \text{ psi}) = 25 \text{ psi}$$

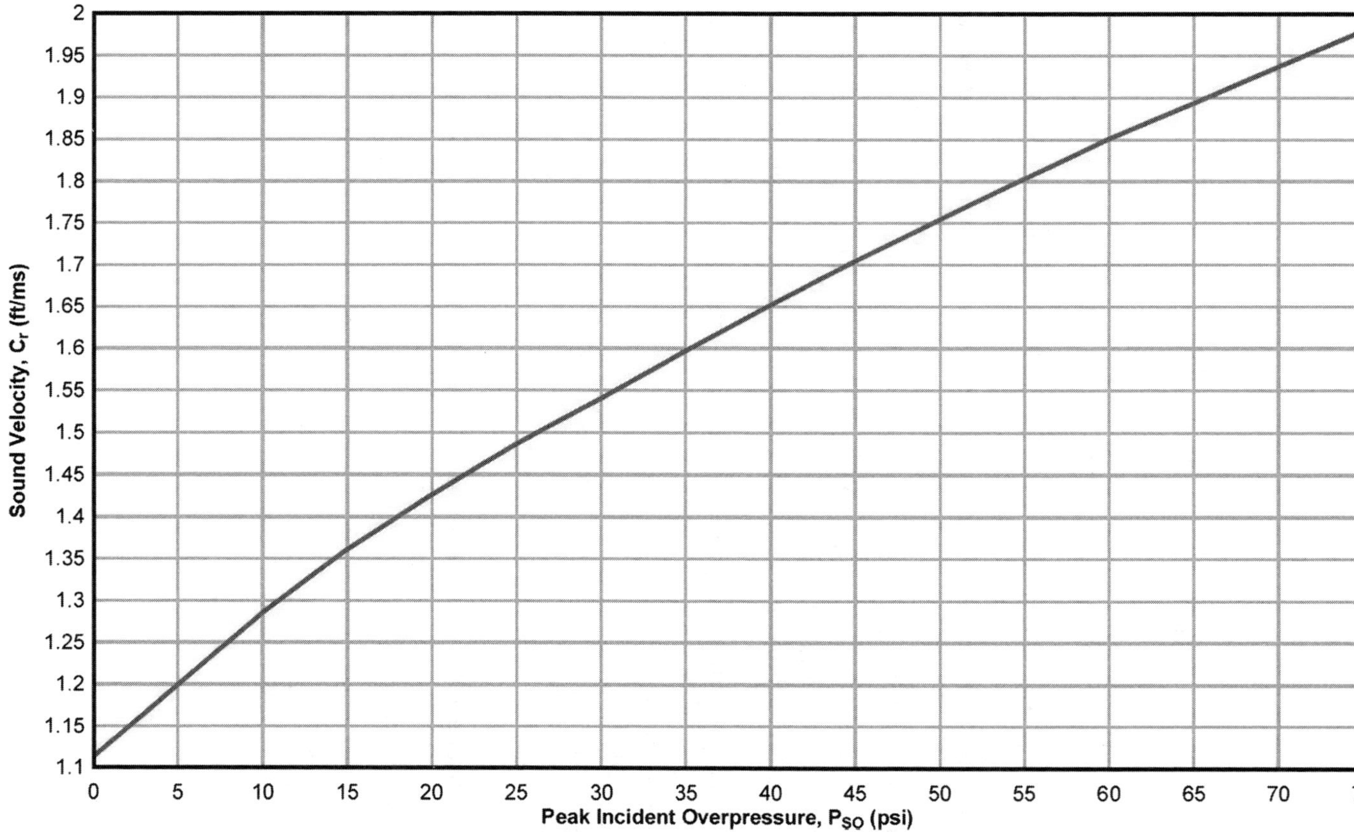

Figure 7.4 Velocity of Sound in Reflected Overpressure Region versus Peak Incident Overpressure (UFC 3-340-02)

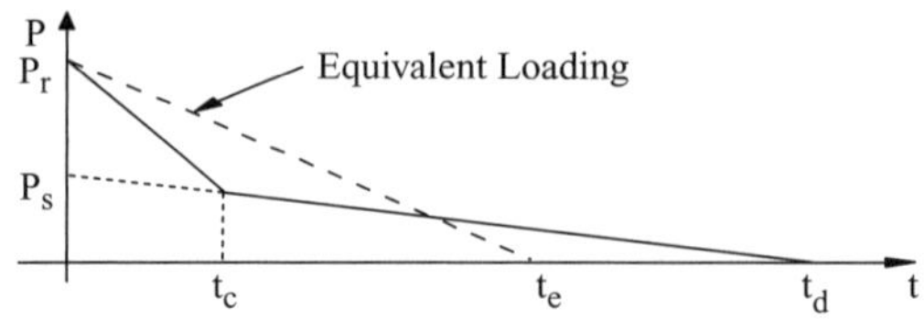

Figure 7.5 Front Wall Loading Diagram

Alternatively, use Equation 7.2 to compute the reflected overpressure,

$$P_r = [2 + 0.05\ (P_{so})]P_{so} = [2 + 0.05(10 \text{ psi})](10 \text{ psi}) = 25 \text{ psi}$$

To compute the reflected pressure clearing time, two parameters are needed,

$S =$ minimum of B_H or B_W = minimum of 20 ft or 150 ft = 20 ft
$G =$ larger of B_H or B_W = larger of 20 ft or 150 ft = 150 ft

With $P_{so} = 10$ psi, use Figure 7.4 to determine the sound velocity in the reflected region, $C_r = 1.28$ ft/ms

Use Equation 7.3 to compute the reflected overpressure clearing time,

$$t_c = 4S/[1 + S/G]C_r = 4(20 \text{ ft})/[1 + (20 \text{ ft})/(150 \text{ ft})](1.28 \text{ ft/ms}) = 55 \text{ ms}$$

This result is nearly equal to t_o, thus $t_c \approx t_o = 56$ ms

Drag coefficient, $C_d = 1.0$

Use Equation 7.4 to calculate the stagnation pressure,

$$P_s = P_{so} + C_d\ (q_o) = (10 \text{ psi}) + (1.0)(2.2 \text{ psi}) = 12.2 \text{ psi}$$

Use Equation 7.5 to calculate the front wall impulse,

$$I_s = 0.5\ (P_r - P_s)t_c + 0.5P_s t_o = 0.5\ [(25 \text{ psi}) - (12.2 \text{ psi})](56 \text{ ms}) + 0.5\ (12.2 \text{ psi})(56 \text{ ms}) = 700 \text{ psi-ms}$$

Use Equation 7.6 to calculate the effective duration,

$$t_e = 2I_s/P_r = 2\ (700 \text{ psi-ms})/(25 \text{ psi}) = 56 \text{ ms}$$

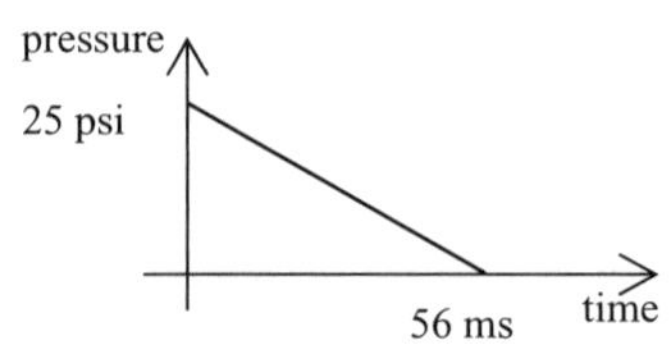

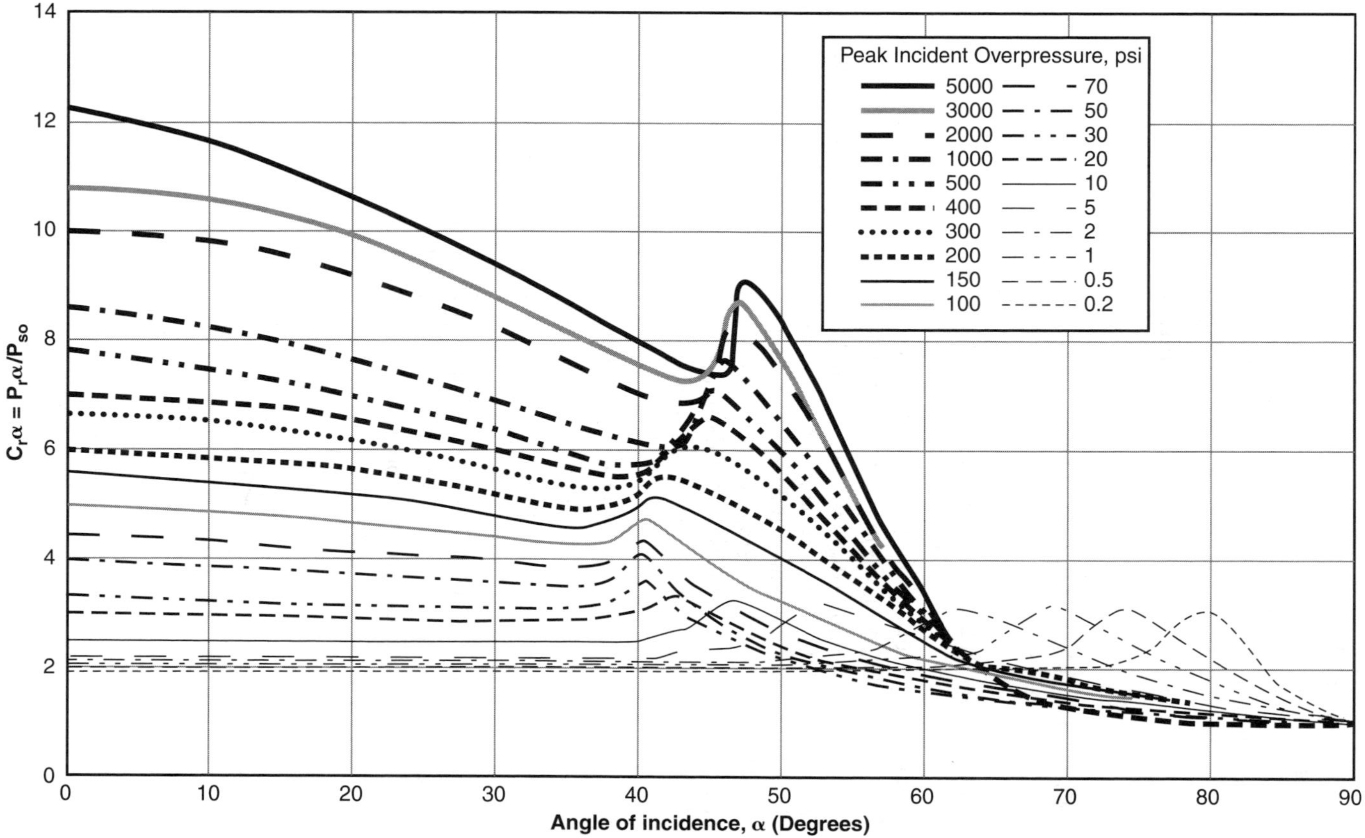

Figure 7.6 Reflected Pressure Coefficient versus Angle of Incidence (UFC 3-340-02)

7.3.2 Empirical Method—Oblique Angle Example

Given angle of incidence, $\alpha = 40$ degrees
Given side-on overpressure, $P_{so} = 10$ psi
Using Figure 7.6, determine the reflected pressure coefficient, $C_{r\alpha} = 2.4$
Using Equation 7.7, calculate the reflected overpressure,

$$P_{r\alpha} = (C_{l\alpha})(P_{so}) = (2.4)(10\text{ psi}) = 24\text{ psi}$$

7.4 SIDE WALL AND ROOF LOADS

Roofs and side walls represent surfaces that are parallel to the path of the advancing blast wave. There is no reflection effect for this situation; however, the average pressure applicable to a specific area, for example a structural element, depends on the length of the blast wave and the length of the structural element. Figure 7.7 shows a diagram of a blast wave traveling across a roof. If a section of the roof tributary to a supporting beam is oriented perpendicular to the traveling wave, then the roof length would be short and the blast wave would have a full effect. If the roof beam were oriented parallel to the traveling wave, then the roof length would be longer, the average pressure would be reduced, and a rise time would become evident. Calculations for these two situations are presented in the side wall and roof examples. Values are obtained as follows:

- Equivalent uniform pressure factor, C_E, is determined from Figure 7.8.
- Scaled uniform pressure rise time, t_r, is determined from Figure 7.9.
- Scaled uniform pressure duration, t_o, is determined from Figure 7.10.

A building's depth could be significant with respect to the distance from the blast source. Thus the peak free-field overpressure for a side or rear surface could be lower than that calculated for the front wall. The calculation of peak free-field

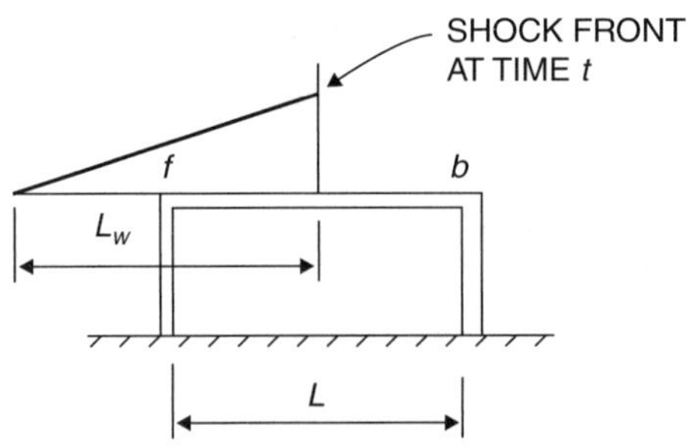

Figure 7.7 Roof Loading Diagram

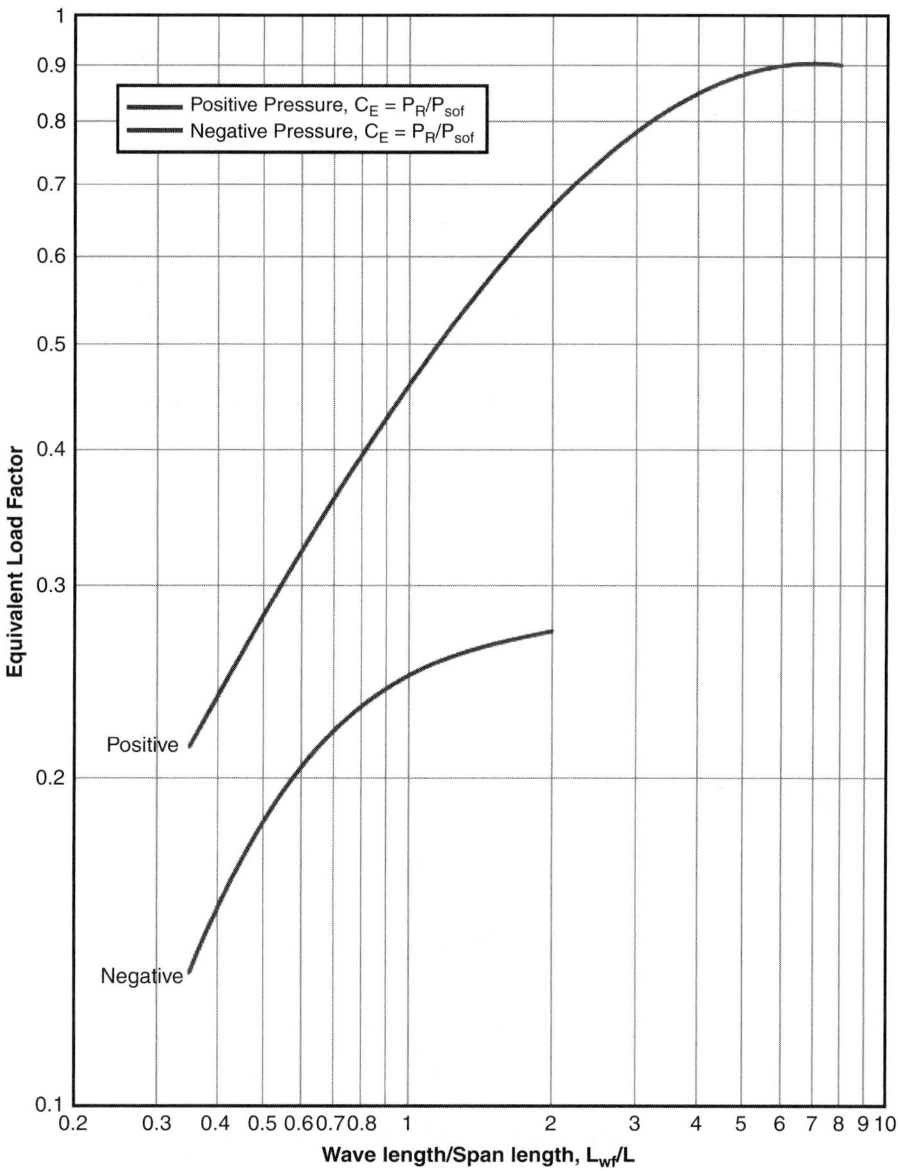

Figure 7.8 Peak Equivalent Uniform Roof Pressure (UFC 3-340-02)

overpressure is determined using the criteria in Chapter 6. In many cases this effect is negligible and is ignored, for simplicity.

For the calculation of roof and side wall dynamic wind pressure, a drag coefficient, C_d, determined from Table 7.1 is used with the q_o value determined from Figure 7.1.

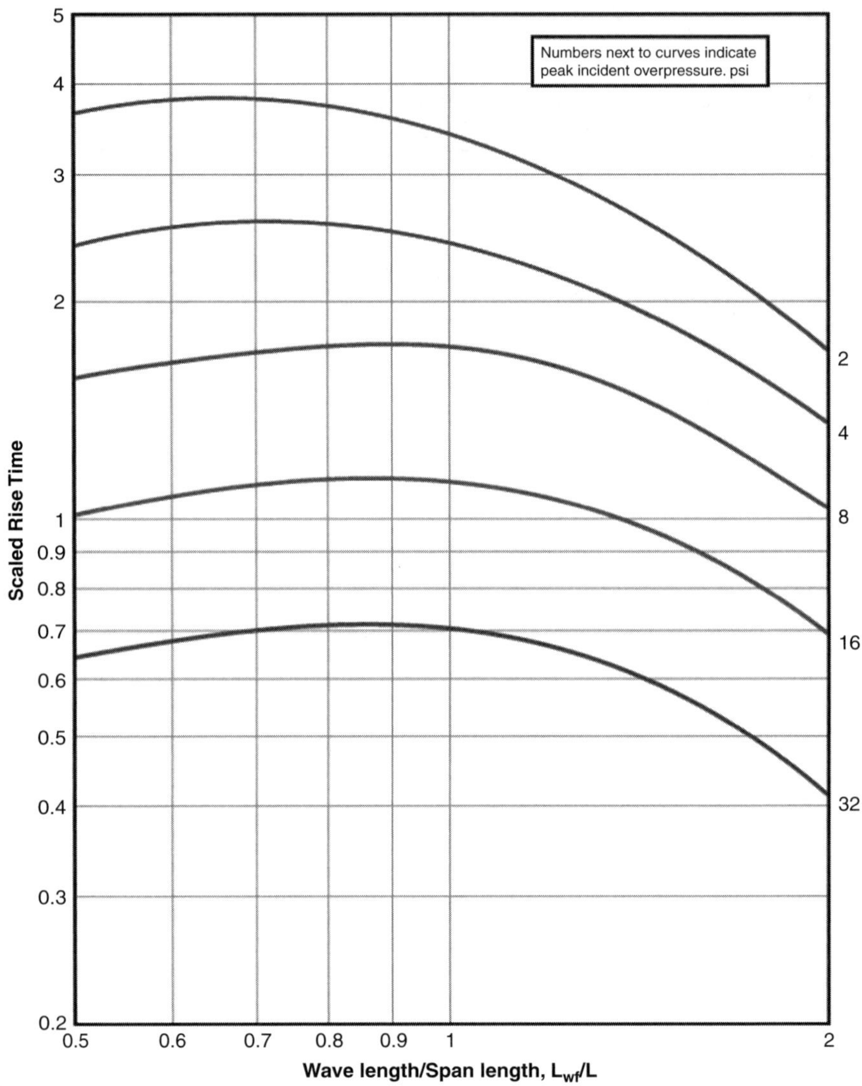

Figure 7.9 Scaled Rise Time of Equivalent Uniform Positive Roof Pressures (UFC 3-340-02)

The total effective pressure on a roof or side wall is determined from the following,

$$P_a = C_E P_{so} + C_d\, q_o \tag{7.8}$$

7.4.1 Empirical Method—Side Wall Loading Example

The side wall is the same as the front wall, spanning vertically from foundation to roof. This calculation will be for a unit width wall segment, $L = 1$ ft.

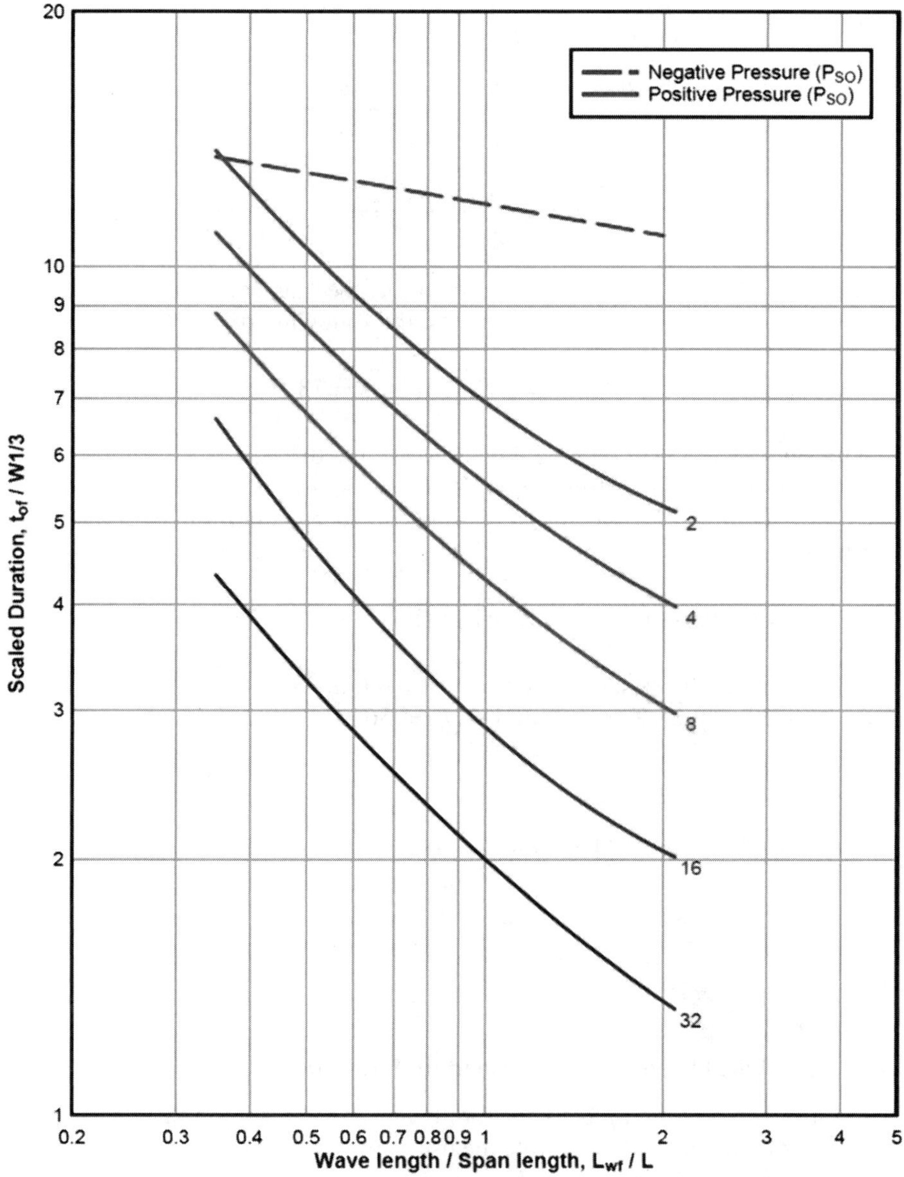

Figure 7.10 Scaled Duration of Equivalent Uniform Roof Pressures (UFC 3-340-02)

Table 7.1 Roof, Side Wall, and Rear Wall Drag Coefficients

Peak dynamic pressure	Drag coefficient
0–25 psi	−0.40
25–50 psi	−0.30
50–130 psi	−0.20

With $q_o = 2.2$ psi, use Table 7.1 to determine the drag coefficient, $C_d = -0.4$
To determine the equivalent load values, the following parameter is needed,

$$L_w/L = (78 \text{ ft})/(1 \text{ ft}) = 78$$

With such a short wall segment in comparison to the blast wave length, it isn't necessary to obtain values from Figures 7.8, 7.9, and 7.10.

By inspection, $C_E = 1.0$, use Equation 7.8 to calculate the effective peak overpressure,

$$P_a = C_E P_{so} + C_d\, q_o = (1.0)(10 \text{ psi}) + (-0.4)(2.2 \text{ psi}) = 9.1 \text{ psi}$$

By inspection, the effective rise time, $t_r = 0.0$ ms
By inspection, the effective duration, $t_d = 56$ ms

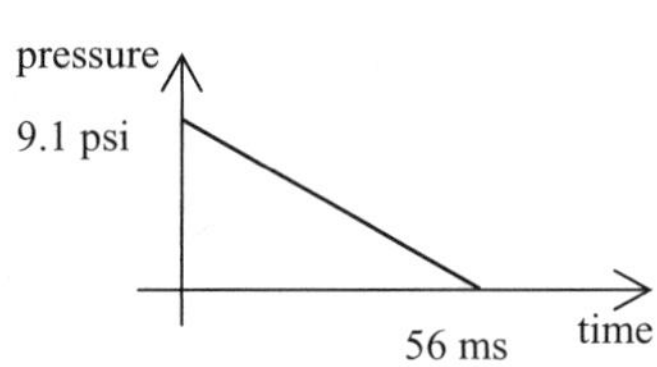

7.4.2 Empirical Method—Roof Loading Example

The roof is a slab spanning between roof beams, and an appropriate span should be used for L. However, to illustrate the use of the figures, a section spanning the entire building will be used, thus $L = 50$ ft

With $q_o = 2.2$ psi, use Table 7.1 to determine the drag coefficient, $C_d = -0.4$
To determine the equivalent load values, the following parameter is needed,

$$L_w/L = (78 \text{ ft})/(50 \text{ ft}) = 1.6$$

With $L_w/L = 1.6$, use Figure 7.8 to determine the effective uniform pressure factor, $C_E \approx 0.60$

Use Equation 7.8 to calculate the effective peak overpressure,

$$P_a = C_E P_{so} + C_d\, q_o = (0.60)(10 \text{ psi}) + (-0.4)(2.2 \text{ psi}) = 5.1 \text{ psi}$$

With $L_w/L = 1.6$, use Figure 7.9 to determine the effective scaled rise time,

$$t_r/W^{1/3} \approx 1.4 \text{ ms/lb}^{1/3}$$

Rise time, $t_r = (t_r/W^{1/3})(W^{1/3}) = (1.4 \text{ ms/lb}^{1/3})(10{,}000 \text{ lb})^{1/3} = 30$ ms
With $L_w/L = 1.6$, use Figure 7.10 to determine the effective scaled duration,

$$t_d/W^{1/3} \approx 3.0$$

Duration, $t_d = (t_d/W^{1/3}(W^{1/3}) = (3.0 \text{ ms/lb}^{1/3})(10{,}000 \text{ lb})^{1/3} = 65$ ms
Total positive phase duration

$$t_o = t_r + t_d = (30 \text{ ms}) + (65 \text{ ms}) = 95 \text{ ms}$$

7.5 REAR WALL LOADS

The rear wall is the wall facing directly away from the blast source, as illustrated in Figure 7.11. The calculation of rear wall blast loads is similar to that for the side walls or roof. In fact, UFC 3-340-02 treats the rear wall as a roof extension.

For the calculation of rear wall dynamic wind pressure, a drag coefficient, C_d, determined from Table 7.1, is used with the q_o value determined from Figure 7.1.

7.5.1 Empirical Method—Rear Wall Loading Example

The example will be for a wall segment 1 foot wide, L = building height = 20 ft

With $q_o = 2.2$ psi, use Table 7.1 to determine the drag coefficient, $C_d = -0.4$

To determine the equivalent load values, the following parameter is needed,

$$L_w/L = (78 \text{ ft})/(20 \text{ ft}) = 3.9$$

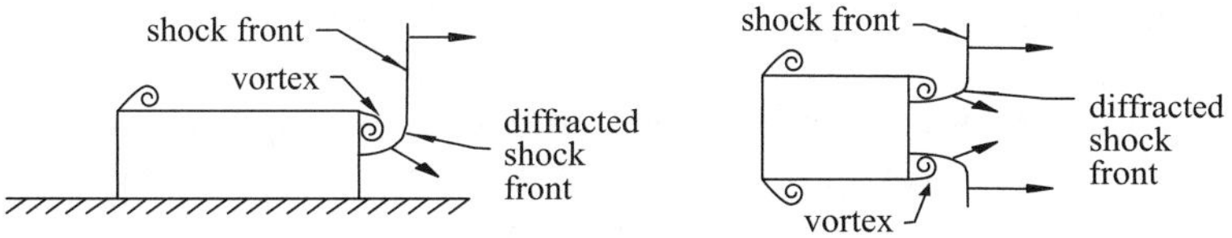

Figure 7.11 Rear Wall (TNO Green Book)

With $L_w/L = 3.9$, use Figure 7.8 to determine the effective uniform pressure factor, $C_E \approx 0.85$

Use Equation 7.8 to calculate the effective peak overpressure,

$$P_a = C_E P_{so} + C_d\, q_o = (0.85)(10 \text{ psi}) + (-0.4)(2.2 \text{ psi}) = 7.6 \text{ psi}$$

With $L_w/L = 3.9$, extrapolate from Figure 7.9 to approximate the effective scaled rise time, $t_r/W^{1/3} \approx 0.5$ ms/lb$^{1/3}$

$$\text{rise time, } t_r = (t_r/W^{1/3})(W^{1/3}) = (0.5 \text{ ms/lb}^{1/3})(10{,}000 \text{ lb})^{1/3} = 11 \text{ ms}$$

Note that an alternative rise time calculation, based on the time for the shock front to travel across the element, would be $t_r = L/U_s =$ (20 ft)/(1.4 ft/ms) = 14 ms

With $L_w/L = 3.9$, extrapolate from Figure 7.10 to approximate the effective scaled duration, $t_d/W^{1/3} \approx 2.1$

Duration, $t_d = (t_d/W^{1/3}(W^{1/3}) = (2.1 \text{ ms/lb}^{1/3})(10{,}000 \text{ lb})^{1/3} = 45$ ms

Note that the duration approaches the free field duration as L_w/L gets large.

Total positive phase duration,

$$t_o = t_r + t_d = (11 \text{ ms}) + (45 \text{ ms}) = 56 \text{ ms}$$

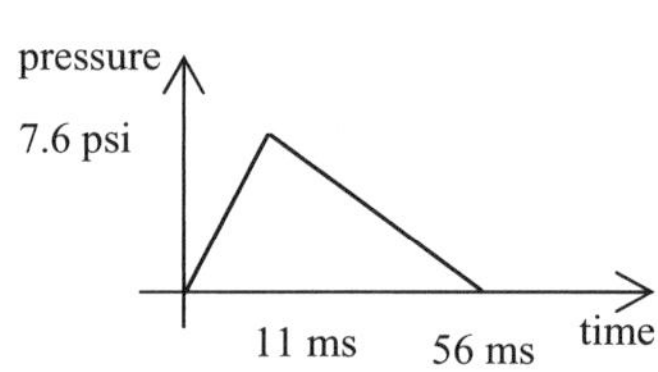

7.6 CONFINED EXPLOSIONS

As discussed in Chapter 6 an interior explosion involves more phenomena than an external explosion, and the loading (Figure 7.12) changes accordingly. In the internal explosion, there is first the directly incident shock as in an external explosion. This is followed by multiple shock reflections off the other surfaces that are confining the explosion. Finally, there is a longer duration gas pressure throughout the interior as the gaseous products of the detonation come to thermodynamic equilibrium.

In practice, it is often convenient to approximate this typical loading with a suddenly applied bilinear pulse, as shown in Figure 7.13. The first part of this loading is the directly incident shock as computed in Section 7.3 for front wall loads. The second part is a triangular approximation to the gas pressure, which generally has a lower peak but longer duration than the shock loading.

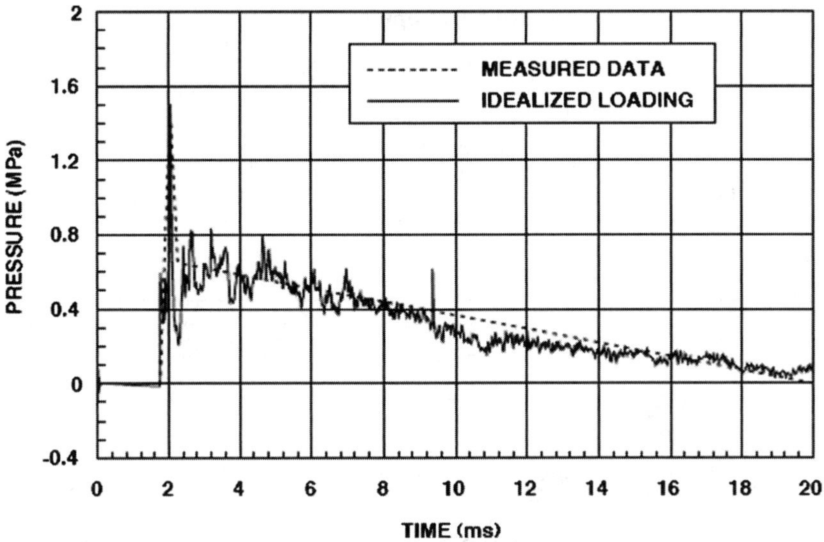

Figure 7.12 Loading from Confined Explosion

To calculate this gas pressure loading, the peak gas pressure, P_g, is obtained from Figure 6.7, as described in the previous chapter. Then the scaled gas impulse, $i_g/W^{1/3}$, is computed from Figures 7.14 through 7.25 as a function of loading density, W/V_f, the scaled reflected shock impulse, $i_r/W^{1/3}$, and the scaled mass, $W_F/W^{1/3}$, of the frangible openings in the confined volume. Table 7.2 indicates which of the Figures 7.14 through 7.25 pertains to various

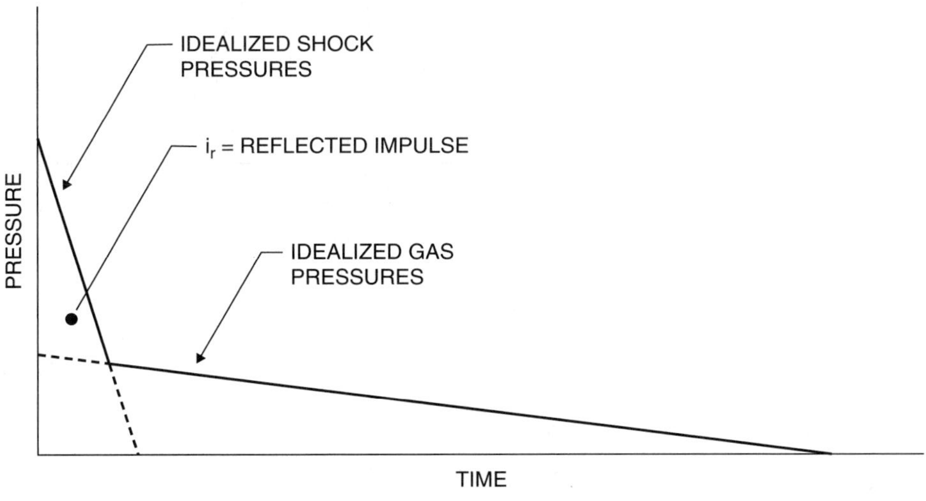

Figure 7.13 Approximate Loading from Confined Explosion

Table 7.2 Figures for Scaled Gas Impulse $ig/W^{1/3}$

	$i_r/W^{1/3} = 20$	$i_r/W^{1/3} = 100$	$i_r/W^{1/3} = 600$	$i_r/W^{1/3} = 2{,}000$
$W/V_f = 0.002$	Fig. 7.14	Fig. 7.15	Fig. 7.16	
$W/V_f = 0.015$	Fig. 7.17	Fig. 7.18	Fig. 7.19	
$W/V_f = 0.15$	Fig. 7.20	Fig. 7.21	Fig. 7.22	
$W/V_f = 1.0$		Fig. 7.23	Fig. 7.24	Fig. 7.25

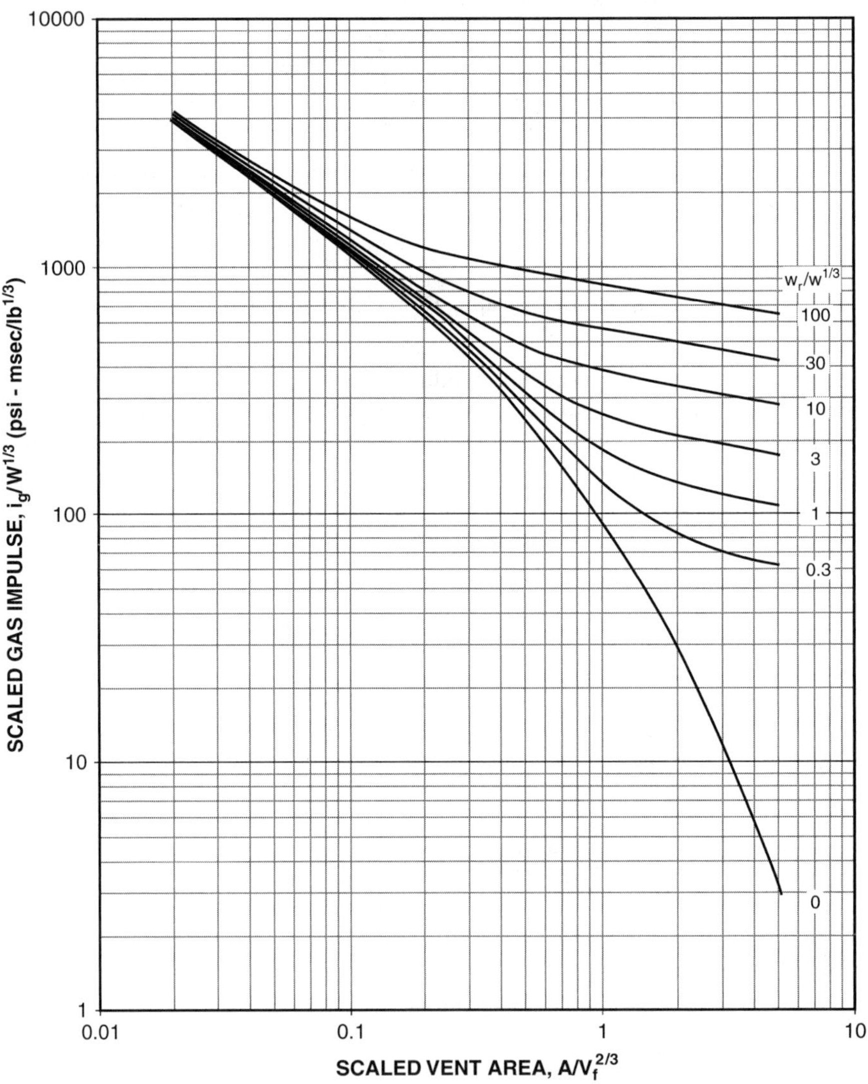

Figure 7.14 Scaled Gas Impulse for $W/V_f = 0.002$ and $i_r/W^{1/3} = 20$

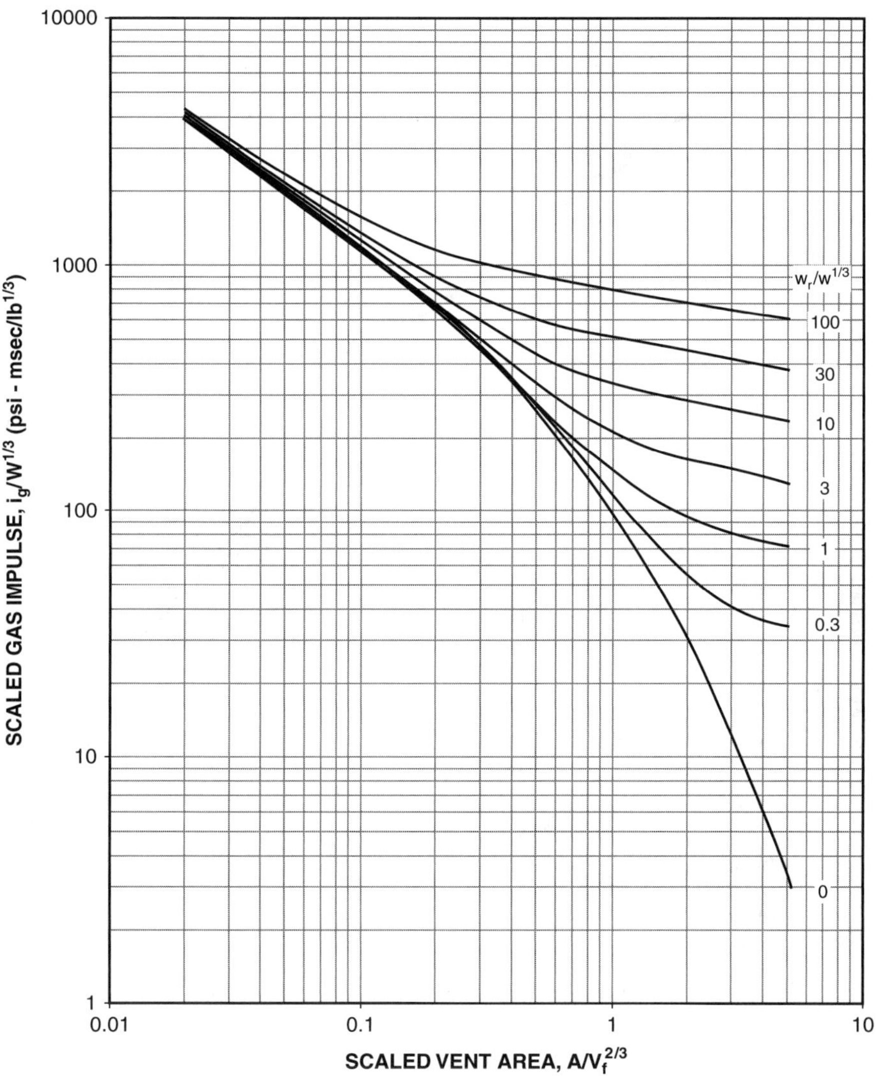

Figure 7.15 Scaled Gas Impulse for $W/V_f = 0.002$ and $i_r/W^{1/3} = 100$

values of W/V_f and $i_r/W^{1/3}$. For values between those given in this table, the scaled gas impulse may be found by interpolation. Once the scaled gas impulse is obtained, the actual impulse follows by unscaling, i.e., multiplication by $W^{1/3}$. The equivalent triangular duration of the gas loading is then $t_g = 2i_r/P_g$.

For example, let us estimate the loading from the detonation of $W = 16$ lbs of TNT at the center of the floor a cubicle that measures 20 ft on a side. Suppose

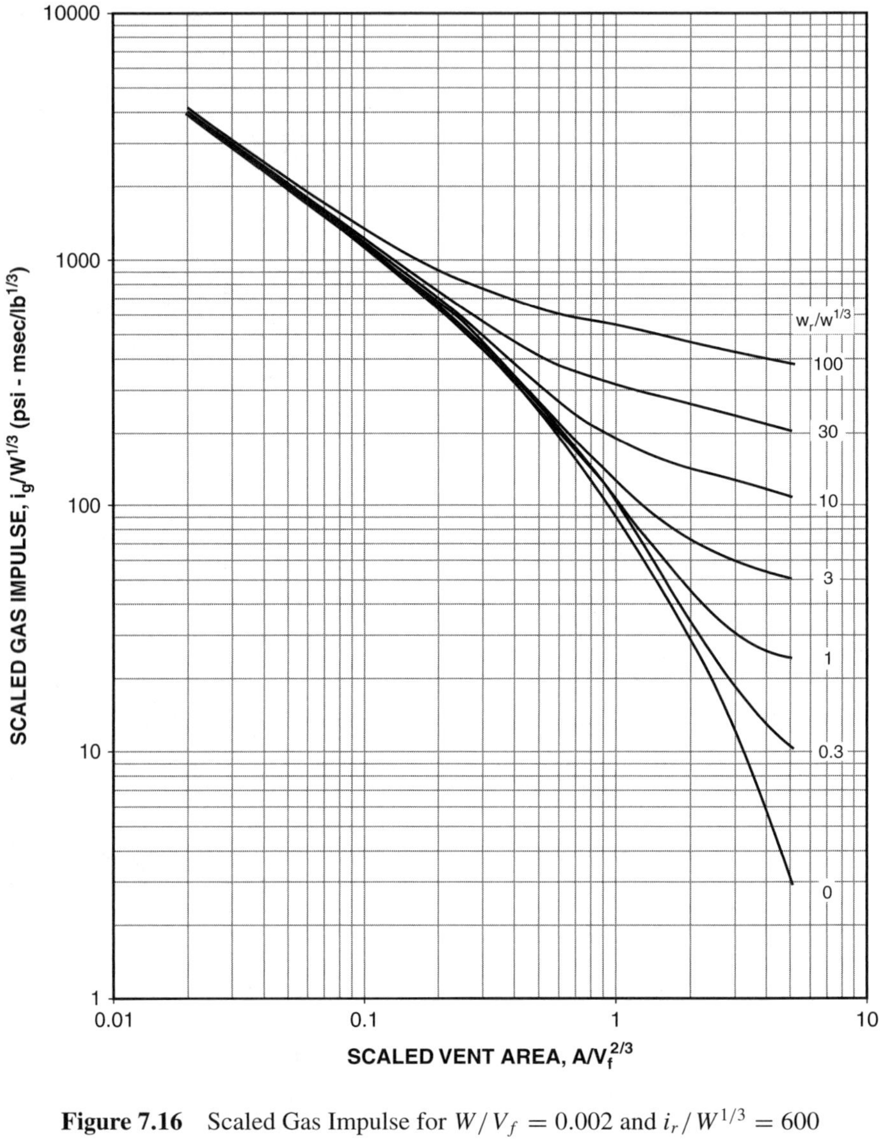

Figure 7.16 Scaled Gas Impulse for $W/V_f = 0.002$ and $i_r/W^{1/3} = 600$

that the total vent area is $A = 20$ ft^2 and the weight of the frangible surfaces is $W_F = 25$ lbs.

Using the method described in Section 7.3, the directly incident shock is described by

$$P_r = 317 \text{ psi}$$
$$t_o = 0.868 \text{ msec}$$

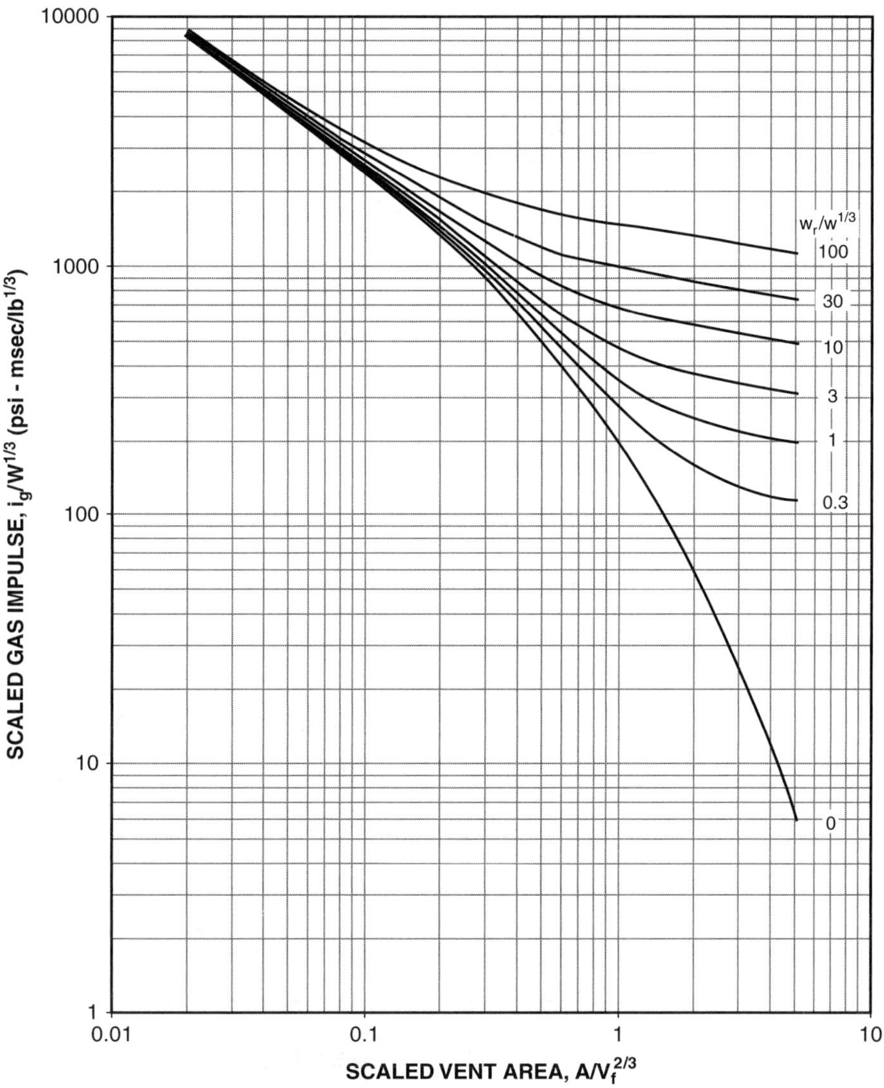

Figure 7.17 Scaled Gas Impulse for $W/V_f = 0.015$ and $i_r/W^{1/3} = 20$

For the gas pressure, the loading density is

$$W/V_f = 16/20^3 = 0.002 \text{ lbs/ft}^3$$

From Figure 6.7, the corresponding peak pressure is

$$P_g = 25.5 \text{ psi}$$

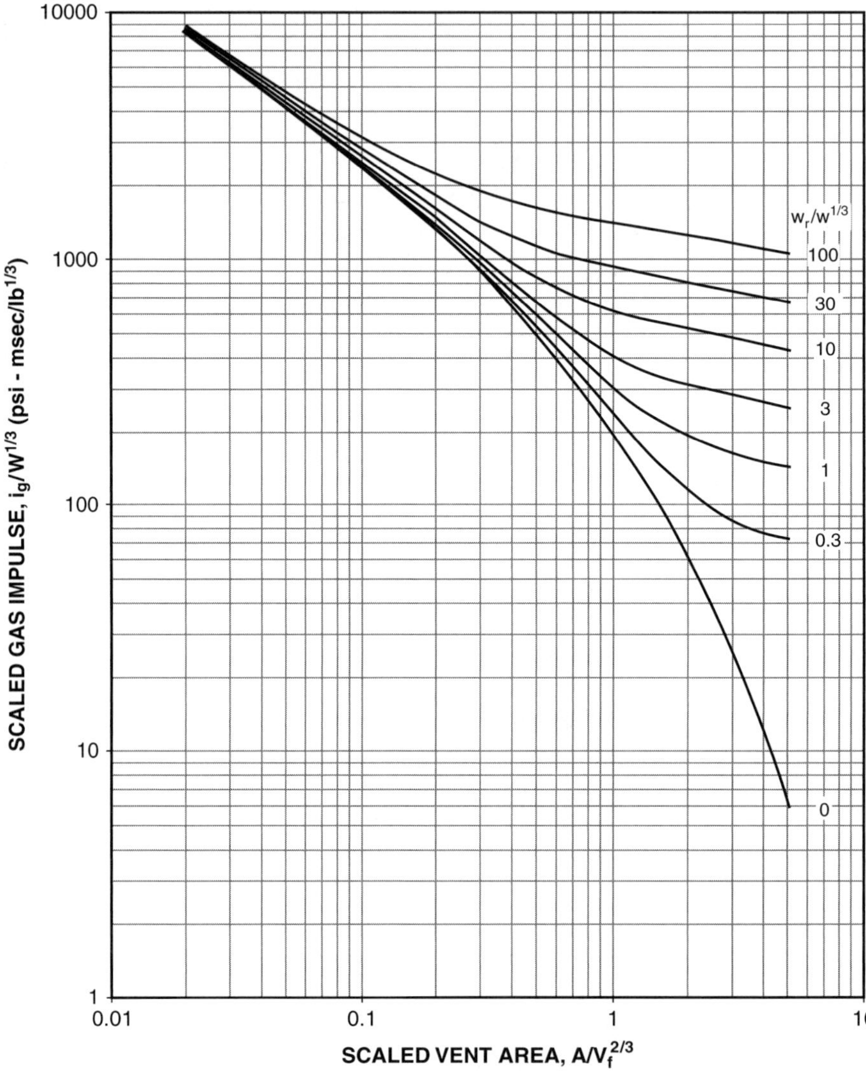

Figure 7.18 Scaled Gas Impulse for $W/V_f = 0.015$ and $i_r/W^{1/3} = 100$

For the scaled reflected shock impulse, $i_r/W^{1/3} = 54.6$ psi $\times$ msec/lb$^{1/3}$
Interpolation between Figures 7.14 and 7.15 gives the scaled gas impulse $i_g/W^{1/3} = 1990$ psi $\times$ msec/lb$^{1/3}$
The actual gas impulse is then

$$i_g = 1990 \times 16^{1/3} = 5010 \text{ psi} \times \text{msec}$$

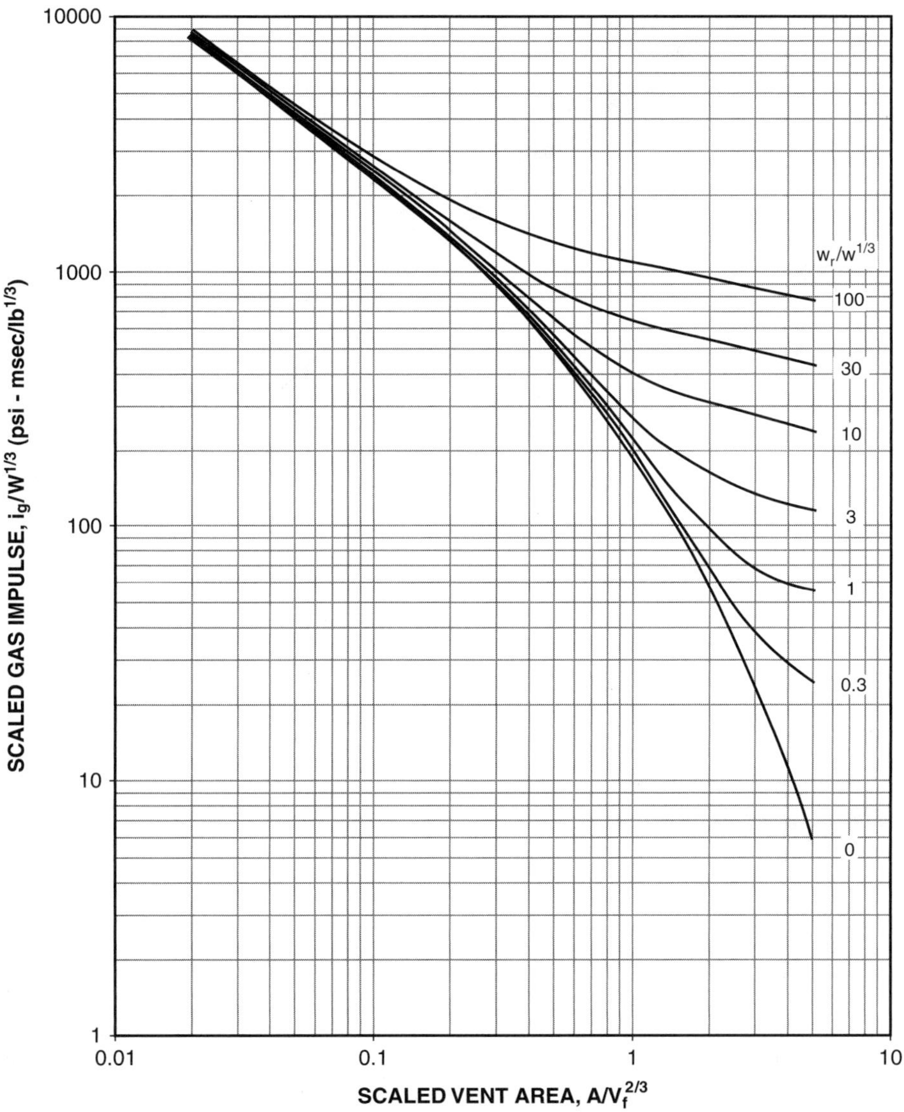

Figure 7.19 Scaled Gas Impulse for $W/V_f = 0.015$ and $i_r/W^{1/3} = 600$

and the equivalent triangular duration is

$$t_g = 2 \times 5010/25.5 = 393 \text{ msec}$$

This dynamic loading is summarized in Figure 7.26. As $t_o \ll t_g$ in this case, the shock portion appears very close to the vertical axis.

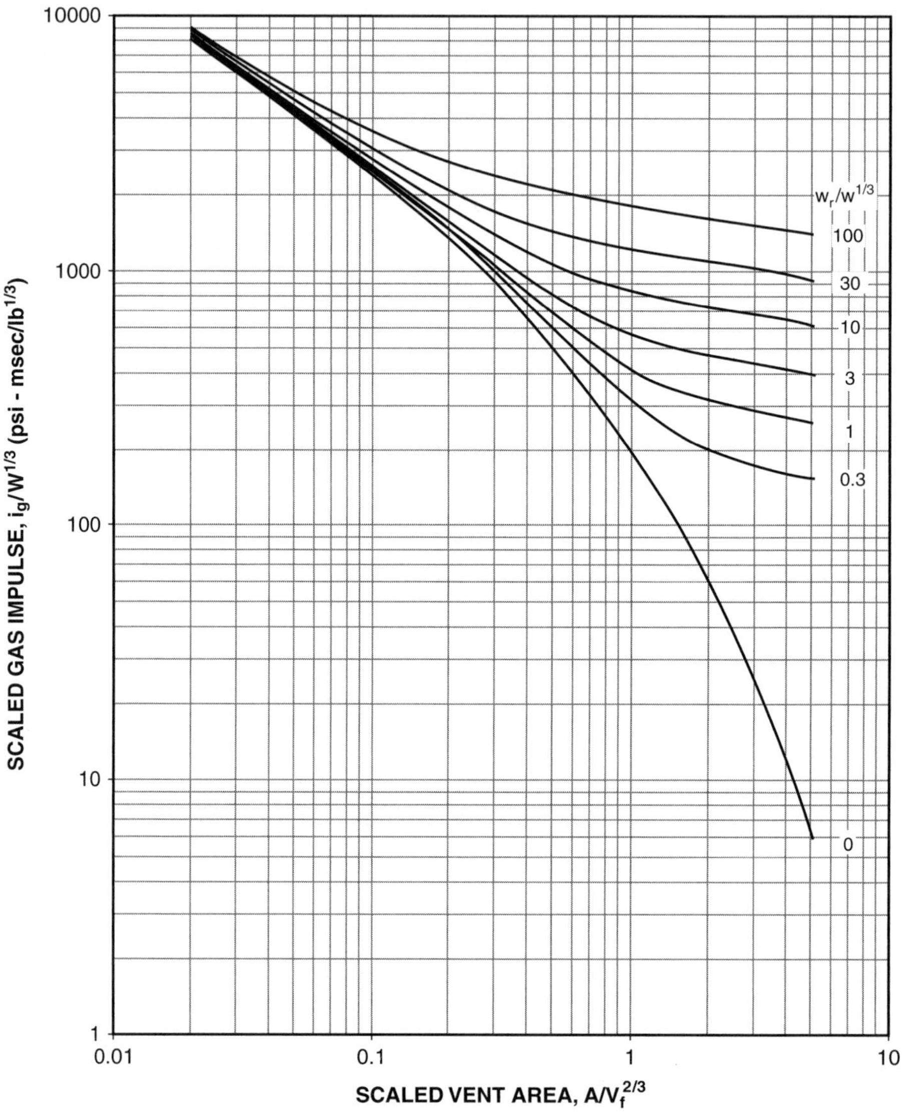

Figure 7.20 Scaled Gas Impulse for $W/V_f = 0.15$ and $i_r/W^{1/3} = 20$

7.7 LEAKAGE

If the openings in the exterior of a structure are not designed for an expected blast loading, there will be a leakage of pressure into the interior. Idealized pressure-time series for this loading are discussed in UFC 3-340-02 as a function

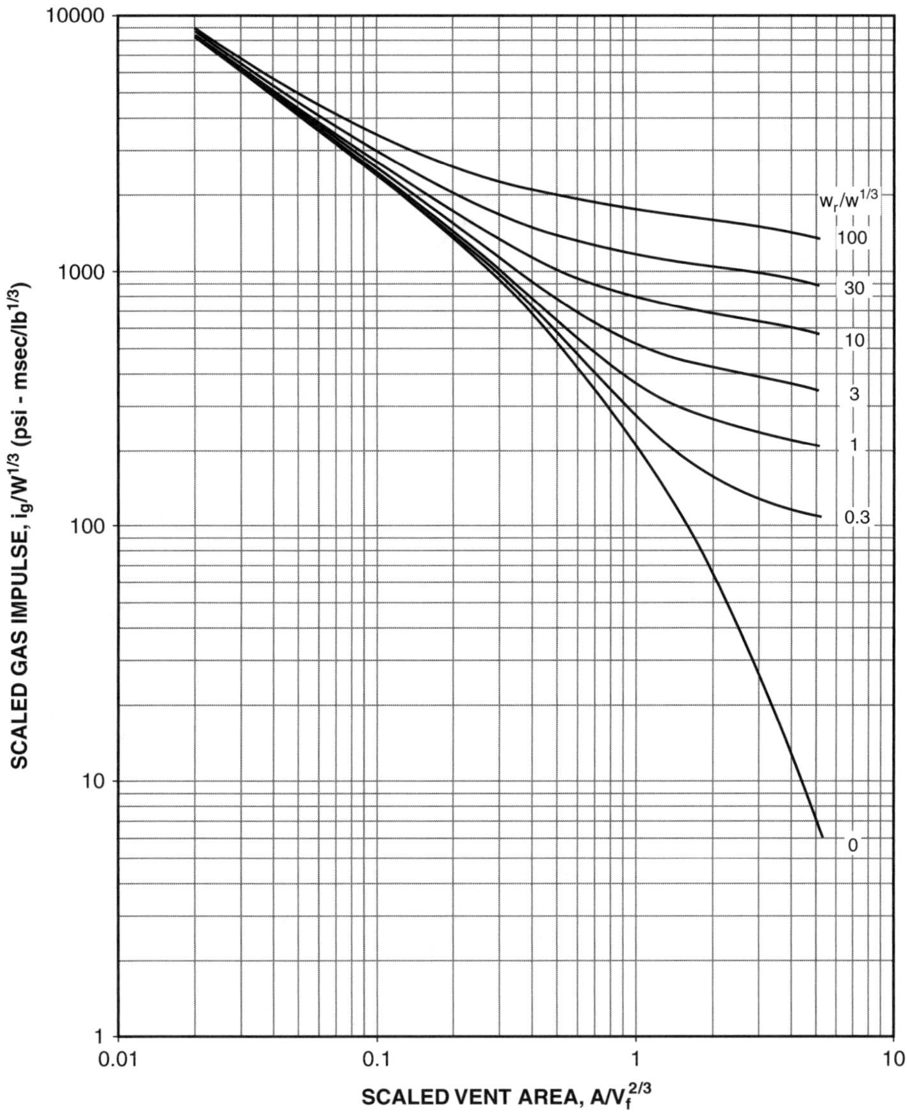

Figure 7.21 Scaled Gas Impulse for $W/V_f = 0.15$ and $i_r/W^{1/3} = 100$

of the structural geometry and shock wave characteristics. Different curves are developed for the exterior front wall, interior front wall, interior side wall and roof, and interior back wall. This manual also discusses closures that are deliberately designed to resist the blast but have small openings for vents and ducts that admit some overpressure.

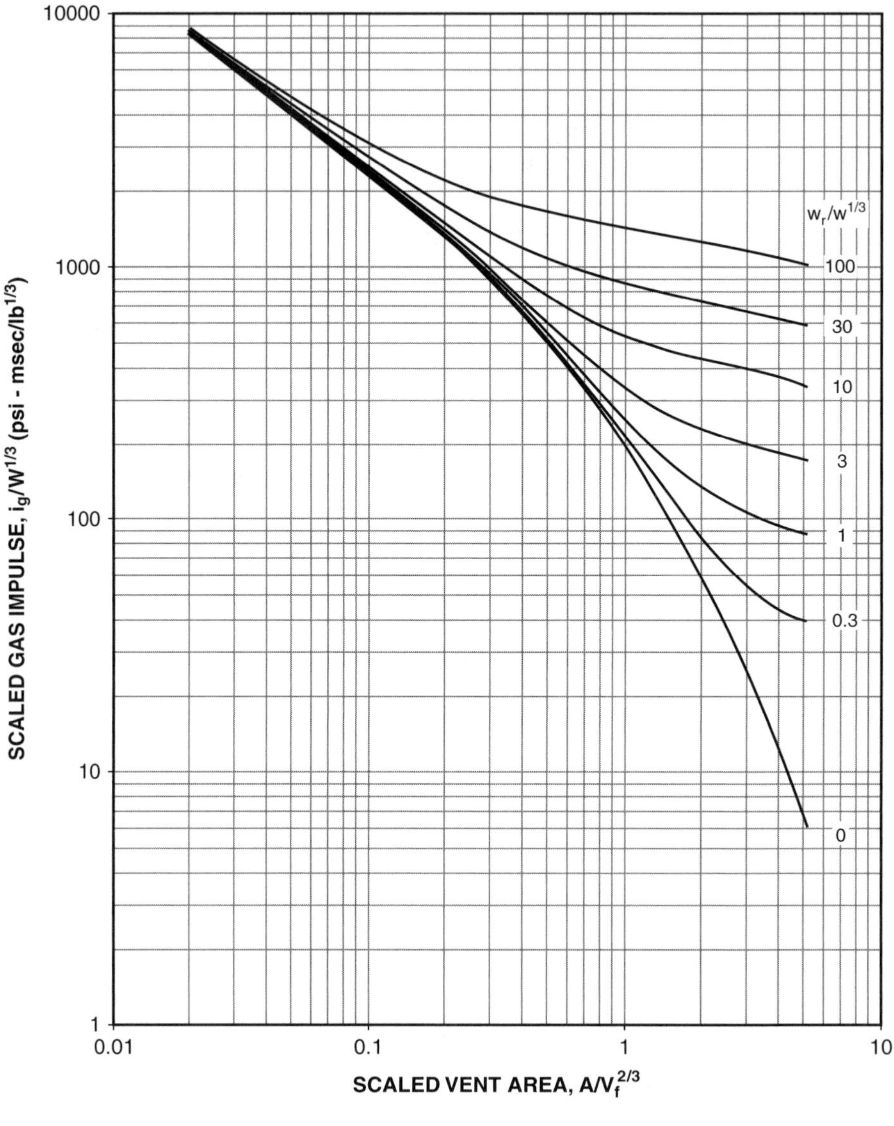

Figure 7.22 Scaled Gas Impulse for $W/V_f = 0.15$ and $i_r/W^{1/3} = 600$

7.8 RAY-TRACING PROCEDURES

Ray-tracing algorithms have been developed to provide a higher resolution of the blast loading than provided by the previously discussed empirical methods in complex geometrical situations. These procedures generally work backwards from each location of interest on a structure to capture a finite number

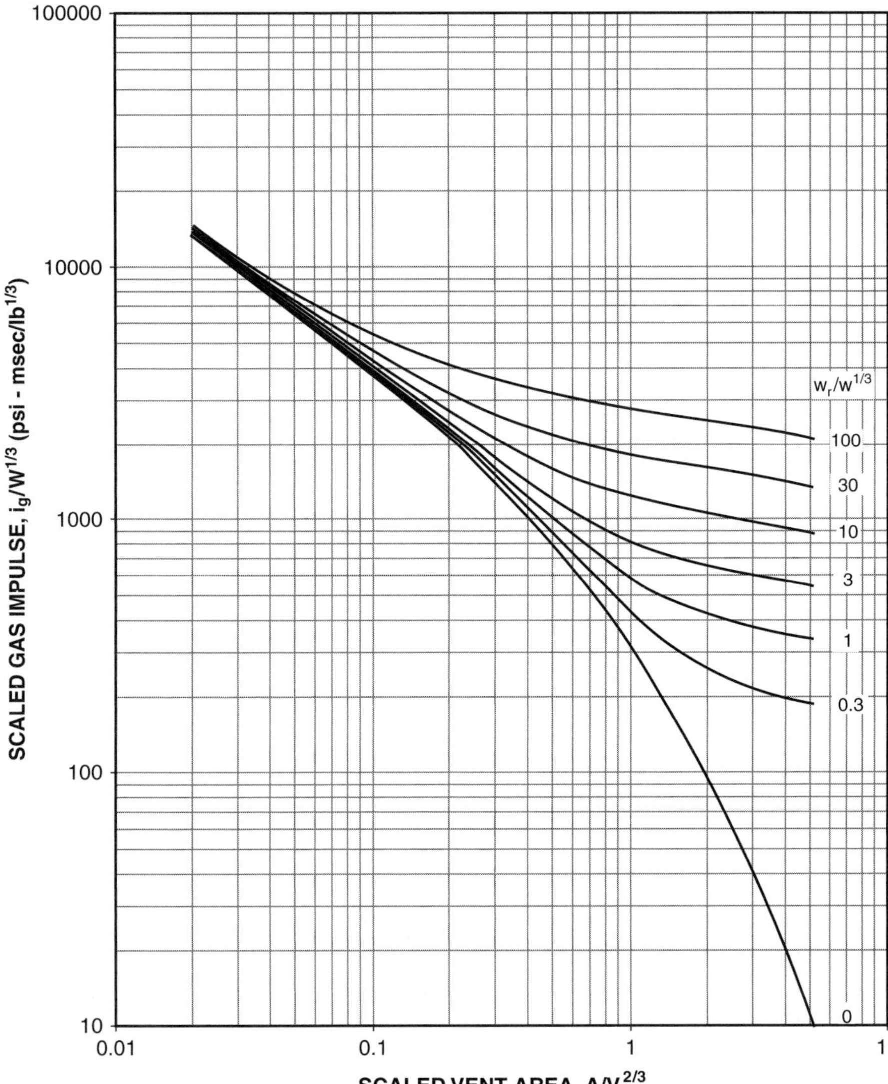

Figure 7.23 Scaled Gas Impulse for $W/V_f = 1.0$ and $i_r/W^{1/3} = 100$

of the possible multiple reflections at each location. As more reflections are considered, the fidelity increases, but so does the computational effort. In practice, the coefficients of such physics-based models are "tuned" to match test data. Accordingly, these are often termed "semi-empirical" models. Their applicability is limited to configurations and charge weight ratios for which data are available. Some semi-empirical codes have the capability to approximate shock

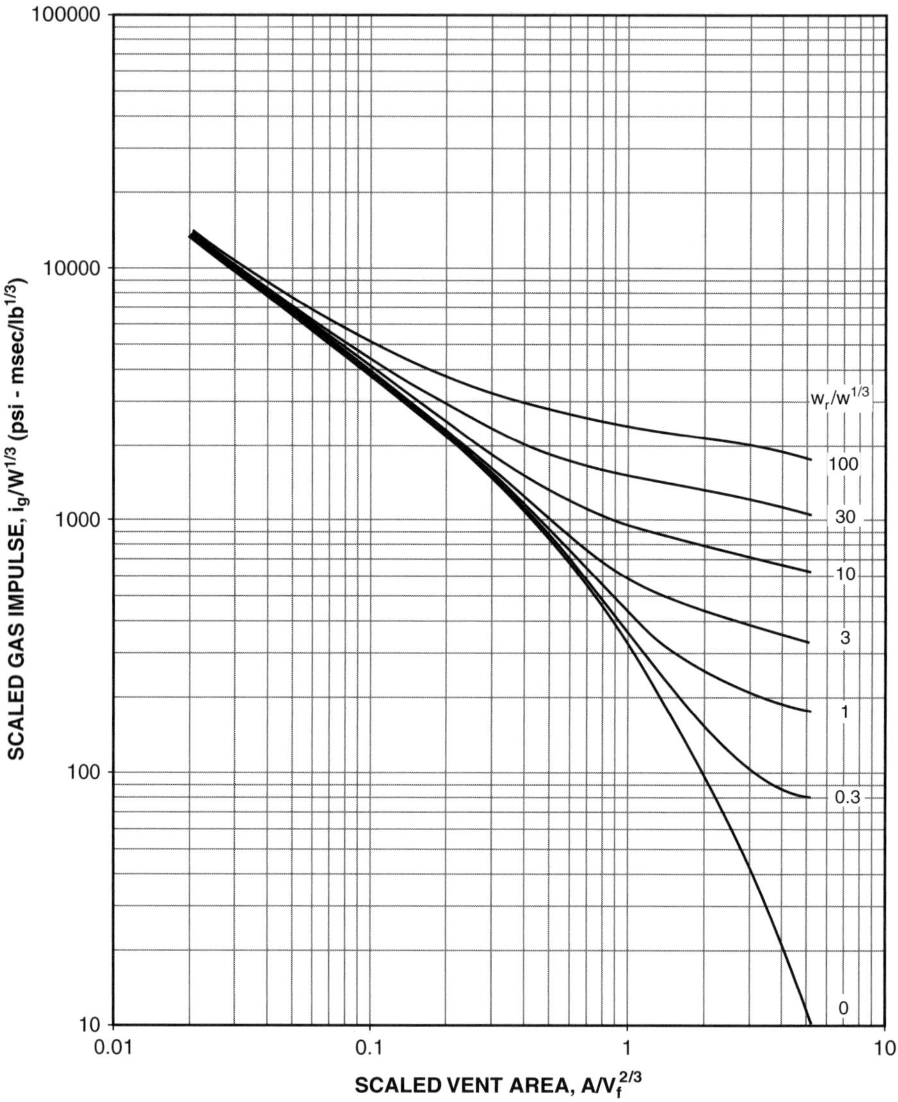

Figure 7.24 Scaled Gas Impulse for $W/V_f = 1.0$ and $i_r/W^{1/3} = 600$

diffraction, shielding, and reflection. In both fidelity and computational effort, these methods lie between the previously discussed fully empirical procedures and fully physics-based computational fluid dynamics. These semi-empirical methods have been developed primarily by defense-related agencies and are restricted in distribution to the government and its contractors.

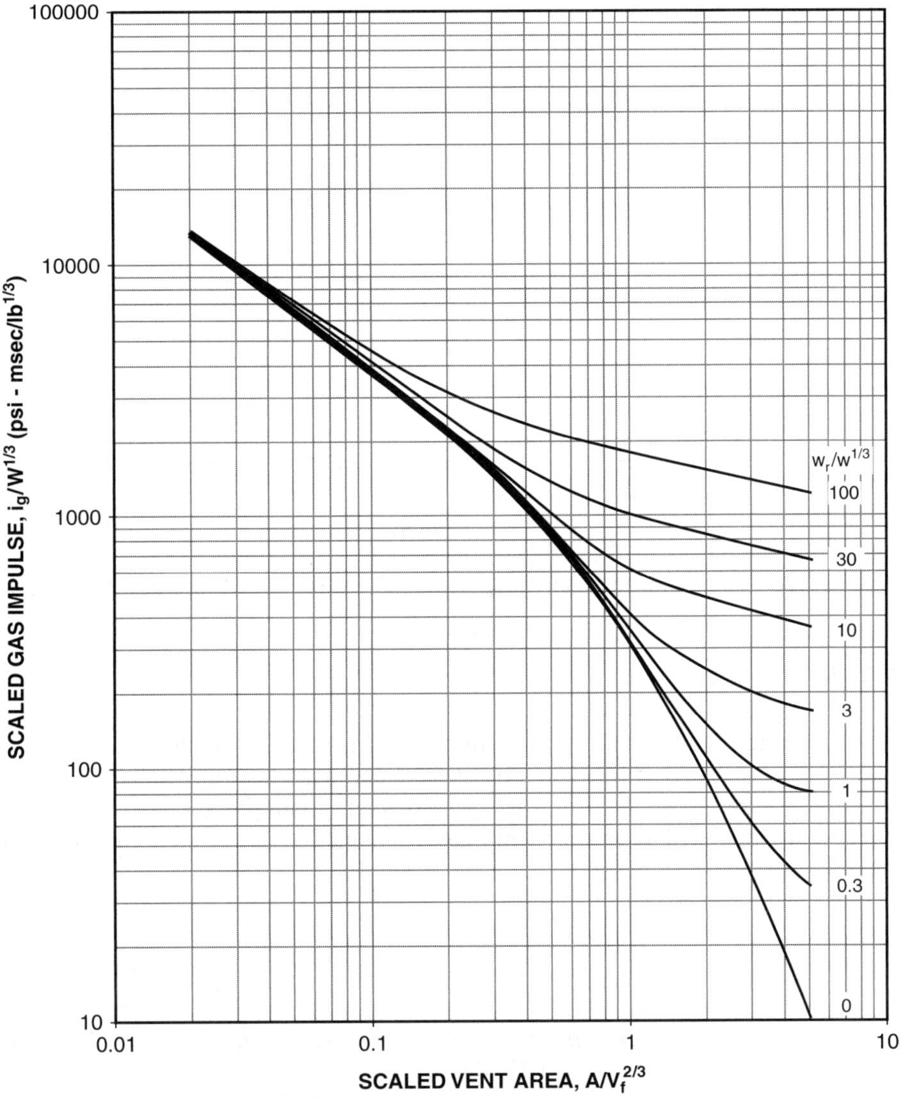

Figure 7.25 Scaled Gas Impulse for $W/V_f = 1.0$ and $i_r/W^{1/3} = 2{,}000$

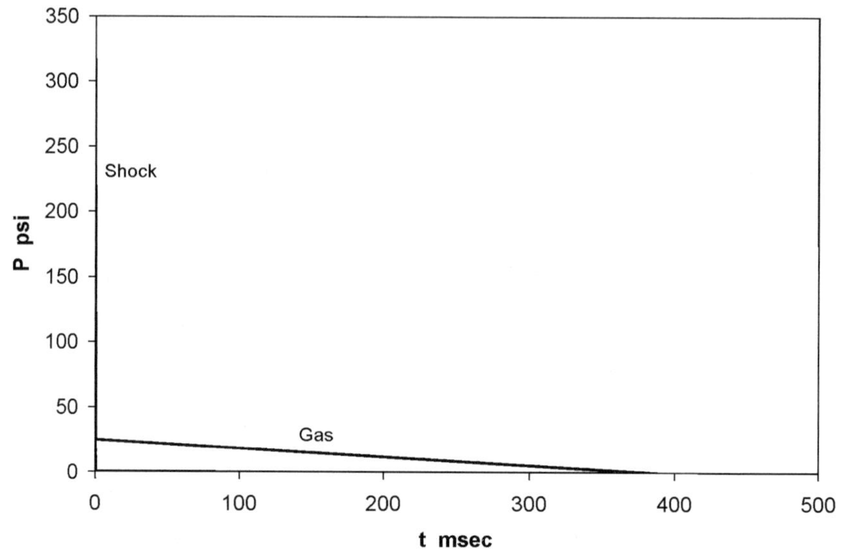

Figure 7.26 Approximate Loading for Detonation of 16 lbs of TNT in a 20-ft Cubicle

7.9 SUMMARY

In this chapter, methods for determining the dynamic blast loading on structural elements from the characteristics of blast phenomena have been discussed and illustrated with examples. Emphasis was placed on simple approximate procedures that apply widely in practical blast-resistant design. More comprehensive methods were also discussed.

REFERENCES

ASCE. 1985. *Design of Structures to Resist Nuclear Weapons Effects* (Manual No. 42). Reston, VA: American Society of Civil Engineers.

ASCE Petrochemical. 1997. *Design of Blast Resistant Buildings in Petrochemical Facilities.* Reston, VA: Task Committee on Blast Resistant Design, American Society of Civil Engineers.

ASCE Physical Security. 1999. *Structural Design for Physical Security*, Reston, VA: American Society of Civil Engineers.

Biggs J. M. 1964. *Introduction to Structural Dynamics*. New York: McGraw-Hill Book Company.

Committee for the Prevention of Disasters Due to Dangerous Substances. 1992. Method for the determination of possible damage to people and objects resulting from releases of hazardous materials (CPR 16E) *TNO Green Book*. The Hague, The Netherlands: The Director-General of Labour.

Mays, G. C. and P. D. Smith, eds. 1995. *Blast Effects on Buildings*. London: Thomas Tedford Publications.

Newmark, Nathan M. 1956. An engineering approach to blast-resistant design (Paper 2786). *ASCE Transactions* **121**: 45–64. Reston, VA: American Society of Civil Engineers.

U.S. Department of Defense. 1965. *Design of Structures to Resist the Effects of Atomic Weapons* (TM 5-856). Washington, DC: U.S. Department of Defense.

________. 1990. *Structures to Resist the Effects of Accidental Explosions* (TM 5-1300). Washington, DC: U.S. Department of Defense.

________. 2008. *Structures to Resist the Effects of Accidental Explosions* (UFC 3-340-02). Washington, DC: U.S. Department of Defense.

8 Fragmentation

Kim King

8.1 INTRODUCTION

Injury and structural damage during an explosion may result not only from the direct pressure and impulse of an air blast, but also from the impact of flying objects called debris and fragmentation. Military weapons are typically explosive charges with some type of metal casing. Upon detonation, this case is ruptured and expelled as fragmentation at high velocity. Similarly, terrorist devices may be embedded with objects such as ball bearings or nails that will be ejected at high velocity. Other fragmentation may occur as objects interact with the blast wave created from a detonation and become airborne. Fragments can range in size from small irregularly shaped objects (fractions of an inch) to large objects such as portions of buildings or vehicles. Two types of fragments may be generated during an explosion: primary and secondary fragments.

Fragmentation generated during an explosion can damage structures by penetrating walls as well as impacting structural elements with sufficient impulse to cause material failure or global structural failure of a structural element. This chapter discusses fragmentation resulting from an explosion and established methods of predicting the hazards of airborne fragments.

8.2 DEBRIS

Debris generically refers to the broken remains of a destroyed object. Debris from a destroyed structure near an explosion can be thrown as secondary fragmentation.

8.3 LOADINGS

Fragmentation can typically be separated into two categories: primary and secondary fragmentation. Primary fragments are typically produced from the container or casing of an explosive charge. Primary fragments can also be generated in manufacturing processes where equipment or tools are in contact with explosive material. Examples of primary fragments include the casing of conventional munitions, metal containers such as kettles and hoppers used in the manufacture of explosives, the steel casing of pipe bombs, and the metal housing of rocket engines. For high explosives, the casing will typically shatter into many small fragments traveling at very high initial velocities up to several thousand feet per second. Primary fragments are typically small, on the order of 1 gram. They typically have "chunky" or irregular geometry with linear dimensions all of the same order. However, some casings are scored or formed to fail in a specific manner to generate a specific fragment pattern.

Secondary fragments are produced when the shock wave encounters objects or structures located near the source of detonation. Secondary fragments are also known as "secondary debris" or "secondary missiles." Secondary fragments are generally larger than primary fragments and have lower initial velocities. Examples of secondary fragments include broken glass from window systems, debris from exterior walls, and dislodged architectural ornamentation. Secondary fragmentation can also include items located near (but not in contact with) the explosive source, such as parts of an automobile near a terrorist vehicle bomb. Damage from secondary fragments is not easy to accurately predict because of the variation in size, shape, and initial velocity.

8.3.1 Primary Fragmentation

Much work has been done to characterize the type of fragments and corresponding fragment velocity of military munitions. Many factors can affect the primary fragment shape, mass, and velocity, including the explosive type, shape of the explosive charge, arrangement of casing, and so forth. For simplification, the most common shape of a cased explosive (a cylinder) with uniform case thickness will be considered.

Primary Fragment Shape The shape of primary fragments may be predictable if, for example, the casing is scored to produce a particular shape, or a specific munition, with a known fragment shape, is known. But a standard shape for primary fragments is typically assumed for design purposes where a specific shape is not known. The typical fragment shape is shown in Figure 8.1.

Primary Fragment Mass For the most common device shape, a uniform cylinder, and a given confidence level ($C_L \leq 0.999$), the design fragment weight can be estimated with the following procedure:

$$W_{df} = M_A^2 \cdot \ln^2(1 - C_L) \tag{8.1}$$

where: W_{df} = design fragment weight (ounces)
M_A = fragment distribution factor
C_L = confidence level

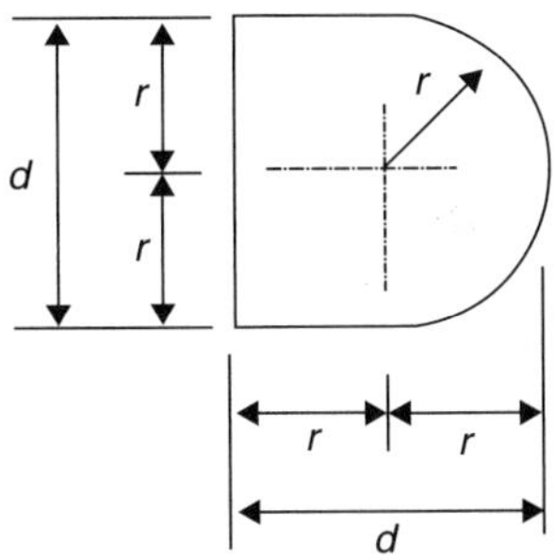

d = Diameter of cylindrical portion of fragment

r = Radius of hemispherical portion of fragment ($d/2$)

W_f = Fragment weight, $0.645 \cdot d^3/\gamma = 0 \cdot 186 \cdot d^3$

D = Caliber density = 0.186.lb/in^3

N = Nose shape factor = $0.72 + 0.25 \cdot (n - 0.25)^{1/2} = 0.845$

n = Caliber radius of the tangent ogive of the fragment nose = $r/d = 0.5$

Figure 8.1 Standard Primary Fragment Shape

$$M_A = B \cdot t_c^{5/6} \cdot d_i^{1/3} \cdot (1 + t_c/d_i) \tag{8.2}$$

where: B = explosive constant (see Table 8.1)
t_c = average casing thickness (inch)
d_i = average inside diameter of casing (inch)

Table 8.1 Mott Scaling Constants for Mild Steel Casings and Various Explosives (Mott 1947)

Explosive	A (oz$^{1/2}$ inches$^{-3/2}$)	B (oz$^{1/2}$ inches$^{-7/6}$)
Baratol	—	0.512
Composition A-3	—	0.22
Composition B	0.214	0.222
Cyclotol (75/25)	—	0.197
H-6	—	0.276
HBX-1	—	0.256
HBX-3	—	0.323
Pentolite (50/50)	0.238	0.248
PTX-1	—	0.222
PTX-2	—	0.227
RDX	0.205	0.212
Tetryl	0.265	0.272
TNT	0.302	0.312

Table 8.2 Gurney Energy

Explosive	Specific Weight (lb/inch^3)	$\sqrt{2 \cdot E'}$
Composition B	0.0621	9100
Composition C-3	0.0578	8800
HMX	0.0682	9750
Nitromethane	0.0411	7900
PBX-9404	0.0664	9500
PETN	0.0635	9600
RDX	0.0639	9600
TACOT	0.0581	7000
Tetryl	0.0585	8200
TNT	0.0588	8000
Trimonite No. 1	0.0397	3400
Tritonal (TNT/AL - 80/20)	0.0621	7600

Primary Fragmentation Velocity The most common technique for calculating the initial velocity of primary fragmentation in contact with an explosive charge, the Gurney method, is provided here (Gurney 1947). The initial velocity of primary fragments produced from a cylindrical cased explosive is a function of the explosive output and the ratio of the explosive charge weight to casing weight. The initial velocity of primary fragments resulting from a cylindrical cased explosive with evenly distributed explosives and casing is expressed as:

$$V_{op} = FS \cdot \sqrt{2 \cdot E'} \cdot \sqrt{\frac{W/W_c}{1 + 0.5 \cdot W/W_c}} \tag{8.3}$$

where: V_{op} = initial velocity of primary fragment (feet/sec)
$\sqrt{2 \cdot E'}$ = Gurney Energy Constant (see Table 8.2)
FS = factor of safety (1.2)
W = design charge weight = $FS \cdot W_{act}$ (feet/sec)
W_{act} = actual quantity of explosive material (lb)
W_c = weight of casing (lb)

8.3.2 Secondary Fragmentation

Secondary fragments generated during an explosion can represent a wide variety of sizes, shapes, and initial velocities. An array of scenarios must typically be considered to establish the design fragment (worst-case loading on the structure). Varying factors such as fragment size, shape, structural constraint, and orientation must be evaluated to establish the controlling fragment scenario that produces the most severe response of the structure in question.

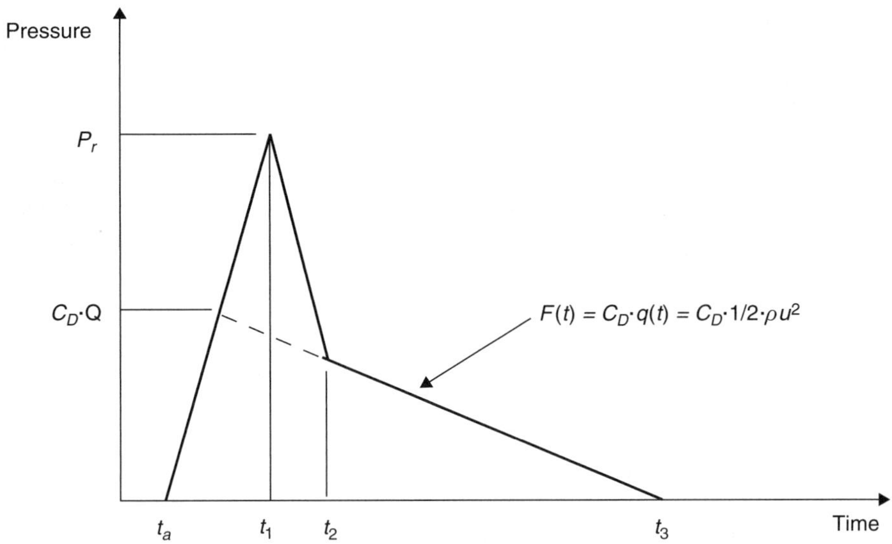

Figure 8.2 Idealized Time Loading on an Irregular Fragment

Initial steps for identifying the hazard associated with secondary fragmentation are listed below.

1. Define blast loading (see Chapter 7 for blast loading).
2. Determine distance from the center of the explosive to the potential fragment (refer to structural details of the equipment and/or structure in consideration).
3. Determine the size, shape, and connection of the potential fragment (see structural details for the equipment and/or structure in consideration).
4. Determine potential fragment velocity using the methods detailed in this chapter.

The blast load applied to the object can be simplified to a three-step function that represents the rise to peak pressure, initial pressure decay that is affected by diffraction of the wave around the object and rarefaction waves that form across the front of the object, and finally the drag phase loading (see Figure 8.2).

Unconstrained Fragments Unconstrained fragments are generated when objects that are not attached interact with the blast wave and are thrown from the incident origin. Predictions for unconstrained fragment velocity assume that the object is not attached (i.e., no energy lost breaking the object free), the object will not deform, and gravity will not affect velocity during acceleration.

Unconstrained Fragments Far from the Charge Potential fragments are defined as far from the charge if the objects are located at a distance more than 20 times the radius of the explosive. For potential fragments located far from the charge, the initial fragment velocity can be calculated using the following methods (U.S. Department of Defense 2008).

The basic equation of motion describes the object acceleration, where the time-dependent pressure applied to the area of the object facing the blast is equal to the mass of the object times acceleration. This relationship can be integrated and rearranged to yield:

$$V(t_d) = \int_0^{t_d} \frac{A}{M} i_d \tag{8.4}$$

where: V = velocity (feet/sec)
t_d = load duration (msec)
A = area of object facing blast (inches^2)
M = mass of object (lb-msec^2/inch)
i_d = total drag and diffraction (psi-msec/lb$^{1/3}$)

Equation 8.4 can be explicitly resolved if the loading history can be mathematically defined in a function. However, Figure 8.3 can be used to graphically solve for the nondimensional velocity and then solve for the object initial velocity, by solving for the quantity P_{so}/P_o (on the ordinate) and the following value on the abscissa:

$$\frac{12 \cdot C_D \cdot i_s \cdot a_o}{1000 \cdot P_{so} \cdot (K \cdot H + X)} \tag{8.5}$$

Establish the nondimensional object velocity using the various curves. The nondimensional velocity ($\bar{v}$) is equal to:

$$\bar{v} = \frac{144 \cdot v_o \cdot M \cdot a_o}{10^6 \cdot P_o \cdot (K \cdot H + X)} \tag{8.6}$$

Rearranging Equation 8.6 will result in the initial object velocity.

where: P_{so} = peak incident overpressure (psi)
P_o = atmospheric pressure (psi)
C_d = drag coefficient (see Figure 8.4)
i_s = incident-specific impulse (psi-msec)
a_o = velocity of sound in air (feet/sec)
K = constant (see Table 8.3)
H = minimum transverse dimension of the mean presented area of object (inch)

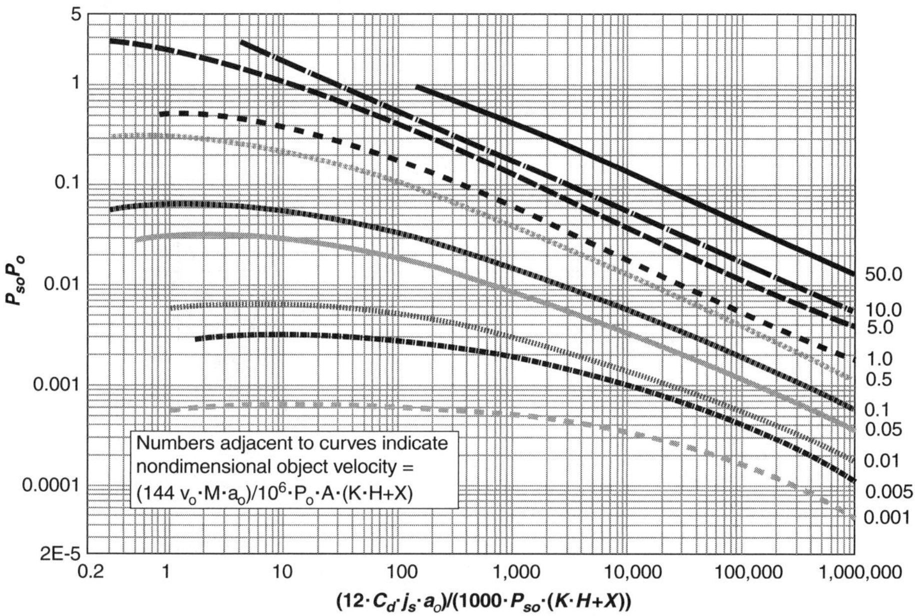

Figure 8.3 Nondimensional Object Velocity

X = distance from the front of the object to the location of its largest cross section normal to the plane of the shock front (inch)

v_o = initial velocity of object (feet/sec)

Unconstrained Fragments Near the Charge Potential fragments are defined as near the charge if the objects are located at a distance less than or equal to 20 times the radius of explosive. Initial fragment velocity for potential fragments can be calculated using the following methods (U.S. Department of Defense 2008).

The first step to determine the velocity of a close-in unconstrained fragment is to determine the specific impulse. Two charge configurations are considered, spherical and cylindrical. For spherical charges with $R/R_e \leq 5.07$:

$$\frac{i}{\beta \cdot R_{eff}} = \left[\frac{R_e}{R_t}\right]^{0.158} = 38{,}000 \cdot \left[\frac{R_e}{R}\right]^{1.4} \tag{8.7}$$

and $R_{eff} = R_e$

For cylindrical charges with $R/R_e \leq 5.25$:

$$\frac{i}{\beta \cdot R_{eff}} = \left[\frac{R_e}{R_t}\right]^{0.158} = 46{,}500 \cdot \left[\frac{R_e}{R}\right] \tag{8.8}$$

Shape	Sketch	C_O
Circular cylinder (Long rod), side-on	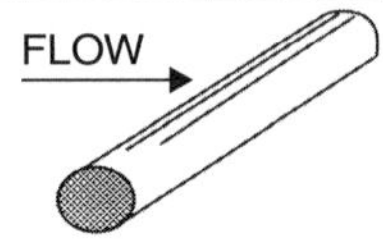	1.20
Sphere	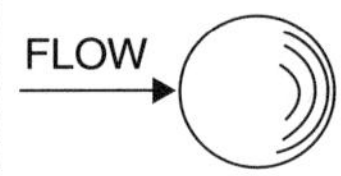	0.47
Rod, end-on	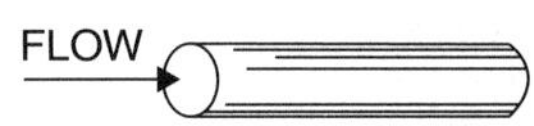	0.82
Disc, face-on	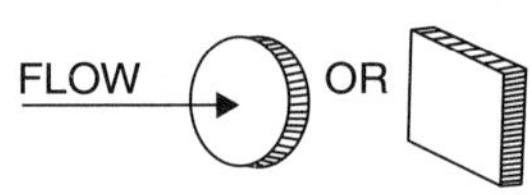	1.17
Cube, face-on	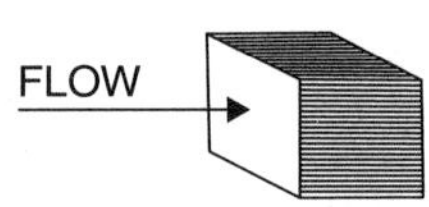	1.05
Cube, edge-on	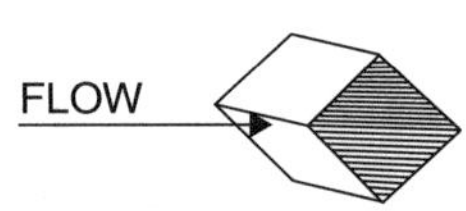	0.80
Long rectangular member, face-on	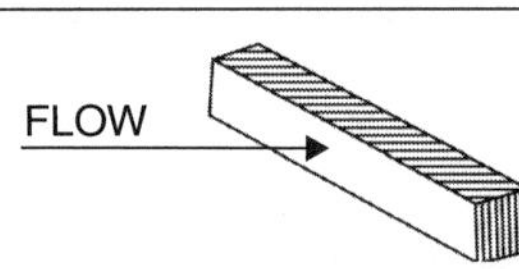	2.05
Long rectangular member, edge-on	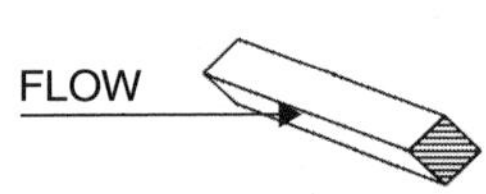	1.55
Narrow strip, face-on	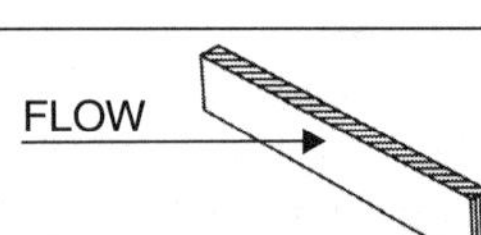	1.98

Figure 8.4 Drag Coefficients (Baker et al. 1977)

Table 8.3 K Constant

Steel	K-Value
Object on reflecting surface (i.e., on ground)	4
Object in air	2

For cylindrical charges with $5.25 < R/R_e \leq 10$:

$$\frac{i}{\beta \cdot R_{eff}} = \left[\frac{R_e}{R_t}\right]^{0.158} = 161{,}700 \cdot \left[\frac{R_e}{R}\right]^{1.75} \tag{8.9}$$

and

$$R_{eff} = 0.909 \cdot \left[\frac{L_e}{R_e}\right]^{0.333} \cdot R_e \tag{8.10}$$

where: i = specific acquired impulse (psi-msec)
β = nondimensional shape factor of the target from Figure 8.5
R_{eff} = effective radius (feet)
R_e = radius of the explosive (feet)
R = standoff distance (feet)
R_t = target radius (feet)
L_e = length of cylindrical explosive (inch)

Figure 8.6 graphically shows specific acquired impulse for spheres and cylinders. The data for this relationship were experimentally derived and should not be used beyond the distances shown in the figure. When these standoff distances are greater than plotted values, the normal reflected impulse may be used as an approximation for specific acquired impulse.

Because the target is so close to the charge, the actual load history is unimportant, and the impulse acting on the target is equal to the applied momentum. Therefore, $i = M \cdot V/A$ and the velocity of a specific target is:

$$v_o = \frac{1000 \cdot i \cdot \beta \cdot A}{12 \cdot M} \tag{8.11}$$

Constrained Fragments A constrained secondary fragment is generated when an object is torn loose from its moorings and thrown from the explosion source. The velocity of a constrained secondary fragment is dependent on the specific impulse applied to the object minus the impulse required to free the object from the support. Therefore, the following is true:

$$I - I_{st} = M \cdot v_o \tag{8.12}$$

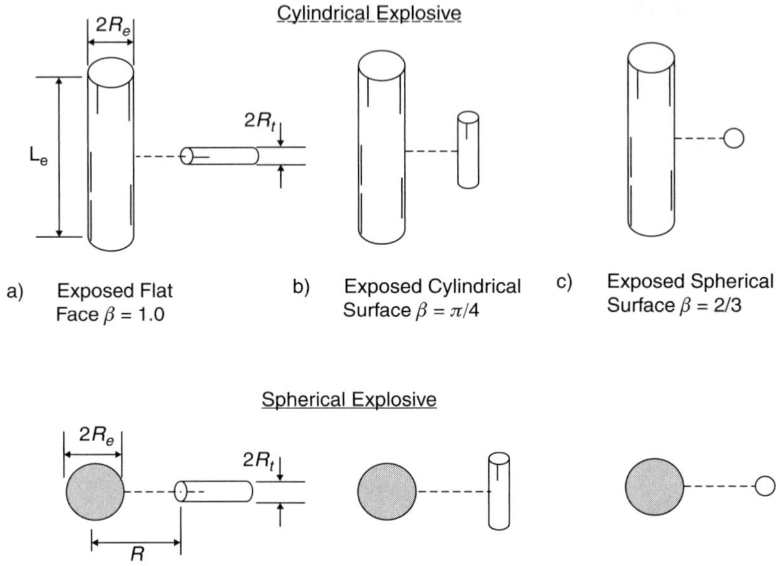

Figure 8.5 Nondimensional Shape Factor (U.S. Department of Defense 2008)

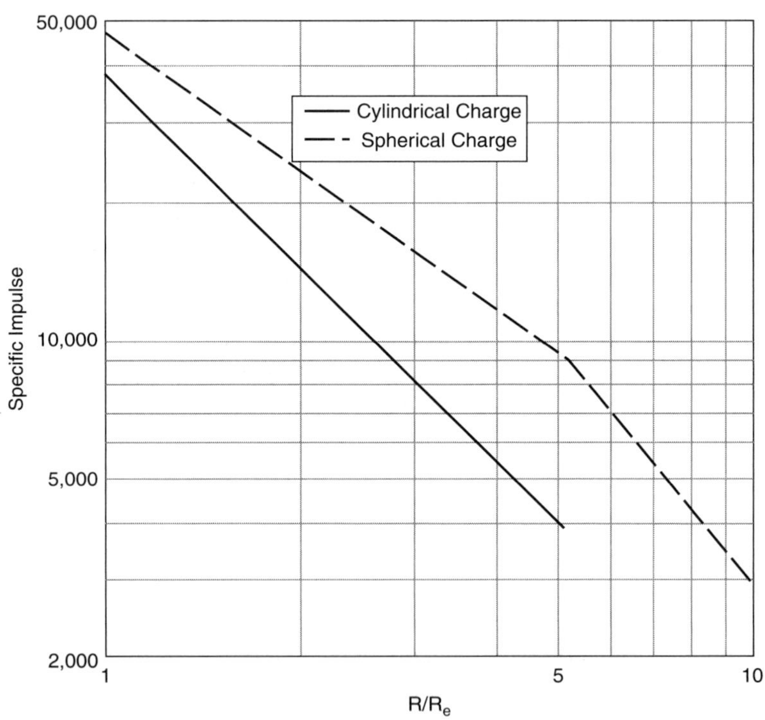

Figure 8.6 Specific Acquired Impulse

Table 8.4 Material Toughness

Steel	Toughness inches-lbf/inches^3
ASTM A36	12,000
ASTM A441	15,000
ASTM A514 Grade F	19,000

where: I = total blast impulse applied to the fragment (psi-msec)
I_{st} = impulse required to free the fragment from the support (psi-msec)

Empirical methods have been produced to estimate the initial fragment velocity from a cantilever or a fixed-fixed beam (clamped-clamped) (U.S. Department of Defense 2008). If the following relationship is true, the object velocity is zero because the applied impulse is insufficient to disengage the object.

If:

$$\frac{i \cdot b_f}{\sqrt{\rho_f \cdot T} \cdot A_b \cdot \left(\dfrac{2 \cdot L_b}{b_f}\right)^{0.3}} \leq 0.602, \tag{8.13}$$

where: i = unit impulse applied to the object (psi-msec)
b_f = width of fragment exposed to blast (inch)
ρ_f = mass density of the fragment lb-msec^2/inch^4)
A_b = smallest cross-sectional area of the fragment (inch^2)
L_b = length of fragment exposed to blast (inch)
T = toughness of material (see Table 8.4)

Then, $v_o = 0 \cdot ft/s$ [m/s]; otherwise:

$$v_o = \frac{1000}{12} \cdot \left(\sqrt{T/\rho_f}\right) \cdot \left[K_1 + K_2 \cdot \left(\frac{i \cdot b_f}{\sqrt{\rho_f \cdot T} \cdot A_b \cdot \left(\dfrac{2 \cdot L_b}{b_f}\right)^{0.3}}\right)\right] \tag{8.14}$$

where: K_1 = constant (see Table 8.5)
K_2 = constant (see Table 8.5)

Table 8.5 Constants K_1 and K_2

Support Condition	K_1	K_2
ASTM A36	−0.2369	+0.3931
ASTM A441	−0.6498	+0.4358

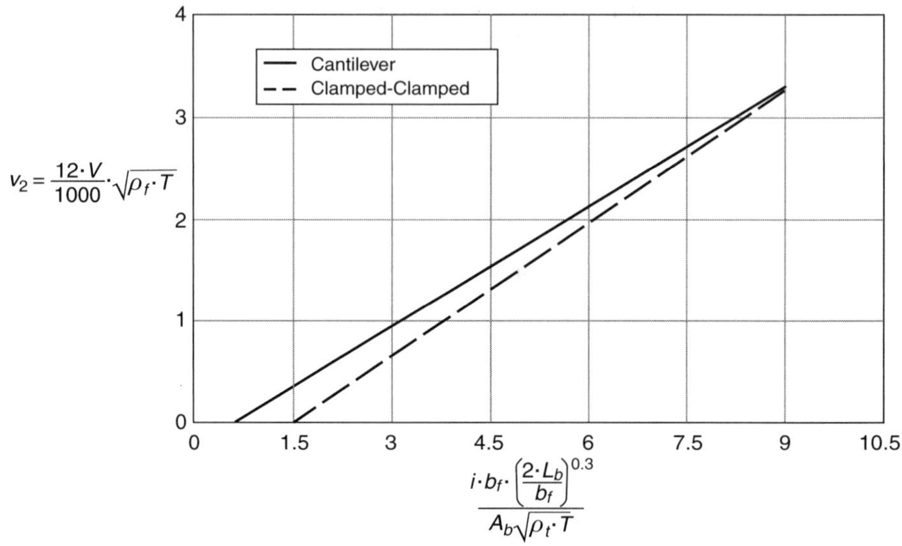

Figure 8.7 Scaled Fragment Velocities for Constrained Objects

Figure 8.7 shows Equation 8.14 graphically for both cantilever and fixed-fixed support conditions.

8.4 DESIGN FRAGMENT PARAMETERS

As a fragment is thrown from its initial position, several factors influence the impact the fragment will have on the target. Those factors include the fragment trajectory, the velocity upon impact, and the angle at which the fragment actually impacts the target (referred to as angle of obliquity). A zero angle of obliquity is the most conservative and typically assumed for design, since the actual angle of obliquity is rarely known.

8.4.1 Fragment Final Velocity

Fragment velocities vary with distance due primarily to drag forces. As the fragment travels away from the event, the initial velocity will decay until impact. It is assumed the initial velocity is achieved instantaneously. If a target is located close to an explosive event (less than 20 feet [U.S. Department of Defense 2008]), the decay in velocity is minimal and can be ignored. For targets farther from the event, the following method can estimate the final fragment velocity.

$$V_s = v_o \cdot e^{-[12 \cdot k_v \cdot R_f]} \tag{8.15}$$

where: V_s = fragment velocity at distance of interest (R_f)
R_f = distance from the center of detonation
k_v = velocity decay coefficient

and:

$$k_v = \frac{A}{W_f} \cdot \rho_a \cdot C_D \tag{8.16}$$

where: W_f = fragment weight (ounces)
A/W_f = fragment form factor
ρ_a = specific density of air (ounces/inch^3)

For primary fragments, some basic assumptions simplify the equation for final fragment velocity.

Assume:

$$\begin{aligned} A/W_f &= 0.78/W_f^{1/3} \text{(for mild steel fragments)} \\ \rho_a &= 0.00071\text{–ounces/inch}^3 \\ C_D &= 0.6 \end{aligned}$$

Resulting in:

$$V_s = v_o \cdot e^{-\left[0.004 \cdot \frac{R_f}{W_f{}^{1/3}}\right]} \tag{8.17}$$

8.4.2 Fragment Trajectory

Once the initial velocity of primary or secondary fragments is determined, it will be necessary to establish the potential range of the accelerated fragments, to establish the hazard associated with the fragments. The range is established by solving the equations that define the trajectory the fragment will follow until it strikes the ground or a target. Because fragments generated during an explosion are typically irregular in shape and mass, determining the fluid dynamic forces acting on the fragments to influence the trajectory can be difficult, and simplified fluid dynamic force prediction methods are typically adopted. Baker describes the acceleration in the X and Y directions based on both lift and drag forces (Baker et al. 1983). However, explosively produced primary and secondary fragments are typically "chunky" or drag-controlled fragments. Thus, most methods ignore the lift forces and evaluate the accelerations based on drag forces only (U.S. Department of Defense 2008).

The equations that define the accelerations and velocities in the X and Y directions can be solved simultaneously to determine the distance traveled by the fragment. Figure 8.8 represents a series of calculations for the resulting fragment range given a variety of initial conditions. A number of initial trajectories were

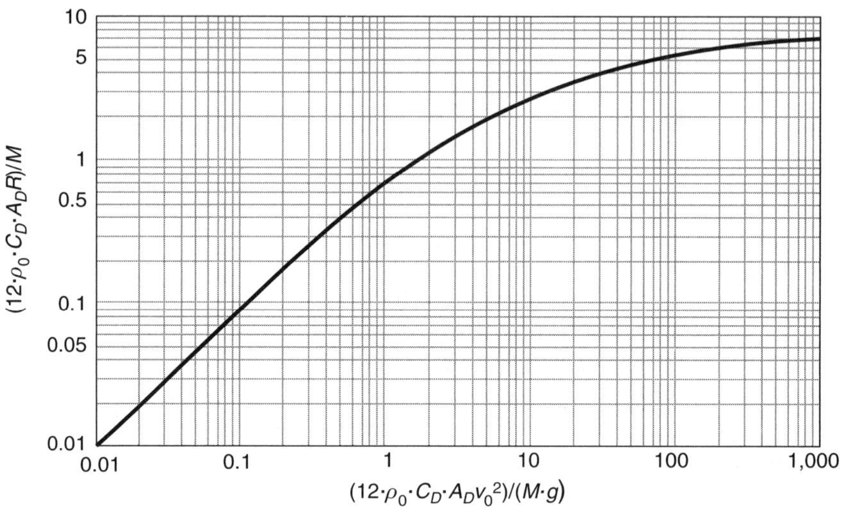

Figure 8.8 Fragment Range Prediction

considered to determine the maximum range; thus, the initial trajectory does not have to be established to use Figure 8.8. Calculate the nondimensional velocity [$\overline{v}$], shown on the abscissa in Figure 8.8, with the following:

$$\overline{v} = \frac{12 \cdot \rho_o \cdot C_d \cdot A_D \cdot v_o^2}{M \cdot g} \tag{8.18}$$

where: ρ_o = mass density of the medium through which the object travels (lb-msec^2/inch^4)
A_D = drag area (inch^2)
g = gravity force (32.2 feet/sec^2)

From Figure 8.8 determine the dimensional range [$\overline{R}$], shown on the ordinate. Rearrange $\overline{R}$ to solve for the maximum range [R].

$$R = \frac{M \cdot \overline{R}}{12 \cdot \rho_o \cdot C_d \cdot A_D} \tag{8.19}$$

8.5 FRAGMENT IMPACT DAMAGE

Almost any type of structure is susceptible to damage from explosively driven fragmentation. Damage can range from cosmetic damage such as bent or cracked architectural features to perforation (where a fragment passes through a wall or structural element), penetration (where a fragment strikes and lodges in a structural element), or even global collapse (when a larger fragment strikes a target with enough momentum to cause a primary element to fail).

Detonation of explosives can result in the formation of primary and secondary fragmentation. These fragments can range from small primary fragments with initial velocities in the order of thousands of feet per second to larger secondary fragments with initial velocities of hundreds of feet per second. When a fragment strikes a target, it will perforate the target, become embedded in the target (i.e., penetrate the target either with or without spalling), or be completely deflected by the target. Factors that influence the fragment/target response include the initial fragment velocity, the distance between the explosion and the target, the angle at which the fragment strikes the target (angle of obliquity), and the physical properties of the fragment (mass, shape, and material strength) and the target (strength and thickness).

8.5.1 Fragment Penetration into Miscellaneous Materials (THOR Equation)

Empirical THOR equations were developed based on testing for metallic and nonmetallic materials (Ballistic Analysis Laboratory 1961, Ballistic Research Laboratory 1963). All testing was performed to simulate primary fragments, and the resulting data should be applied accordingly. The test data were assembled into the THOR reports by Greenspon in 1976 (Greenspon 1976). These data were gathered for military purposes and as such should be used in commercial and industrial structures with specific limitations in mind. The method assumes small fragment sizes into specific targets. The equation for minimum thickness of a plate to resist perforation by a mild steel fragment is:

$$t = \frac{10 - \frac{C_1+7C_2+3C_3}{C^2}}{A_f} \cdot m_f^{\left(-\frac{C_3}{C_2}\right)} \cdot V_s^{\left(\frac{1-C_5}{C_2}\right)} \cdot (\cos\phi)^{\frac{C_4}{C_2}} \tag{8.20}$$

where:
t = fragment penetration (m)
m_f = fragment mass (kg)
A_f = presented area of fragment (m^2)
V_s = impact velocity (m/sec)
ϕ = impact angle relative to normal from tangent (rad)
$C_1 - C_{10}$ = empirical constants (see Table 8.6)

Other useful THOR equations for application to fragment impact on structures are residual velocity and residual fragment mass equations. The THOR equation for residual velocity is:

$$V_r = V_s - 10^{(C_1+7C_2+3C_3)} \cdot \left(t \cdot A_f\right)^{C_2} \cdot m_f^{C_3} \cdot (1/\cos\phi)^{C_4} \cdot V_s^{C_5} \tag{8.21}$$

where: V_r = residual velocity of fragment after perforation (m/sec)

Table 8.6 THOR Constants

Target Material	C_1	C_2	C_3	C_4	C_5	C_6	C_7	C_8	C_9	C_{10}
Aluminum (2024-T3)	2.907	1.029	−1.072	1.250	−0.139	−6.548	0.227	0.694	−0.361	1.901
Cast iron	1.037	1.042	−1.051	1.028	0.523	−9.052	0.162	0.673	2.091	2.710
Copper	0.314	0.678	−0.730	0.846	0.802	−5.904	0.340	0.568	1.422	1.650
Doron	3.431	1.021	−1.014	0.917	−0.362	−10.161	0.251	0.343	0.706	2.906
Glass, bullet-resistant	1.046	0.705	−0.723	0.690	0.465	−6.341	0.305	0.429	0.747	1.819
Lead	0.203	0.499	−0.502	0.655	0.818	−3.267	0.506	0.350	0.777	0.934
Lexan	0.328	0.720	−1.657	0.773	0.603	−7.063	0.480	0.465	1.171	1.765
Magnesium	2.534	1.092	−1.117	1.050	−0.087	−6.026	0.285	0.803	−0.172	1.519
Nylon, bonded	0.674	1.144	−0.968	0.743	0.392	−12.165	0.035	0.775	0.045	3.451
Nylon, unbonded	2.590	0.835	−0.654	0.990	−0.162	−6.619	−0.067	0.903	−0.351	1.717
Plexiglass, as cast	1.310	1.044	−1.035	1.073	0.242	−6.115	1.402	−0.137	0.674	1.324
Plexiglass, stretched	−0.093	1.112	−0.903	0.715	0.686	−6.431	0.437	0.169	0.620	1.683
Steel, face hardened	1.631	0.674	−0.791	0.989	0.434	−1.768	0.234	0.744	0.469	0.483
Steel, hard homogeneous	2.877	0.889	−0.945	1.262	0.019	−3.017	0.346	0.629	0.327	0.880
Steel, mild homogeneous	2.801	0.889	−0.945	1.262	0.019	−2.616	0.138	0.835	0.143	0.761
Titanium	2.118	1.103	−1.095	1.369	0.167	−1.927	1.086	−0.748	1.327	0.459
Tuballoy	0.441	0.583	−0.603	0.865	0.828	−4.564	0.560	0.447	0.640	1.381

The THOR equation for residual fragment weight is:

$$m_r = m_f - 10^{(C_6+7C_7+3C_8-3)} \cdot \left(t \cdot A_f\right)^{C_7} \cdot m_f^{C_8} \cdot (1/\cos\phi)^{C_9} \cdot V_s^{C_{10}} \tag{8.22}$$

where: m_r = residual mass (kg)

The range for each variable in the THOR equations is given in Table 8.7.

8.5.2 Steel

Limit Velocity of "Chunky" Nondeforming Fragments for a Thin Metal Target Fragment penetration into thin metal structural elements such as typical metal buildings can be calculated based on the metal target material and fragment properties. A fragment velocity greater than the limit velocity (V_{50}) will penetrate the metal target. The limit velocity for a thin metal target can be calculated using:

$$V_{50} = \frac{\sqrt{\sigma_t \cdot \rho_t} \cdot V_{50n}}{\rho_f} \tag{8.23}$$

where: σ_t = target yield stress
ρ_t = target density
ρ_f = fragment density
V_{50n} = nondimensional limit velocity (calculate h/a and determine V_{50n} from Figure 8.9)
h = target thickness
a = fragment radius

Fragment Penetration into Mild Steel Targets Fragment penetration calculations are based on some simplifying assumptions. The primary assumption is that the fragment shape is a normal cylinder with a rounded leading edge (see Figure 8.1). Additionally, the steel material is assumed to have a Brinell hardness of less than 150. Results will be conservative for materials with increased hardness. Depth of penetration for mild steel targets impacted by armor-piercing steel fragments is:

$$x = 0.30 \cdot W_f^{0.33} \cdot V_s^{1.22} \tag{8.24}$$

Depth of penetration for mild steel targets impacted by mild steel fragments is:

$$x = 0.21 \cdot W_f^{0.33} \cdot V_s^{1.22} \tag{8.25}$$

where: x = depth of penetration (inches)
W_f = fragment weight (ounces)
V^s = striking velocity (feet/sec)

Table 8.7 Range of Variables in THOR Equations

Target Material	Target Thickness Range t mm	Striking Velocity Range V_s m/sec	Fragment Size Range m_f kg (GRAINS)
Aluminum (2024-T 3)	0.5 to 51.0	365.8 to 3,353	3.2×10^{-4} to 1.60×10^{-2} (5 to 240)
Cast iron	4.8 to 14.0	335.0 to 1,859	9.7×10^{-4} to 1.60×10^{-2}(15 to 240)
Copper	1.5 to 25.0	335.4 to 3,475	9.7×10^{-4} to 1.60×10^{-2} (15 to 240)
Doron	1.3 to 38.0	152.4 to 3,353	2.3×10^{-8} to 5.6×10^{-6} (3.57×10^{-4} to 8.57×10^{-2})
Glass, bullet-resistant	5.0 to 42.0	61.0 to 3,048	1.4×10^{-7} to 4.4×10^{-6} (2.14×10^{-3} to 6.79×10^{-2})
Lead	1.8 to 25.0	152.4 to 3,170	9.7×10^{-4} to 1.60×10^{-2} (15 to 240)
Lexan	3.2 to 25.0	304.8 to 3,505	4.6×10^{-8} to 2.2×10^{-6}(7.14×10^{-4} to 3.43×10^{-2})
Magnesium alloy	1.3 to 76.0	152.4 to 3,200	9.7×10^{-4} to 1.60×10^{-2} (15 to 240)
Nylon, bonded	11.0 to 51.0	304.8 to 3,658	4.6×10^{-8} to 7.6×10^{-6} (7.14×10^{-4} to 1.18×10^{-1})
Nylon, unbonded	0.5 to 76.0	91.4 to 3,048	4.6×10^{-8} to 1.9×10^{-6}(7.14×10^{-4} to 2.96×10^{-2})
Plexiglass, as cast	5.0 to 28.0	61.0 to 2,897	4.6×10^{-8} to 4.4×10^{-6} (7.14×10^{-4} to 6.79×10^{-2})
Plexiglass, stretched	1.3 to 25.0	152.4 to 3,353	4.6×10^{-8} to 4.4×10^{-6} (7.14×10^{-4} to 6.79×10^{-2})
Steel, face hardened	3.6 to 13.0	762 to 2,987	9.7×10^{-4} to 1.60×10^{-2}(1 5 to 240)
Steel, homogeneous	8.0 to 25.0	182.9 to 3,658	3.2×10^{-4} to 5.30×10^{-2} (5 to 825)
Titanium alloy	1.0 to 30.0	213.4 to 3,170	1.9×10^{-3} to 1.60×10^{-2} (30 to 240)
Tuballoy	2.5 to 5.0	1,372 to 3,078	1.9×10^{-3} to 3.10×10^{-2} (30 to 475)

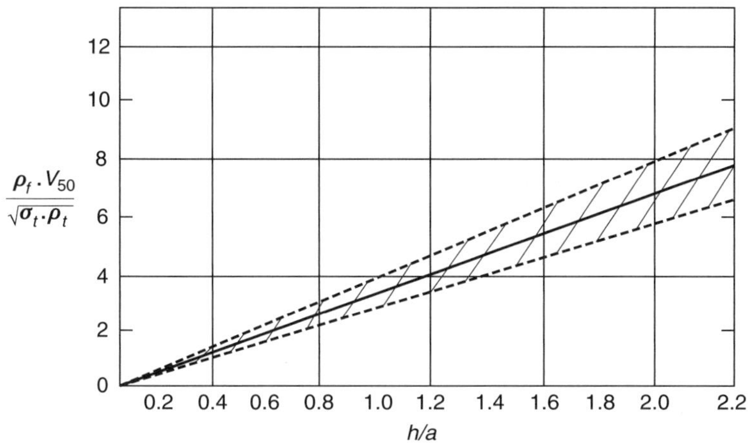

Figure 8.9 Nondimensional Limit Velocity versus Nondimensional Thickness for "Chunky" Nondeforming Fragments (U.S. Department of Energy 1992)

Most fragments generated during an explosion will not resemble a simplified bullet—the majority of primary fragments are more blunt; thus, the bullet-shaped fragment is conservative. Certainly some sharper-edged fragments are generated; however, the likelihood that these fragments will strike the target in the appropriate orientation to control the design is minimal.

8.5.3 Fragment Penetration into Concrete Targets

When a fragment strikes a concrete target, the impulse imparted by the fragment on the concrete will cause a crater of irregular size. The depth of the crater grows with impact velocity for a given fragment size (or with a reduction in fragment cross-sectional area for a given velocity). If the velocity is high enough (typically greater than 1,000 feet/second [U.S. Department of Defense 2008]), the fragment will penetrate beyond the crater. If the fragment deforms upon impact, the crater size will decrease or may not form at all.

In addition to the crater on the impact side, a crater may form (generating concrete spall) on the opposite side of the target adjacent to impact. This crater is the result of a compression wave passing through the concrete material and reflecting off the free surface of the opposite concrete face. The reflection causes tension stresses at the surface of the concrete element. If the tension stress exceeds the compressive strength of the concrete, spalling of the concrete target will occur. If the impulse imparted on the target is sufficient, the spall crater may extend to the reinforcing steel. As impact velocity increases, the impact and spall craters increase, and the fragment will become lodged in the target or pass through the target (perforate).

Armor-Piercing Fragment Penetration into Concrete Targets The maximum penetration of an armor-piercing fragment into a massive concrete target is defined as:

For $X_f \leq 2{\cdot}d$:

$$X_f = 4.0 \cdot 10^{-3} \cdot \sqrt{K \cdot N \cdot D} \cdot d^{1.1} \cdot V_s^{0.9} \tag{8.26}$$

For $X_f > 2{\cdot}d$:

$$X_f = 4.0 \cdot 10^{-6} \cdot (K \cdot N \cdot D) \cdot d^{1.2} \cdot V_s^{1.8} + d \tag{8.27}$$

where: X_f = maximum penetration by armor-piercing fragment (inch)
K = penetration constant $= K = 12.91/\sqrt{f_c'}$
N = nose shape factor (see Figure 8.1)
D = caliber density (see Figure 8.1)
d = fragment diameter (inch)

For a standard primary fragment and concrete strength (f'_c) of 4,000 psi, Equations 8.26 and 8.27 can be expressed in terms of fragment diameter (inches) as:

For $X_f \leq 2{\cdot}d$:

$$X_f = 2.86 \cdot 10^{-3} \cdot d^{1.1} \cdot V_s^{0.9} \tag{8.28}$$

For $X_f > 2{\cdot}d$:

$$X_f = 2.04 \cdot 10^{-6} \cdot d^{1.2} \cdot V_s^{1.8} + d \tag{8.29}$$

Or in terms of fragment weight:

For $X_f \leq 2{\cdot}d$:

$$X_f = 1.92 \cdot 10^{-3} \cdot W_f^{0.37} \cdot V_s^{0.9} \tag{8.30}$$

For $X_f > 2{\cdot}d$:

$$X_f = 1.32 \cdot 10^{-6} \cdot W_f^{0.4} \cdot V_s^{1.8} + 0.695 \cdot W_f^{0.33} \tag{8.31}$$

For concrete strengths other than 4,000 psi, the penetration depth can be estimated by multiplying the results by the square root of the ratio of the concrete strengths:

$$X_f' = X_f \cdot \sqrt{\frac{4000}{f_c'}} \tag{8.32}$$

where: X'_f = maximum penetration by armor-piercing fragment into concrete with compressive strength of f'_c (inch)

Table 8.8 Fragment Penetration Factors, K_3

Type of Material	K_3
Armorpiercing steel	1.00
Mild steel	0.70
Lead	0.50
Aluminum	0.15

Non-Armor-Piercing Fragment Penetration into Concrete Targets For non-armor-piercing fragments, the depth of penetration into a concrete target can be calculated by taking the fragment metal hardness into account:

$$X_f = K_3 \cdot X_f \tag{8.33}$$

where: X''_f = maximum penetration by non-armor-piercing fragment (inch)
K_3 = constant to account for fragment metal hardness (see Table 8.8)

8.5.4 Fragment Perforation of Concrete Targets

The equations presented in Section 8.5.3 assume the concrete is of infinite thickness. Of course the material thickness of a concrete slab of finite thickness required to prevent perforation is greater than the maximum penetration depth into an infinite thickness (due to confinement and compressive resistance of a massive concrete structure). The minimum concrete thickness required to prevent perforation as a function of maximum penetration into a massive concrete structure is:

$$T_{pf} = 1.13 \cdot X_f \cdot d^{0.1} + 1.311 \cdot d \tag{8.34}$$

where: T_{pf} = minimum concrete thickness to prevent perforation by design fragment (inch)
X_f = maximum fragment penetration into a massive concrete structure as described in Section 8.5.3

If a fragment perforates a concrete target, it can pose a threat to personnel and structural elements behind the concrete target. The residual velocity of the fragment can be calculated using equations that define the fragment velocity as a function of penetration depth:

For $X_f \leq 2 \cdot d$:

$$\frac{V_r}{V_s} = \left[1 - \cdot \left(\frac{T_c}{T_{pf}}\right)^2\right]^{0.555} \tag{8.35}$$

For $X_f > 2 \cdot d$:

$$\frac{V_r}{V_s} = \left[1 - \cdot \left(\frac{T_c}{T_{pf}}\right)\right]^{0.555} \tag{8.36}$$

where: T_c = concrete thickness up to T_{pf} *(inch)*
V_r = residual velocity of fragment as it leaves the element (feet/sec)

8.5.5 Fragment Spalling of Concrete Targets

As discussed in Section 8.5.3, concrete spall occurs when the impacting fragment imparts enough impulse to generate a compression wave and a resulting tension wave in the concrete that exceeds the compressive strength of the concrete. The spall crater typically does not exceed the reinforcement depth. Concrete spalling is a function of fragment penetration depth. Therefore, the minimum concrete thickness required to prevent spalling as a function of maximum penetration into a massive concrete structure can be expressed as:

$$T_{sp} = 1.215 \cdot X_f \cdot d^{0.1} + 2.12 \cdot d \tag{8.37}$$

where: T_{sp} = minimum concrete thickness to prevent spall by design fragment (inch)
X_f = maximum fragment penetration into a massive concrete structure as described in Section 8.5.3 (inch)

If spalling occurs, it is likely the concrete spall will generate a secondary fragment that is hazardous to personnel and equipment behind the concrete target. While the velocity produced when the concrete material fails is likely to be relatively low, the spall fragment velocity will also be affected by the wall motion in response to the blast loading. Therefore, it is typically accepted that concrete spall should be prevented by designing the concrete thickness to be sufficient to avoid spall, or by providing an additional cover for the concrete surface (i.e., a catch system or fiber-reinforced laminate).

8.5.6 Roofing Materials

Most roofing materials are relatively lightweight and have low thresholds of serious damage. However, the angle of obliquity will likely reduce the effect of fragment damage. Lower limits for fragment damage to miscellaneous lightweight roofing materials based on fragment momentum are shown in Table 8.9.

Table 8.9 Fragment Impact Damage for Roofing Materials (Baker et al. 1977)

Roofing Material	Minimum Fragment Momentum (mV) for Serious Damage (lbf-sec)	Comments
Asphalt shingles	0.159	CracksShingle
	1.370	Damage deck
Built-up roof	<0.159	Crack tar flood coat
	0.451	Crack surface of conventional built-up roof (BUR) without top layer of stones
	>0.991	Crack surface of conventional BUR with 2.867 lb/feet^2 (137 Pa) top layer of stones
Miscellaneous		
1/8-inch asbestos cement shingles	0.159	
1/4-inch asbestos cement shingles	0.285	
1/4-inch green slate	0.285	
1/4-inch gray slate	0.159	
1.2-inch cedar shingles	0.159	
3/4-inch red clay tile	0.285	
Standing seam terne metal	0.991	Plywood deck cracked

8.5.7 Other Materials

A number of other materials have been evaluated for fragment penetration, including sand, various rock and other geological materials, and combinations of sand and concrete panels. While they are not presented here, they are covered in varying detail in *Structures to Resist the Effects of Accidental Explosions* (U.S. Department of Defense 2008), *Explosion Hazards and Evaluation* (Baker et al. 1983), and *A Manual for the Prediction of Blast and Fragment Loadings on Structures* (Department of Energy 1992), as well as the Department of the Army Historical Summary (DAHS) manual *Fundamentals of Protective Structure Design for Conventional Weapons* (Department of the Army 1998).

REFERENCES

Baker W. E., P. A. Cox, P. S. Westine, J. J. Kulesz, and R. A. Strehlow. 1983. *Explosion Hazards and Evaluation*. Fundamental Studies in Engineering 5. Amsterdam: Elsevier Scientific Publishing Company.

Baker W. E., J. J. Kulesz, R. E. Ricker, R. L. Bessey, P. S. Westine, v. b. Parr, and g. a. Oldham. 1977. *Workbook for Predicting Pressure Wave and Fragmentation Effects of Exploding Propellant Tanks and Gas Storage Vessels*, NASA CR-134906, NASA Lewis Research Center.

Ballistic Analysis Laboratory, Johns Hopkins University. 1961. *The Resistance of Various Metallic Materials to Perforation by Steel Fragments; Empirical Relationships for Fragments Residual Velocity and Residual Weight* (Project THOR Technical Report 47). Baltimore, MD: Johns Hopkins University, Institute for Cooperative Research.

Ballistic Research Laboratory. 1963. *The Resistance of Various Non-Metallic Materials to Perforation by Steel Fragments, Empirical Relationships for Fragment Residual Velocity and Residual Weight* (TR 51). Aberdeen Proving Ground, MD: Ballistic Research Laboratory.

Department of the Army. 1998. *Fundamentals of Protective Structure Design for Conventional Weapons* (TM 5-855-1, Air Force AFPAM 32-1147[I], Navy P-1080, DSWA DAHSCWEMAN-97). For Official Use Only. Washington, DC: Department of the Army.

Greenspon J. E. 1976. *An Approximate Non-Dimensional Representation of the THOR Equations* (U.S. Army Materiel Systems Analysis Activity TR #173).

Gurney R. W. 1947. *The Initial Velocities of Fragments from Bombs, Shells, and Grenades* (Report No. 648). Aberdeen Proving Ground, MD: Ballistic Research Laboratory.

Mott R. I. 1943. *A Theoretical Formula for the Distribution of Weights of Fragments* (Ministry of Supply, AC-3642).

U.S. Department of Defense (DoD). 2008. *Structures to Resist the Effects of Accidental Explosions* (UFC 3-340-02). Washington, DC: U.S. Department of Defense.

U.S. Department of Energy. 1992. *A Manual for the Prediction of Blast and Fragment Loadings on Structures* (DOE/TIC-11268). San Antonio, TX: Southwest Research Institute.

III System Analysis and Design

9 Structural Systems Design

Robert Smilowitz and Darren Tennant

9.1 GENERAL DISCUSSION

In order to protect the occupants of a building that is either the target of an explosive attack or the unintended victim of collateral damage resulting from an attack against a neighboring property, attention must be paid to the performance of the structural system. Although the structure may require extensive repair following an incident, the primary goal of protective design is to prevent structural collapse and to minimize debris. This goal is accomplished through the design and detailing of structural components to sustain specified intensities of blast loading, and the design and detailing of structural systems to provide ductile and redundant load paths, regardless of the cause or event. This combination of threat-specific and threat-independent design provides the most robust structural system that is able to survive both the anticipated and the unimaginable catastrophic events. As with the performance-based design of structures to resist strong ground motions, the design of structures to resist blast loads considers successive levels of damage and deformation in response to increasingly severe events. Also similar to seismic design, the analytical methods that are required to demonstrate the performance of structural components and structural systems in response to extreme blast loading are necessarily dynamic and inelastic.

9.1.1 Seismic versus Blast

Performance-based methods are used to design structures that must resist strong ground motions and structures that must resist blast loading. In both cases, different extents of acceptable damage are specified for events of different intensity. This approach acknowledges the financial burden associated with comprehensive protection in response to the most extreme loading conditions and rationally accepts successively greater risk for successively more severe events. This risk-based approach starts with a risk assessment that considers the function, criticality, occupancy, site conditions, and design features. However, unlike natural hazards, man-made events are entirely unpredicable. As a result, risk associated with blast loading can at best be quantified in a relative sense, in order to compare different levels of protection or different protective measures.

Since there are not enough data to justify the absolute quantification of risk, the choice of design basis threat is inherently subjective and is often based on past experience.

The analytical methods that are used to evaluate the performance of structures in response to these forms of extreme loading involve the solution of the equations of motion. Since ground motions are typically assumed to be applied uniformly over the foundations of most structures, the spectral approach allows the designer to envelop the expected response without having to consider a specific input motion time history. However, blast loads and blast effects are typically localized to the immediate vicinity of the detonation, and these enveloping methods are not applicable. Localized blast loads will deform the exterior bays of the structure to a much greater extent than the interior bays, and, in the time frame of the applied loading, the story shears may not necessarily be distributed through the diaphragms in proportion to the framing stiffness. Typically, the ductility demands at the blast-loaded face of the structure far exceed the demands elsewhere throughout the remainder of the structure.

Some types of construction are inherently more ductile than others, and these are better suited for seismic and blast-resistant design. While cast-in-place concrete can be detailed to provide sufficient ductility, post-tensioned two-way flat-slab structures are problematic. Punching shear failures at the slab joint must be avoided under lateral sway deformations, and these systems possess limited lateral deformation capability. Although a two-way flat-slab system might perform satisfactorily in response to strong ground motion when it is used in combination with a stiffer shear wall system that controls the ultimate lateral sway of the structure and prevents destabilizing P-Delta effects, the punching shear failure vulnerability is a function of the slab performance in response to blast infill pressures and is not improved with the addition of shear walls.

The fundamental difference between seismic excitations and blast-resistant design is the distribution of ductility demand throughout the structure. Strong ground motions engage the entire lateral resisting system, while resistance to blast loading is more concentrated in the structure in the vicinity of the explosion. Moment frames may provide high ductility and energy absorption when the plastic hinging of beams is spread throughout the structure. In order for inelastic deformation to be distributed throughout the structure, the beam-column connections must develop the flexural hinge in the beam. The elements must also be proportioned to prevent plastic story mechanisms that might limit the inelastic energy absorption to isolated regions of the frame. This is achieved through the "strong-column, weak-beam" (SCWB) approaches for both seismic and blast design. Since blast loading is more localized than seismic excitations, the resulting plastic hinging is more localized, with more severe rotations than those that result from strong ground motions. Structural systems that effectively resist the lateral forces resulting from strong ground motions, such as concentrically braced frames (CBF) and eccentrically braced frames (EBF), must be considered on a case-by-case basis for use in blast-resistant design.

However, the manner in which seismic and blast forces are distributed throughout the building is strongly affected by the size, shape, and location of structural elements. Although seismic forces are proportional to the mass of the building, inertial resistance reduces the response to blast loading. Structures that are designed to resist both blast forces and strong ground motion benefit from low height-to-base ratios, balanced resistance, symmetrical plans, uniform sections and elevations, the placement of shear walls and lateral bracing to maximize torsional resistance, short spans, direct load paths, and uniform floor heights. Although the benefits of employing these design elements may be similar, the reasons for incorporating these design features may differ. For example, seismic excitations may induce torsional response modes in structures with reentrant corners, whereas these conditions provide pockets where blast pressures may reflect off adjacent walls and amplify the blast effects. Similarly, first-floor arcades that produce overhangs or reentrant corners create localized concentrations of blast pressure and expose areas of the floor slab to uplift blast pressures. Adjacent structures may impact one another as they respond to base motions, while adjacent structures may be subjected to multiple reflections of blast waves. The geology of the site influences the characteristics of the ground motion that load the structure, and the geology influences the size of the blast crater and the reflectivity of the blast waves off the ground surface.

The ductility demands for both seismic and blast-resistant design require attention to details. Confinement of the concrete core prevents premature buckling of the rebar and maintains load-carrying capacity, despite the significant cracking associated with large inelastic deformation. Closely spaced ties and spiral reinforcement are particularly effective in increasing the ductility of a concrete compression element. Except for near-contact detonations, carbon fiber wraps and steel jacket retrofits provide comparable confinement to existing structures; however, carbon fiber materials may be damaged when the standoff distances are too short. Steel column splices must be located away from regions of plastic hinging or must be detailed to develop the full moment capacity of the section. Closely spaced stiffeners or concrete encasement will prevent local flange buckling of steel sections. Reinforced concrete beam sections require equal resistance to positive and negative bending moments. In addition to the effects of load reversals and rebound, doubly reinforced sections possess greater ductility than singly reinforced counterparts. Steel beams may be constructed composite with the concrete deck in order to increase the ultimate capacity of the section; however, this increase is not equally effective for both positive and negative moments. While the composite slab may brace the top flange of the steel section, the bottom flange is vulnerable to buckling. Tube sections and concrete encasement are particularly effective in preventing flange buckling under load reversals.

9.1.2 Analytical Methods

A wide variety of analytical approaches may be used to determine the performance of structures in response to explosive loading. The various methods include empirical explosive test data that can be expressed as prescriptive requirements or pressure-impulse charts, single-element response analysis, structural system response analysis, and detailed finite element analyses. These dynamic response methods all characterize the behavior of the structural system in response to a short-duration but high-intensity dynamic loading. The dynamic analysis methods must be able to capture the frequency of the loading function and the dominant frequencies of the structural system, and represent the governing failure mechanisms so as not to overlook critical failure mechanisms that may precipitate collapse or hazardous debris. The designer must therefore anticipate the likely failure mechanism before selecting the appropriate analytical approach.

9.2 MODELING

Long before complex finite element analysis methods were readily available, the design of structures to resist blast loading relied extensively on single-degree-of-freedom (SDOF) methods. These methods were generally conservative as long as the designer anticipated the system response and made sure the analytical model captured all failure mechanisms and response characteristics. Advanced finite element methods may be used where structural complexity or spatial distribution of loading requires more detailed analyses. These methods can represent complex geometries, connectivity, and material properties. While simplified models are prone to problems associated with too little information, complex models need to be tailored to available resources and time constraints. The optimal model therefore contains enough information to capture the response of interest without becoming so excessively complex that it cannot solve the equations of motion within a reasonable solution time. Accepted practice requires that enough resolution be included in the model to accurately capture response modes.

When fast-running simplified methods are deemed inadequate, detailed analytical models can often be constructed to calculate structural performance in response to blast and impact. The current state of the art relies on explicit dynamic inelastic finite element software such as NLFlex, LS-DYNA, AUTODYN, Abaqus, Pronto, and the like. When modeling a significant section of a building, such as a couple of bays, the required level of resolution or element size to calculate an accurate response on current-generation computer hardware limits the extent of the structure that can be analyzed. For these 3D models, the required computer RAM increases by the cube of the resolution, and the computational time increases by the 4th power. As a result, the analysis of larger models will become feasible as computer hardware advances.

At this resolution level, each rebar is typically represented explicitly with beam or bar elements, and structural steel is typically represented explicitly with shell or plate elements. Concrete, soil, and rock are represented with continuum elements. In view of this background, macroscopic phenomenological constitutive models are the state of the art for common construction materials such as steel and concrete. Research efforts on more fundamental meso- and micromechanical approaches fill the literature; however, these are currently too computationally demanding for production use.

The goal of a phenomenological model is to accurately capture the important characteristics of material response in the expected loading regime, while obeying the laws of thermodynamics. The theory must be implemented in a robust algorithm that is efficient in terms of both RAM and CPU usage. Important material characteristics for concrete include strength dependence on pressure, energy absorption or hysteresis due to pore compaction, increase of resistance with loading rate, and softening due to fracture or crushing. Concrete also transitions from a brittle material that fractures at low pressures to a more ductile material that flows at higher pressures. Important considerations for steel include hardening after yield, strength dependence on loading rate, and fracture dependence on confining pressure. A realistic representation of fracture or material softening is very important in blast applications. Once softening initiates, material damage localizes into discrete cracks or narrow bands. Maintaining accurate simulations during localization requires some sort of regularization. At currently achievable resolution levels, this usually requires a fracture energy-based approach whereby the softening stress-strain response is a function of element dimension.

Theories of plasticity, viscoplasticity, and/or continuum damage mechanics are a few of the popular theoretically sound frameworks for phenomenological models. A well-formulated model will use parameter identification from a small number of lab tests to reasonably characterize material response over a very wide range of loading conditions.

9.2.1 Systems

Several of the most common structural systems will be discussed, including the significant features that must be represented in the analytical models. While this list represents a sampling of structural systems that may be analyzed, it is not intended to limit the types of structural systems that may be used for blast-resistant design.

Concrete and Steel Moment Frames Moment frame buildings may be analyzed one component at a time, modeling each component as an SDOF system with dynamic blast loads either applied tributary to the loaded surface area or collected through the reaction forces from secondary members as they frame into the primary structure. The appropriate performance limits for the calculated inelastic deformations are typically conservatively specified, particularly for axial members, where P-Delta effects may initiate structural instability. Furthermore, simplified SDOF methods are incapable of capturing localized steel flange deformations, steel splice performance, concrete panel zone deformations, and other localized response characteristics. These SDOF approaches are only appropriate for standoff detonations that apply fairly uniform loads to the member. The overall frame response may similarly be determined using an SDOF model of the building's lateral resisting system. These blast-induced base shears recognize the flexibility of the frame and a characteristic extent of frame ductility; however, since blast damage is typically localized in proximity to the point of detonation, these overall ductility levels are approximate and represent overall energy dissipation of the framing system. Fundamental to the modeling and analysis of the frame using these methods is the appropriate detailing of the connections so the structural members will be capable of developing the calculated extent of rotation and lateral drift.

Steel-Braced Frames and Concrete Frames with Concrete Shear Walls Similar to the response of concrete and steel moment frames, the response of steel-braced frames or concrete frames with concrete shear walls may be analyzed one component at a time, to determine local response in proximity to the detonation; however, the global response of the structure will be characterized by a stiffer lateral resisting system with a higher fundamental frequency of response. Not only will the period of the structural system be shorter than that of a comparable moment frame structure, but the amount of energy dissipation will be less. This type of structural system will typically develop larger blast-induced base shears, and the floor diaphragms must be capable of distributing the collected lateral loads to the various lateral resisting components.

Precast Tilt-Up with Concrete Shear Walls Precast tilt-up load-bearing walls may be modeled as single-degree-of-freedom systems, provided the detonation is sufficiently far from the building to produce a uniform loading over the floor height. The flexural deformations of these wall systems must be limited in order to prevent P-Delta effects that may precipitate instability, and the diaphragms must be sufficiently tied to the wall in order to transfer the large lateral loads and rebound forces. Subsequent diaphragm models are required in order to demonstrate their ability to distribute loads to the concrete shear walls.

Reinforced Masonry Walls with Rigid Diaphragms Reinforced masonry walls may be analyzed as single-degree-of-freedom systems; however, pressure-impulse diagrams that represent empirical data and the results of analytical studies may be used to evaluate their performance in response to blast loading. As with precast tilt-up construction, analyses must be performed in order to demonstrate that the diaphragms are capable of distributing the lateral loads to the shear walls and are able to restrain the large rebound forces.

9.2.2 Materials

Structural materials are affected by extremely high loading rates, high confining pressures, and large inelastic deformations in response to blast loading. Simplified elastic-plastic resistance functions are often used to represent ductile materials with dynamic enhancement factors that account for the strain rate loading. Explicit dynamic inelastic finite element analyses require more sophisticated material models to represent the behavior of structural material.

Materials such as concrete and steel are typically stronger than the minimum values specified on the construction documents. Although ASTM specifications and codes define the minimum properties for materials, the actual materials being installed typically have strengths that exceed these minimum requirements. For the analysis and design of reinforced concrete and structural steel elements for flexure, it is generally acceptable to increase the steel material's minimum yield strength by 10%. Ignoring the material's average strength and dynamic strength increase generally results in increased factors of safety for bending elements; however, underestimating the effects of material overstrength is nonconservative for shear and connection design.

Structural elements also exhibit higher strengths when subjected to dynamic blast loads than they exhibit in response to static loads. Material testing demonstrates an increase in strength in response to high-strain-rate dynamic loadings. As with the use of average strength factors, underestimating the high-strain-rate effects produces greater factors of safety for flexural element design but is nonconservative for shear and connection design. However, the factored yield strength that will be used for determining the flexural resistance function should not exceed the ultimate dynamic strength of the material.

Threat-Dependent Effects Explosive threat can be grouped into three general categories based on the explosive weight and standoff, contact/near contact, and near field and far field. Contact and near contact place the explosive threat in close proximity to the structure, generally at scaled ranges less than 0.5, scaled standoff range ($Z = R/W^{1/3}$), where R is standoff, and W is explosive weight. At this standoff distance, the explosive effects are highly localized, and the air-blast loads have extremely high-pressure short-duration pulses. At these scaled ranges, use of simplified tools is questionable. Scaled ranges between 0.5 and 2 fall into the near-field range. If charge shape is not an issue and the structure surface is flat and regular, simplified air-blast loading functions are applicable. Far-field threats generally exist at scaled ranges greater than 2.0. Pressure loading is more uniform across the structure with lower peak pressures and long duration pulses. Air-blast clearing can be a significant issue for far-field threats.

Near Range versus Far Range The amplitude of peak pressures for explosive detonations that are in close proximity to a structure or structural element is extremely high. At these short distances the blast loading is taking place within the explosive by-products, and the reflection factors may be in excess of 10. Although the duration of the applied loads is short, particularly where complete venting of the explosion occurs, the corresponding impulses are also extremely high. Fragments associated with the high-pressure range usually consist of high-velocity primary fragments from the explosive casing breakup or the acceleration of objects positioned close to the explosion. Simplified analytical tools that are based on the modified Friedlander equations (Hyde 1992, Kingery and Bulmash

1984) become less reliable for scaled ranges that are less than 0.37 to 0.45. Advanced analysis and testing show that when charge shapes other than spheres or hemispheres are considered, simplified methods lose accuracy at scaled ranges as high as 2.0. Advanced analytical methods based on computational fluid dynamics (CFD) are required to calculate accurate loading for near contact or for complicated surfaces.

For near contact, the durations of the applied loads are much shorter than the time it takes the structural elements to reach maximum deflections; they can often be analyzed for the impulse rather than the peak pressure associated with longer-duration blast pressures. Also associated with the "close-in" effects of a high-pressure design range is the possible occurrence of spalling and scabbing of concrete elements and post-failure fragments. Spalling and scabbing result from the disengagement of the concrete cover over the reinforcement on the back side of a concrete structure (the side away from the explosion). Spalling is caused by the high-intensity-blast shock wave reflecting off the back surface as a tension wave, and scabbing is associated with the disengagement of concrete as it undergoes large displacements. Secondary fragments that result from spalling and scabbing of structural materials can be a hazard to people and equipment.

Structures that are vulnerable to close-in detonations are vulnerable to breach and direct shear. Protective and containment structures adjacent to occupied areas should be constructed with double walls separated by a void of sufficient size to allow for the deflection of the outer sacrificial wall. The relative strength of the two walls and the required space between them may be determined on a case-by-case basis. Although simplified approximations may provide design guidance, explicit dynamic nonlinear finite element analyses are required to accurately determine the potential for breach and spall.

Where scaled ranges are large, the duration of the blast loading may exceed the response time of the structure. Dynamic or quasi-static response characteristics may be observed for this type of loading. Simplified single-degree-of-freedom methods may accurately determine flexure and flexural shear failures that may result from far-range loading.

9.2.3 Members

This section will discuss the modeling of structural components, such as columns, beams, slabs, and walls, to determine their behavior in response to blast loading. Included in this discussion will be the evaluation of deformation and reaction forces associated with flexure, shear, and axial effects. Also discussed will be issues such as breach, spall, and fracture.

In a numerical model, the more accurate the representation of a member, the better the results will be. Modeling of structural components, such as columns, beams, slabs, and walls, can be done on the component level, as an assembly or as a complete structure. When they are modeled on the component level, care must be taken to prevent the results from being adversely impacted by the assumed boundary conditions. Generally boundary conditions will be either free, fixed, pinned, or a combination. It is good practice to vary the boundary conditions to see the impact each choice has on the response of the component. Fixing the axial response at both ends of a member will result in a stiffer section than a member with a fixed and a free end. Application of boundary conditions directly to the ends of a member can result in additional confinement to pressure-sensitive materials, such as concrete. This additional confinement can prevent shear failure and force a member into a flexural response. Single-component models allow the analyst the opportunity to perform grid refinement studies to determine the impact of resolution on the accuracy of the solution. While single-component models are the simplest to construct and the fastest to run, the results may not provide the required accuracy.

The next step up from a single-component model is an assembly model. Examples would be a column with the addition of bracing beams, or a wall with beams and columns surrounding. The increased model size and run time for assembly models can improve accuracy due to more realistic boundary representation. The boundary conditions are further removed from the component of interest, resulting in a better representation of the physical system. The increased complexity of assembly models necessitates the introduction of connection details. This is especially true in modeling of steel-frame structures.

Full-structure models provide the greatest accuracy in modeling of a structure if the level of refinement can be maintained. The trade-offs among model size, run time, and level of refinement usually result in models of a full structure having a lower level of refinement. The reduction in refinement is traded off versus the more accurate boundary conditions and the addition of the overall response and flexibility of the full structure.

Advanced finite element methods using accurate material models are capable of capturing the response of members through failure. Level of damage can be obtained directly from the analysis results by evaluating deformations, velocities, strains levels, and material damage metrics. Simplified analysis approaches rely on guidelines based on comparison with testing or advanced analysis to relate displacements or rotations with level of damage. These guidelines, which can be found in military manuals and design references, are generally conservative (Unified Facilities Criteria 2002, Unified Facilities Criteria 2008b). This will better enable a structure to survive the applied load. The available test databases are not extensive enough to cover all possible design configurations; as a result, the accuracy of predicted damage can be affected.

Shear and Flexural Reaction Forces For most structural members, their response when subjected to a blast loading is a combination of shear and flexure. How the member responds is a function of the structural component's properties and the intensity of the loading. Long, slender members are more likely to respond in flexure than short, thick sections. The intensity of the load also impacts the type of response. A more impulsive loading is more likely to produce a shear response, as compared to a longer-duration load. As a result, a column that responds primarily in bending due to an explosive threat at a large standoff distance may respond in shear due to a near-contact blast.

Reaction forces collected at the boundaries of component models can be extreme. Reactions from shear deformation modes are usually of short duration with a high peak force. Reactions from primarily bending modes have a longer duration on the order of the period of the member. Whether a section deforms in shear or flexure depends on the properties of the section and the applied load.

Inclusion of supporting members or modeling of a region extending beyond the component of interest will result in more accurate representation of reaction forces. When the compliance of bracing members is included in the modeling, the reaction at the column-beam intersection will be reduced. This allows for a more efficient and cost-effective design of a connection.

Axial Effects Accounting for large deformation effects such as arching, stress stiffening, tension membrane, and P-Delta buckling is critical to the accurate evaluation of structures subjected to blast. Simplified modeling approaches, such as SDOF, do not account for large deformation effects in their resistance functions. As a result, the prediction of response is usually based on rotations or displacements, and the level of damage is evaluated based on guidance from calibration of the models to test results or an advanced analysis database. Extrapolation outside of the available test database can result in inaccurate assessments of structural damage. Advanced finite element methods are formulated to account for geometric nonlinear and large deformation effects and are better suited for evaluating response over the full range of damage through collapse.

The additional confinement provided by axial preload can influence the response of concrete structural components such as columns or load-bearing walls subjected to blast loading. Strength dependence on confining pressure can increase the capacity of concrete materials subjected to blast loading. In general a vertical structural component loaded with a lateral blast pressure will initially experience a compression membrane, followed by a tension membrane as the deformations increase. During compression membrane behavior, the structural resistance to lateral deformation may increase. However, as the section deforms due to blast loading and the lateral deformation grows, the ability to resist deformation transitions from compression arching to tension resistance. Post-blast damage stability is dependent on the damaged section's ability to continue to carry axial load. Geometric P-Delta effects can result in collapse under axial loads.

The response of steel columns subjected to blast is less sensitive to initial axial load. The addition of an axial preload does have an impact on the post-damage failure of the section or collapse due to P-Delta effects. A common method for evaluating the response of axially loaded members is to break the response up into load phases. Initially the member is preloaded with the axial force. The top axial degree of freedom is fixed, locking in the preload. The member is then subjected to the lateral blast load, and the damage to the column due to blast is evaluated. The axial degree of freedom is then released to allow either the axial force on the column to stabilize or the damaged column to collapse under the axial load.

Fracture Capturing fracture of a member requires the use of high-fidelity modeling. Simplified methods such as SDOF models can be calibrated to testing or high-fidelity modeling results to give indications of when fracture occurs. Displacement or rotation limits can be used to predict failure, but these are usually conservative. Advanced finite element methods, using constitutive models that capture the response of materials undergoing large strain and softening, are required to investigate fracture of members. The evaluation of breach and spall is impacted by the model's ability to calculate fracture in the material. If fracture is not accurately modeled, the ability to model breach and spall will be compromised.

9.2.4 Connections

This section will discuss the modeling of structural connections in order to determine their behavior and performance in response to blast loading. Included in this section will be the effects of shear and flexure associated with small deformations and the axial forces generated by large deformations.

Modeling of connections is critical for accurately predicting the response of structures subjected to blast and the potential for progressive collapse following a blast. Evaluation of connections is best accomplished using advanced modeling approaches. The complicated nature of connections requires a refined treatment of the details of their construction. For concrete structures, the additional confinement provided by the reinforcing steel in the joint can increase capacity and ductility. Models of a concrete joint need to include the detailed geometry and accurate material models that can account for the strength increase resulting from this confinement. Steel connections can be complicated and require the modeling of contact surfaces between members and connection plates and angles. Failure of welds or bolts can dominate the response of joints, and detailed models are required to address possible failure modes including fracture of welds or surrounding steel, bolt failure, and bearing failure of bolt holes.

Large Deformations The response of a connection changes as it deforms. Rotations can become large, and the resulting forces at the joint change. Initially at small deformations the forces in a joint can be dominated by shear and compression. As the joint deforms more, the method of force transfer through the joint can change to tension. The forces in the connection are impacted by a number of effects, such as prying, twisting, fracture, failure in bearing, and local buckling. These complicated geometric and material effects are difficult to capture accurately. Details of connection construction are difficult for simplified tools to model accurately.

Shear versus Flexural Effects During blast loading, a connection can experience direct blast loads and loads transferred from neighboring members. As adjacent members fail, additional forces can be transferred through connections from surrounding members. These forces can be transferred by shear and flexure. Bending of neighboring members can result in lateral forces due to tension or compression of the member. Prying at the joint can also induce lateral loads.

9.3 ANALYTICAL APPROACHES

The analytical method that may be used to evaluate a structure's response to short-duration but high-intensity dynamic loading depends on the characteristics of the loading and the anticipated failure mechanism. These approaches range from simplified table lookups to multi-degree-of freedom finite element calculations, and each approach has inherent constraints or limiting assumptions that must be appropriate for a specific application. For example, blast-resistant glass thickness may be selected using ASTM F2248, *Standard Practice for Specifying an Equivalent 3-Second Duration Design Loading for Blast Resistant Glazing Fabricated with Laminated Glass* (ASTM 2003), and ASTM E1300, *Standard Practice for Determining Load Resistance of Glass in Buildings* (ASTM 2007). Empirical explosive test data can often be expressed as prescriptive requirements or pressure-impulse charts. Flexural response behavior of structural components can often be determined using single-degree-of freedom methods. The *Component Explosive Damage Assessment Workbook* (CEDAW) (Oswald 2005) and Single-Degree-of-Freedom Blast Effects Design Spreadsheets (SBEDS) (Protective Design Center 2006) were developed by the U.S. Army Corps of Engineers (ACE) for use in blast effects assessments and utilize the pressure-impulse (P-I) and single-degree-of-freedom (SDOF) response methodologies. Detailed finite element analyses are computationally intensive and require complex modeling of the structure and material properties. The choice of the most efficient analytical approach requires a full understanding of the benefits and limitations.

In all cases, the analytical methods must be numerically robust in order to characterize the transient blast loading pulse and the dominant frequencies of response associated with the postulated failure mechanisms. Care must be taken so as not to neglect the critical failure mechanisms that may precipitate collapse or hazardous debris. Since it is important that the selected analytical approach be capable of representing the structure's likely failure mechanism, the designer must be able to anticipate the response when using simplified analytical methods. This is not the case for a finite element approach in which the model should capture all the failure modes.

9.3.1 P-I Diagrams

Pressure-impulse (P-I) diagrams summarize the performance characteristics of a type of structural or nonstructural element in response to a range of explosive loading. The explosive loading parameters are expressed in terms of peak pressure and impulse, and the corresponding P-I curves indicate the threshold of the

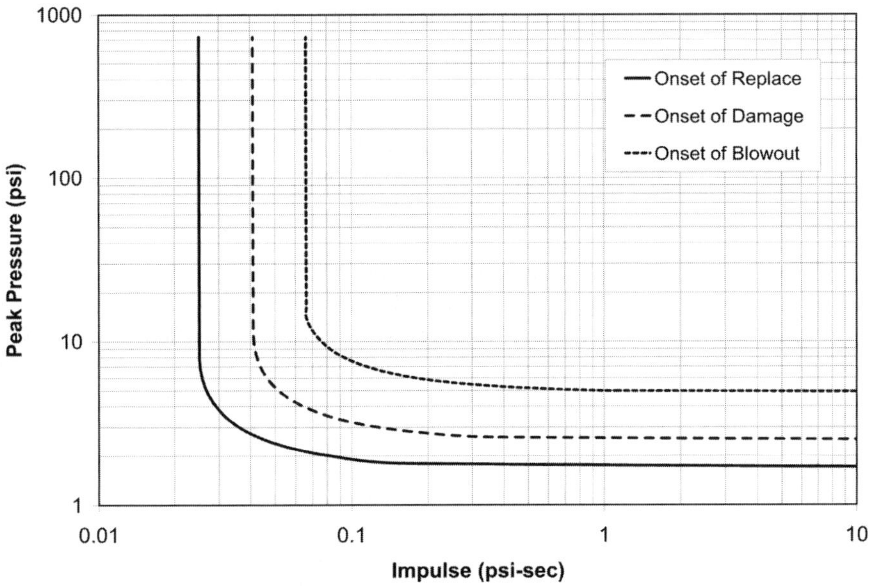

Figure 9.1 Pressure-Impulse (P-I) Diagram for Different Levels of Damage

different levels of protection or hazard. See Figure 9.1. Typically, one end of the concave P-I curves transitions to a pressure asymptote, while the other end of the curve transitions to an impulse asymptote. These two asymptotes represent the minimum impulse for all greater peak pressures or minimum peak pressure for all greater impulses that correspond to a specified performance threshold. All combinations of peak pressure and impulse that lie below or to the left of a P-I curve correspond to a better performance condition, while all combinations of peak pressure and impulse that lie above or to the right of a P-I curve correspond to a worse performance condition. The P-I curves either may be derived from an analytical database of performance or may be fit from available test data. Often, the parameters are presented in nondimensional form where the peak pressure and impulse axes are normalized so that a given set of curves and plots may be used to represent the performance of a greater range of element responses to a wide variety of explosive threat conditions.

9.3.2 Single-Element Analyses

Individual elements may be analyzed independently by means of either single-degree-of-freedom (SDOF) or multi-degree-of-freedom (MDOF) inelastic dynamic methods. These analytical methods require the use of explicit time-step numerical integration algorithms, ranging from constant velocity to linear acceleration techniques, along with the appropriate constitutive relations that represent both the elastic and inelastic behavior of the element. SDOF models are inherently simple, and the accuracy of the response calculation depends on the approximations that are used to characterize the dynamic response of the element. MDOF models may be significantly more complex and may be developed using finite element methods (FEM) to derive the mass and stiffness matrices. While it is permissible to consider damping effects, these effects are generally small in comparison to the large inelastic response that may result from blast loading and are often neglected.

SDOF SDOF methods are adequately described in a variety of structural dynamics textbooks (Biggs 1964) and U.S. Army Technical Manuals (Unified Facilities Criteria Progam 2008b). This approach is based on solving the equation of motion for a spring-mass system with a nonlinear resistance function using numerical integration. The approximate design methods are based on a Rayleigh-Ritz approach for which representative shape functions for the flexural mode are assumed and used to calculate the effective mass, effective stiffness, and effective loading function. When the loading pulse is very short relative to the period of the dynamic system, the initial velocity may be approximated as impulse divided by the effective mass, and the kinetic energy of the system can be equated to the area under the force-displacement resistance function. This provides a closed formed relationship between the required ultimate resistance (R_m), the applied impulse (I), the frequency of the dynamic system (ω) and the allowable ductility (μ).

$$R_m = I\omega/(2\mu - 1)^{0.5}$$

The accuracy of these approximate approaches depends on selecting the appropriate SDOF system to represent the governing failure mechanism of the element. While this approach is most commonly used for the design and analysis of elements that are subjected to blast loading, it requires considerable experience to make sure a critical failure mechanism is not overlooked. Typically, flexural modes of response are represented by the SDOF models; however, direct shear and instability may prove to be more critical response characteristics for short-standoff detonations and for axially loaded columns. The dynamic loads that may be applied to SDOF analyses of individual structural elements may be based on tributary areas or may be derived from the dynamic reactions of the secondary elements that frame into a primary element. The application of dynamic reactions from subsidiary elements accounts for the frequency of response and limiting capacity of the subsidiary elements, but must also include the appropriate amount of mass associated with the subsidiary element response. The dynamic reaction forces from a series of subsidiary element analyses may be converted into an equivalent dynamic load by equating the work performed by the individual reactions through a representative deformation to the work performed by the equivalent SDOF response motion. Therefore, if SDOF analyses are performed for individual secondary beams or purlins, the calculated reaction forces at every time-step may be multiplied by the shape function amplitude of the primary beam or girder at the location of their connections. If this quantity of work is divided by the integral of the shape function over the span, the resulting quantity is the effective uniform load at the corresponding time-step that is applied over the entire span.

MDOF MDOF analyses may be performed using a variety of dynamic response analysis software programs; however, all analyses must be performed at a sufficiently short time-step to guarantee numerical stability, to resolve the frequency content of the model, and to accurately represent all potential modes of failure. Nonlinear dynamic finite element analyses provide the most comprehensive means of determining the self-consistent stiffness and inertial properties, the limiting plastic limits, and the governing failure mechanisms. Furthermore, nonlinear geometric algorithms provide the means to include the destabilizing P-Delta effects due to the coupling of the large lateral transient deformations and the static gravity loads. MDOF models accurately represent the spatial distribution of dynamic pressures and dynamic reactions from subsidiary element analyses without having to generate an equivalent dynamic load.

9.3.3 Structural Systems Response

SDOF responses and MDOF models of individual elements do not account for the interaction between interconnected elements, the phasing of their responses, or the flexibility of the actual boundary conditions. These simplified approaches rely on engineering judgment for the selection of approximate boundary conditions, such as soft springs and additional mass. As a result, the idealizations represented by SDOF and MDOF analyses of individual structural elements do not accurately account for the dissipation of energy as the entire structure is deformed and suffers damage. Structural system response calculations include the relative flexibility and strength of the interconnected structural elements and provide a more accurate distribution of blast loading. Most importantly, however, structural system analyses consider the phasing of the responses between the different structural elements. Often, the region of greatest interest is modeled in greater detail, while the rest of the structure is more coarsely represented. This allows the analyst to dedicate the greatest computational resources to the response of critical details. However, as with all dynamic response analyses, the transition from coarse representations to more detailed models must be gradual so as not to generate spurious results.

9.3.4 Explicit Dynamic Finite Element Analyses

Explicit dynamic finite element methods require sufficient modeling resolution and short enough time-step to capture the high-frequency characteristics of the shock wave loading and structural response. Explicit formulations of the equations of motion express the displacement at a given time-step t_{j+1} (where the indices j represent the increments or steps in time for the duration of the response analysis) in terms of displacements, velocities, or accelerations at previous time-steps. This approach captures the high-frequency effects of the shock loading and easily allows for both material and geometric nonlinear effects; however, all explicit formulations have a critical time-step, above which the solution becomes numerically unstable. The critical time-step is related to the wave speed of the material and the least dimension of the finite element model. The equation of motion is therefore solved at each node, using the current geometry and material properties at that location and point in time.

An implicit formulation of the equations of motion expresses displacement of each node at a given time-step t_{j+1} in terms of all the displacements, velocities, or accelerations at that time-step t_{j+1}. The equations of motion for the entire system must be solved simultaneously; this is usually done using matrix methods and is a very efficient way of calculating the response of linear elastic systems. For nonlinear dynamic systems, however, an explicit formula (predictor) is usually used to estimate the response at the end of each time-step, and this is followed by one or more corrections to improve the results. Such an approach is sometimes very inefficient. Furthermore, even for elastic analysis, if the shock or high-frequency response of the system is needed, the time-step must be made sufficiently small to capture such effects while avoiding undue numerical damping. Implicit schemes are least efficient in such cases.

Given the complexity of using explicit dynamic finite element methods, great care needs to be exercised to achieve sufficient accuracy. Verification and validation of both the software and the analyst are critical to the successful completion of high-fidelity first-principle calculations. Comparison of analysis methods versus physical test data is one method for validating computational tools and approaches. Benchmarking versus known analytical solutions or against other software packages provides further confidence in software and methods. Unless established methods are used based on validation exercises, finite element methods need to be checked with grid refinement calculations to determine whether the model has sufficient resolution. Fine resolution and detailed models are no guarantee of accurate results. Accurate representation of material response, sufficient physical extent of the computational model, and proper boundary conditions are critical to calculating accurate results.

9.4 PROGRESSIVE COLLAPSE

The protective design of structures is often referred to as threat-dependent or threat-independent. In one case, a structure is designed to provide a specified level of performance in response to a specific design basis threat, while in the other, the structure is designed to be insensitive to local damage resulting from unforeseen extraordinary events. In many cases, the uncertainty of the maximum credible threat or the inability to accurately characterize the performance in response to explosive loading requires both approaches to be applied to the design. While the threat-dependent approach is related to the design basis threats, the threat-independent methods evaluate the potential for an initiating event to precipitate a broader extent of collapse. The prevention of progressive collapse may therefore be achieved through the hardening of key structural elements in order to preclude the initiating event, through the provision of ductile redundant load paths, or through compartmentalization that permits local failure to occur without endangering adjoining structural elements. In all cases, consideration must be given to both global ductility demands that involve all elements of the framing system and local ductility in the vicinity of the initiating damage.

The insensitivity to local failure, often referred to as robustness, is a property of the structure, and is independent of possible causes of initial local failure. Where specific threats are not prescribed, or for structures that are deemed low risk, an assumed extent of initial local failure and assumed extent of damage to the remaining structure may be prescribed. For these structures, element-to-element connectivity, minimum tie forces, the elimination of nonductile failure mechanisms, and enhanced stability requirements may improve the structure's robustness. If the structural members and connections can develop these minimum tie forces, which vary with construction type and location in the structure, the structure will be held together in the event local damage disrupts the primary load paths. The purpose of the horizontal and vertical ties is to enhance continuity and ductility, and to develop alternate load paths in the structure. Internal and peripheral horizontal ties are typically provided, along with ties to external columns and walls. The extent of tie force requirements may either be prescribed or calculated, starting with an assumed damage pattern. The alternate load path requirements, also termed "bridging requirements," may be determined using dynamic nonlinear analytical methods and known material damage limits, or may be approximated using equivalent static elastic analysis methods with nominal material damage criteria. The level of conservatism may be varied depending on the level of approximation or the risk assessment for the structure.

Where the characteristics of the structure make bridging impractical, local hardening of key elements or the introduction of structural fuses may be introduced into the structure to either prevent the initial damage or define a limited zone of collapse. For these types of structures, the level of specific local resistance for key elements or the zones of limited collapse must be defined.

Where a threat and risk assessment quantifies the hazards to which a structure may be subjected, specific initiating events may be prescribed, and the performance of the structure may be calculated. Nonstructural protective measures may be developed to deny access to key structural members, or details and sacrificial elements may be introduced to isolate the critical components. For this performance-based approach, the structural system redundancy and element capacities would be designed to resist these initiating events or resist the progression of damage or collapse. The increased resistance is therefore based on a prescribed set of actions and a permissible extent of resulting damage. The intensity of the initiating actions and the permissible response limits may be specified in the security performance criteria for the structure. The analytical methods that demonstrate the behavior of the structure must accurately represent the dynamic nature of the event and the damage-state behavior of the structural materials.

Regardless of whether local hardening, bridging over a damaged structure, or compartmentalization is used, a post-event stability analysis of the structural system must be performed.

For blast-resistant design, a threat must be identified in order for the initial state of damage and the residual capacity of the structure to be determined. This approach is similar to that of performance-based seismic design codes for which the engineer selects the hazard (event) and the desired performance, such as low levels of damage (immediate occupancy), in response to a small earthquake and life-safety protection (damage short of complete collapse) in response to an extreme event. Either the structural elements must be designed to resist the specified threat and maintain load-carrying capacity, or a finite extent of localized damage may be permitted, as long as the damaged structure is still capable of preventing a progression of collapse. This corresponds to the "extreme event" in response to which the structure must remain standing for life safety and evacuation of the occupants.

9.4.1 European Guidance

In the Euro-Norm EN 1991-1-7 (British Standards Institute 2006), the strategies for preventing collapse consider two different approaches: strategies based on identified accidental actions and strategies based on limiting the extent of localized failure. The strategies based on accidental actions consider designing the structure to have sufficient minimum robustness, protective measures that prevent the action or reduce its effects, or designing the structure to resist the effects of the action. The strategies based on limiting the extent of localized failure address the "threat-independent" prescriptive approaches for robustness, such as enhanced redundancy through alternate load paths, key element designs to withstand notional accidental action, and prescriptive rules for integrity and ductility.

However, the criteria for protection are specific to the project and are determined through consultation with the client and the relevant authority. Since strategies based on unidentified accidental actions cover a wide range of possible events, the means to prevent progression of collapse are typically based on limiting the extent of localized failure. The magnitudes of accidental actions that are identified, as in the case of internal explosions and impact, are specified in the EuroNorms. In some cases where there is no risk to human life and where economic, social, or environmental consequences are negligible, the complete collapse of the structure caused by an extreme event may be acceptable.

9.4.2 U.S. Guidance

All current U.S. Guidance documents are based on the U.K. standards that were based on the partial collapse of the Ronan Point apartment house. These standards extrapolated from the behavior of a panelized precast structure to develop prescriptive and performance-based guidelines for other forms of construction. Similar requirements form the basis for the most recent *Building Code of the*

City of New York (New York City Department of Buildings 2008), the 2008 Supplement to the International Building Code (International Code Council 2008), and the U.S. Department of Defense's UFC 4-023-03 (Unified Facilities Criteria 2008a). Since these approaches all rely on catenary action, the structural detailing must consider connections that are able to undergo significant inelastic rotations while sustaining large tensile axial loads.

Since the alternate path method is inherently dynamic and inelastic, the optimal analytical approach considers material and geometric nonlinearity and inertial effects. While the analysis of dynamic systems may be performed using implicit or explicit numerical integration algorithms, the introduction of material and geometric nonlinearity complicates the analysis. Implicit schemes require the reformulation of the stiffness matrix whenever properties are significantly altered, and this can become computationally inefficient when significant extents of inelasticity or catenary behavior are introduced. Explicit schemes require small enough time-steps to ensure numerical stability, and this too may become computationally inefficient. Furthermore, unless the beam-column joints and connections are explicitly modeled, their behavior must be approximated with equivalent nonlinear rotational and translational springs. The force displacement relations for these equivalent nonlinear springs are best determined by performing comprehensive component analyses of specific joints, as was done for the analysis of the Deutsche Bank Building for the *World Trade Center Building Performance Study* (Federal Emergency Management Agency 2002).

The most recent version of UFC 4-023-03 (UFC, 2008a) proposes a simplified procedure that is based on the approaches established for *Seismic Rehabilitation of Existing Buildings* (American Society of Civil Engineers 2007). A ductile material that develops significant deformation following yield, over which the force is either constant or increasing, is said to exhibit deformation-controlled behavior. Alternatively, a brittle material that experiences a steep reduction in capacity following limited inelastic deformation exhibits force-controlled behavior. The representations of structural behavior that are reported in ASCE 41-06 are considered to be reasonable for the evaluation of damage-state performance of buildings. Although simplifications to the dynamic inelastic response analyses reduce the extent of computational resources needed to calculate the performance of the structure following the removal of a key element, these simplified methods require detailed guidance in order to capture all reasonable failure mechanisms. UFC 4-023-03 offers two simplified static analysis procedures, linear static procedure (LSP) and nonlinear static procedure (NSP), as alternatives to the nonlinear dynamic procedure (NDP) approach. These simplified approaches allow the use of commercially available structural analysis software that is readily available in structural design offices. The simplifying approximations allow the analyst to apply load factors to static analyses in order to account for the missing dynamic effects, and provide the means to interpret the results of linear analyses to account for the material and geometric nonlinearities. Since the NDP is the most accurate representation of the structural response to an initial state of damage, there is little ambiguity to the interpretation of the calculated results. Correspondingly, all approximate methods are necessarily conservative, and a more representative analytical understanding will allow the designer to achieve a specified level of robustness using a wider variety of structural options. The use of the simplified LSP is restricted to buildings with regular framing plans, which are defined as irregularity limitations. Both approximate methods were calibrated to nonlinear dynamic analysis to establish simplified system-dependent factors that account for the anticipated dynamic response.

The LSP may be used wherever structural irregularities and demand capacity ratios conform to prescribed limitations. For the LSP, the structural model is analyzed in response to both deformation and force-controlled actions. The actions applied to the tributary areas above the removed key element are amplified to account for the inertial effects associated with the sudden removal of the column, and a concurrent lateral load is applied individually to each face of the structure to evaluate stability. The LSP may still be used if the structural framing is irregular, provided the calculated demand/capacity ratios (DCR) are less than or equal to 2.0; otherwise the NSP must be used. Once the acceptable use of the LSP is determined, the potential for disproportionate collapse is calculated using component or element demand modifiers (m-factors) in an approach that is similar to that of ASCE 41-06. M-factors are tabulated for different types of structural systems and details for different levels of damage-state performance. These m-factors are used to determine both the Dynamic Increase Factors and the maximum acceptable forces for deformation-controlled actions. The expected material strength and the specified material strength respectively define material properties for deformation-controlled and force-controlled analyses, and the expected component strength (Q_{CE}) and lower bound estimate of the component strength (Q_{CL}) respectively define action capacities for deformation-controlled and force-controlled analyses. Q_{UD} is defined as the deformation-controlled action, from linear static model; Q_{UF} is the force-controlled action, from linear static model; and Φ is the strength-reduction factor from the appropriate material-specific code.

$$\Phi \, \mathrm{m} \, Q_{CE} \geq Q_{UD} \quad \text{deflection-controlled criteria}$$
$$\Phi Q_{CL} \geq Q_{UF} \quad \text{force-controlled criteria}$$

The use of the NSP is not restricted by either irregularity or DCR limitations. The actions applied to the structure for the NSP are the same for calculating displacement-controlled and force-controlled responses; however, the force-displacement behavior of all structural components must be explicitly modeled, including connections wherever they are weaker or less ductile than the connected components or the flexibility of the connection produces more than a 10% change in the connection forces or deformations. Tabulated plastic rotation angles, as in ASCE 41-06, are used in the NSP to determine the Dynamic Increase Factors; the deformation limits determine compliance for deformation-controlled actions, and the lower bound estimate of the component strength determines compliance for force-controlled actions.

While the UFC relies on the best information available, based on seismic research, the justification for the specified factors and procedures requires rigorous experimental and analytical study. Furthermore, the performance of actual structural details for the different structural materials and systems requires extensive analytical research and development. Although simplified procedures must conservatively represent inelastic material properties and large displacement mechanics, there is little justification for overly precise procedures to approximate the effects of dynamics and nonlinear behavior. Furthermore, the correlation between the simplified analytical representation of structural behavior and the actual performance of structural components undergoing large deformations requires significant testing and analysis. Even the use of conventional understrength factors, Φ factors, may be questioned for the regime in which materials will be severely distorted.

REFERENCES

American Society of Civil Engineers. 2007. *Seismic Rehabilitation of Existing Buildings* (ASCE 41-06). New York: American Society of Civil Engineers.

ASTM. 2003. *Standard Practice for Specifying an Equivalent 3-Second Duration Design Loading for Blast Resistant Glazing Fabricated with Laminated Glass* (ASTM F2248). West Conshohocken, PA: ASTM International.

ASTM. 2007. *Standard Practice for Determining Load Resistance of Glass in Buildings* (ASTM E1300). West Conshohocken, PA: ASTM International.

Biggs J. M. 1964. *Introduction to Structural Dynamics*. New York: McGraw-Hill Book Company.

British Standards Institute. 2006. *Eurocode 1: Actions on Structures—Part 1–7: General Actions—Accidental Actions* (EN 1991-1-7:2006). London: British Standards Institute.

Federal Emergency Management Agency. 2002. *World Trade Center Building Performance Study: Data Collection, Preliminary Observations and Recommendations* (FEMA 403). Washington, DC: Federal Emergency Management Agency, Department of Homeland Security.

Hyde D. W. 1992. *CONWEP, Conventional Weapons Effects* (USAEWES/SS-R). Vicksburg, MS: U.S. Army Engineer Waterways Experiment Station.

International Code Council. 2008. Supplement to the International Building Code. Washington, DC: International Code Council.

Kingery C. N. and G. Bulmash. 1984. *Airblast Parameters from TNT Spherical Air Burst and Hemispherical Surface Burst* (ARBRL-TR-02555, AD B082713). Aberdeen Proving Ground, MD: U.S. Armament Research and Development Center, Ballistic Research Laboratory.

New York City Department of Buildings. 2008. Supplement to the Building Code of the City of New York, Local Law No 76 Effective December 6, 1968. New York: Department of Buildings.

Oswald C. J. 2005. *Component Explosive Damage Assessment Workbook (CEDAW) Methodology Manual V1.0* (02-0752-001), San Antonio, TX: Baker Engineering and Risk Consultants, Inc.

Protective Design Center. 2006. *Methodology Manual for the Single-Degree-of-Freedom Blast Effects Design Spreadsheets (SBEDS)* (PDC-TR-06-01), Omaha, NE: U.S. Army Corps of Engineers, Protective Design Center.

Unified Facilities Criteria Program. 2002. *Design and Analysis of Hardened Structures to Conventional Weapons Effects* (UFC 3-340-01). Washington, DC: Department of Defense, Unified Facilities Criteria Program.

________. 2008a. *Design of Buildings to Resist Progressive Collapse* (UFC 4-023-03). Washington, DC: Department of Defense, Unified Facilities Criteria Program.

________. 2008b. *Structures to Resist the Effects of Accidental Explosions* (UFC 3-340-02). Washington, DC: Department of Defense, Unified Facilities Criteria Program.

10 Building Envelope and Glazing

Eve Hinman and Christopher Arnold

For an explosion outside a building, the exterior envelope is the critical line of defense that separates the people, operations, and contents inside the building from the air-blast effects outside the building. Unfortunately, for most standard building types, the building envelope is ill-suited to resisting air-blast loads, which are enormous and which act directly on the surface area of the exterior envelope in the out-of-plane direction.

In this chapter, some of the key concepts needed to optimize the design of the exterior envelope to mitigate explosion effects are presented. A variety of wall and window types are considered, as well as doors and louvers. Roof and foundation systems are also addressed.

10.1 DESIGN INTENT

The primary design objective for the exterior envelope of most standard building types, is to mitigate the hazard of flying debris generated by failed exterior walls, windows, and other components, to reduce casualties and business disruption and facilitate rescue and evacuation efforts.

Other objectives are to design the exterior envelope to:

- Fail in a way that does not initiate progressive collapse
- Keep the air blast outside the building to the best of its ability

10.1.1 Life Safety

Impact of flying debris entering a building causes laceration and blunt trauma injuries, and potentially fatalities. Once the exterior envelope is pierced and the air blast enters the building, injuries such as eardrum rupture, lung collapse, and blunt trauma and concussion due to being thrown against objects are the dominant injury modes. Impacts due to internal nonstructural damage to partitions and ceiling, light fixture, and equipment components and falling furniture are also a hazard.

Although it is cost-prohibitive to eliminate all these hazards, it is possible to reduce the risk to a level that is acceptable through informed design decisions.

10.1.2 Emergency Egress and Facilitating Search and Rescue

The building exterior envelope design may be optimized to make it easier for people to get out and for emergency responders to enter safely.

After an explosive event, the debris created by the failed exterior envelope can exacerbate casualties by reducing the efficiency of evacuation and rescue operations. Some examples of this are:

- Internal debris can cause occupants to become trapped in rubble piles.
- Exterior debris can block emergency exits.
- Debris along egress paths can cause delays and additional injuries.
- Exterior falling fragments post-event trigger safety precautions for emergency operations, which can significantly delay rescue, thereby putting the injured at greater risk.
- Toxic dust or fumes generated by debris may have long-term health ramifications.

10.1.3 Critical Functions (Protecting Equipment and Business Processes)

Even at low pressures, air blast that enters the building can cause extensive nonstructural damage to interior partitions and suspended ceilings. The resulting internal disarray can cause extensive business disruption. Even if the damages are repairable, total cost required to return the building to pre-event functionality may exceed the replacement cost of the building.

Sensitive mechanical parts of critical equipment can be disabled by air blast or flying debris that enters the building through the exterior envelope. An interior baffle wall may be used to dissipate the energy of the air blast by forcing it to change direction two or more times before it reaches the equipment. Some other options are to use storm louvers or forced-entry-resistant louvers, which have a chevron cross section which are able to perform a similar function. Louver attachments into the structure are to be designed to resist the capacity of the louver to ensure that the louver fails before the connections. In some cases, a catch system using a grate securely fastened behind the louver may be used to prevent impact into delicate equipment components.

Critical equipment placed on the roof will be less vulnerable to a ground level explosion than equipment at or below street level. For roof equipment, a hardened wall with roof may be used for protection. However, for this solution, note that if the wall fails, the impact of debris may have a more detrimental effect than using lightweight housing that is not blast-resistant.

It is recommended that the transformer be placed away from vulnerable areas, for instance in a parking lot, near the loading dock, or under the sidewalk. Inside the building in a protected location is preferred. The switchgear room is to be separated to the extent practical from the transformer(s), the generators, and any other backup systems, such as an uninterruptible power supply (UPS) or batteries, to improve the chances of retaining at least emergency functions in the event of an explosive attack. In addition to separation, redundancy of critical systems is highly recommended for higher-risk buildings. An example of this is placing one set of generators on one side of the building, and another set on the other side of the building to create redundancy.

10.2 DESIGN APPROACH

For buildings with a low to moderate risk of explosive attack, the design approach is to first design the building exterior envelope for conventional loads, and then to evaluate the response to explosive loads. If the design team is properly consulted when key decisions are made during the concept phase, then the modifications required to meet explosive load requirements can be addressed efficiently. This approach ensures that the design meets all the requirements for gravity and natural hazards, in addition to air-blast effects, and is effective for buildings where explosion effects are a secondary concern.

Take note that some features that are beneficial for blast resistance diminish performance for other loads, if not properly integrated. For example, using a plastic coating that emits toxic fumes in the event of a fire may not be a good solution for enhancing the blast performance of an exterior wall, since fire is a more common hazard than terrorist attack.

Another example that presents a potential conflict between the seismic design and blast design is that increased mass generally increases the design forces, whereas for explosion loads, mass generally improves response. This is because of the highly impulsive nature of explosive loads. Heavy structural components have longer periods of vibration so that, by the time the mass, M, is mobilized and the structural component begins to move with a velocity, V, the load or impulse, I, has been removed. In mathematical terms, the kinetic energy, $KE = MV^2/2$, or momentum, MV, imparted to the structural component is directly proportional to impulse and inversely proportional to the mass: i.e., $KE = MV^2/2 = (Mv)^2/(2M) = I^2/(2m)$. In this situation, careful coordination between the blast consultant and the structural engineer is needed to provide an optimized response that accounts for both loading cases.

Good antiterrorism design is a multidisciplinary effort, requiring the concerted efforts of the architect, structural engineer, security professional, and the other design team members. For instance the passive protection provided by effective building design needs to be balanced with the active or operational security measures implemented (e.g., guards, cameras, dogs, sensors, magnetometers). It is critical that these two sets of measures complement and support one another. In particular, the size weapon that is considered at the perimeter and at the exterior envelope is a function of the size weapon that is able to be easily detected through operational security measures. Like protective design, security needs to be implemented early in the design to be effective. Lines of sight; lighting and placement of cameras along the exterior envelope; and location, number and type of entrances provided are examples of security issues to be addressed early in design with the architect, since they potentially affect the design threat to be used.

The envelope system is to be designed by keeping in mind the concepts of balanced or capacity design (see Section 10.2.3) and ductile response. Ductile systems are able to absorb energy through inelastic flexural deformation. Systems that fail in a brittle manner through shear are to be avoided. By permitting limited exterior envelope damage that does not significantly increase the hazard to the majority of the occupants, it is possible to design more cost-effective systems that absorb energy through deformation, and transmit lower forces into the connections and supporting structure, thus reducing the potential for more serious structural failures.

Perhaps the simplest analytical method is to use a pressure-impulse, or P-I, diagram, in which various damage levels for a component with a given resistance are expressed as curves, and the axes are peak pressure and impulse. By plotting the pressure and impulse acting on the system, you can estimate the level of damage to the exterior envelope component and compare it with allowable limits. These curves may be based on explosive tests or on the output from a semi-empirical or a single-degree-of-freedom system. The limitation of these tools is that they are based on constant parameters for the component under consideration. If the component that is being investigated differs in any way, some extrapolation or interpolation between charts is needed.

Typically, simplified or single-degree-of-freedom (SDOF) methods used for the design of structural components may be used for the exterior envelope (Unified Facilities Criteria Program 2008). For these systems, the component time-dependent peak deflection, $X(t)$, is evaluated by modeling the component as a lumped mass, M, so that the inertia force is $MA(t)$ where $A(t)$ is the acceleration or second derivative of $X(t)$. The resistance is provided by a spring, $R(x)$, which has a linear elastic flexural stiffness, K, and a ultimate flexural resistance of R_u. The loading is modeled as a time-dependent function, $F(t)$. For a system initially at rest, the resulting equation of motion is:

$$MA(t) + R(x) = P(t)$$

where: $R(x) = KX(t)$ when $X(t)$ is less than or equal to the maximum elastic limit displacement X_e and

$$R(x) = R_u \text{ when } X(t) > X_e$$

In these expressions, M and R, and P are "equivalent" values which represent the portion of the mass, resistance, and loading that is participating in the displacement of the member, and which provide a natural frequency and

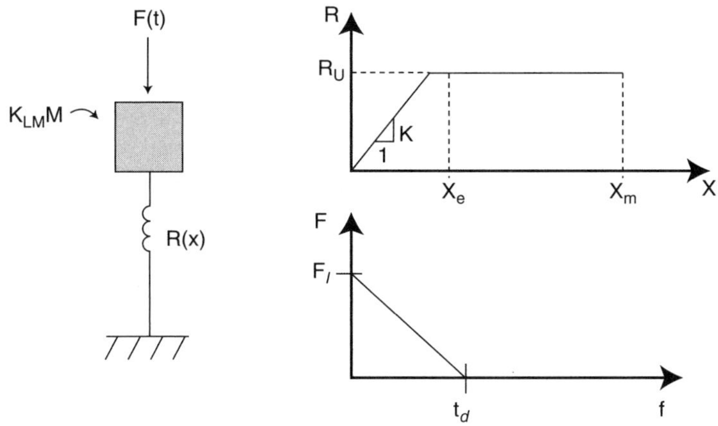

Figure 10.1 Graphical Representation of an SDOF System

displacement that are the same as that of the actual system. Figure 10.1 provides a graphical representation of an SDOF system.

These equations may be easily solved by using numerical methods. The resulting maximum displacement is typically evaluated by comparing the ratio of the maximum displacement *Xm* and the maximum elastic limit displacement, or ductility *Xm/Xe*. Another measure is the support rotation, which is defined as the angle between the undeflected member and the line connecting the support point to the point of peak deflection. For example, for a uniformly loaded, simply supported beam of length *L* the support rotation is equal to the tangent of 2*Xm/L*. Figure 10.2 illustrates the concepts of displacement ductility and rotation. Typically, the maximum allowable ductility may vary from 1 for a brittle material like monolithic glass to about 10 for ductile systems like steel. For heavily damaged systems, a ductility of up to 30 may be allowed. Permitted ductilities are generally about 2 degrees for most systems, but may be permitted to be 4 degrees or more for highly ductile structures where extensive but low-hazard damage is permitted.

Other models that may be used include two or more degrees-of-freedom systems or piecewise linear continuum models. Two-degrees-of-freedom models have been used with some success, in particular for a beam and girder where a portion of the load is transmitted directly to the girder from the beam, and where the beam experiences rigid body as well as deformational displacements. Piecewise linear systems are of particular interest because they are able to represent behavior more accurately for bilinear or trilinear resistance functions and loading functions. This type of analysis was not a practical solution until the recent advent of powerful computational tools such as MATLAB (MathWorks 2006) or Mathcad (MathSoft 2001).

However, for complex exterior envelope systems or air-blast loading scenarios or systems using innovative materials or energy absorption methods, simplified

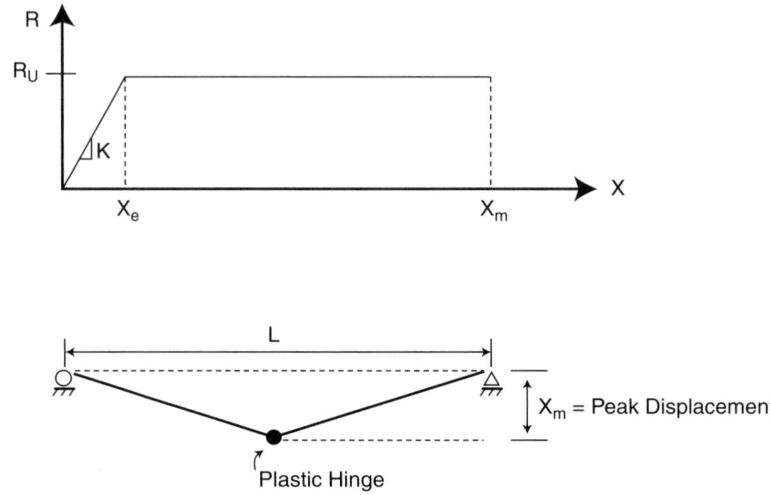

Figure 10.2 Graphical Representation of the Concepts of Displacement Ductility and Rotation

analysis is not sufficient. In these cases, the engineer needs to use finite-element numerical time integration techniques and/or explosive testing to verify response. Again, nonlinear dynamic time history analysis is required to adequately model the system. Material models used for these programs are generally more sophisticated than the linear elastic perfectly plastic models used for simplified analysis. In this case, the material models provide a more realistic representation for large deformation. Examples of computer codes on the market that are capable of performing these types of computations include SAP2000/NonLinear (Computers and Structures 1990) and LS-DYNA (Livermore Software Technology Corp. 1999). There are also codes available to model the expansion of an air blast and its reflection against the structure surface, which may be used to generate the loads applied to the model using a software product such as CONWEP (Hyde 1988) or ATBLAST, a software product prepared by Applied Research Associates.

The time and cost required to perform the analysis cannot be ignored in choosing computational methods. Because the design process may require iteration, the cost of analysis must be justified in terms of benefits to the project and increased confidence in the reliability of the results. In some cases, a simplified approach will be used for the preliminary design and a more sophisticated approach, using finite elements and/or explosive testing, may be used for the final verification of the design.

If testing has been performed for a similar system, it is necessary to obtain the testing report and verify that the assumptions and findings are consistent with the project requirements, prior to accepting testing in lieu of detailed analysis. In particular, check to make sure that the support conditions for the member are accurately represented. Also check to see that both the pressure and impulse are at least as large as what the design requires. If only the pressure or the impulse meets criteria, some additional calculations will be needed to verify response. It is recommended that the testing be supplemented with pre- and post-analysis, using an advanced method such as finite elements.

Often, only a proof of concept design is provided by the blast engineer during design. The final blast design is the responsibility of the subcontractor implementing the design. A detailed set of specifications is strongly recommended to ensure that the design intent is met. Some of the requirements to consider, including in the specifications for the exterior facade subcontractors, are:

- Provide performance-based design requirements for response.
- Put all the blast requirements in a single location in the specifications, rather than sprinkling them through the individual sections. This makes it easier for everyone to keep track of these special requirements.
- Require the blast engineer's approval of the subcontractor based on successful performance on similar projects.
- Require the subcontractor to submit blast calculations for review.

10.2.1 Response Criteria

Levels of damage to the exterior envelope may be described by these terms: minor, moderate, major, catastrophic. Qualitative descriptions of each damage level are given below as they pertain to the exterior envelope. Note that all these descriptions imply that there is little or no structural damage (i.e., all the damage is limited to the exterior envelope).

Minor: There is little if any damage. The exterior envelope may sustain some damage, but virtually no failure. Damage may consist of cracked windows, failed sunshades still attached to the building, or permanent deformation of walls and window frames. Injuries may be expected, and fatalities are possible but unlikely.

Moderate: Failure of the exterior envelope is confined to a localized area and is usually repairable. Hazardous window failure is limited to one face of the building. Injuries and possible fatalities are expected.

Major: Extensive exterior envelope damage and some roof damage are expected. In this case, extensive fatalities are expected. Damage is usually not repairable.

Catastrophic: In this case, all or the majority of the exterior envelope has been removed.

Generally, moderate damage for the design basis threat is a reasonable goal for new construction to meet life-safety objectives. For buildings that need to remain operational during and after an event, or are designated as high risk, minor damage may be the more appropriate damage level.

Quantitative performance is usually evaluated by comparing the ductility (i.e., the peak displacement divided by the elastic limit displacement) and/or support rotation (the angle between the support and the point of peak deflection) to empirically established maximum values that have been established by the military through explosive testing. Maximum permissible values vary depending on the material and the acceptable damage level. Some criteria documents do provide the design values that need to be met.

10.2.2 Static versus Dynamic

A dynamic nonlinear approach is more likely to provide a design that meets the design constraints of the project than a static approach. Elastic static calculations are likely to give overly conservative design solutions if the peak pressure is considered without the effect of load duration. By using dynamic calculations instead of static, we are able to account for the very short duration of the loading. In addition, the inertial effect included in dynamic computations greatly improves response. This is because by the time the mass is mobilized, the loading is greatly diminished, enhancing response. Furthermore, by accepting that damage occurs, we are able to account for the energy absorption of ductile systems that occurs through plastic deformation. Finally, because the loading is so rapid, we are able to engage the enhanced material strength that often occurs with very high strain rates.

10.2.3 Balanced Design

Balanced or capacity design philosophy for a building system refers to designs that are controlled such that the failure of the weakest component in the system results in the least destruction. For a window system, for instance, it refers to the window glass failing at pressure levels that do not exceed those of the frame, anchorage, and supporting wall system. If the glass is stronger than the supporting members, then the window is likely to fail with the whole panel entering into the building as a single unit, possibly with the frame, anchorage, and the wall attached. This failure mode is considered more hazardous than one in which the glass fragments enter the building, provided that the fragments are designed to minimize injuries. By means of a damage-limiting approach, the damage sequence and extent of damage are controlled.

10.2.4 Load Path

Load path in the exterior envelope generally refers to the transmission of load from the member that presents the most surface area to the exterior (i.e., glass pane, panel, wall, slab) into the building structure. Connections and reaction forces, two of the more nontrivial parts of the path, are discussed below.

Connections For the exterior facade, the connections include:

- The connection between the window glass and the window mullion, transom or frame
- The connection of the window mullion or transom and the window frame
- The connection between the window frame and the wall
- The connection between the wall and the structural frame

Connections are typically the least ductile elements in the exterior envelope system and the most prone to failure if not properly designed. The connections are required to enable the supporting elements to reach their ultimate flexural capacity without failure. They also need to remain intact for the loads imposed by large inelastic deformations of the supporting members, which in turn can cause significant deformation of the connections.

Fasteners holding the window frame in the wall tend to be placed into tension and shear as the window undergoes deformation. The wet glazing are put into shear as a laminated glass pane begins to act as a membrane. As the mullion of a unitized system deforms, stresses are placed on the connections holding the two halves of the mullion together.

The failure envelope of connections respondingto load combinations such as tension and shear need to be considered as part of the design.

Although it seems as if window frame connections should be routinely designed to carry the ultimate loads transmitted by the window pane, this is not the case. The window frame is designed to carry the design loads only. The capacity of the glass may be significantly higher than the design loads.

Reaction Loads (*Ru *versus Applied Load) Reaction loads may conservatively be approximated by the static ultimate capacity of the loading element, and by setting the load and strength factors for the connection design to 1.0. The load factor is set to 1 because we are assuming that the supported member has been pushed to its capacity. The reaction load cannot be higher than the capacity of the member. Similarly, the strength factor is set to 1, again in recognition of the fact that we are letting the supporting member reach its ultimate capacity. Although this may not seem conservative, it is, because we neglected the mitigating effects of ductility and inertia on the system.

This approach is consistent with capacity or balanced design principles and is the best and most reliable. However, it can lead to designs that are not constructible or affordable. If this is the case there are several approaches which may be considered:

- Use the average applied load instead of the peak load, because the edge load along a panel transfers a load that varies with its peak value at the center point and decreases towards the ends. Often the peak value is given, but the average load along the length is less and is more representative of the loading on the anchors.

- Increase the strength factor for flexural or tensile response, to account for strain-hardening effects. For impulsively applied loads, the strength of the material is increased. This effect is referred to as strain hardening.
- Permit the loss of a secondary member (for instance, permit the loss of a beam but not a girder). By enabling the major element in the load path to be designed to the capacity of the elements it supports, we are improving performance while maintaining the cost-effectiveness of the design. This is done often for roof systems where the floor slab capacity load is greatly in excess of the design loads.
- Use an element with a lower ultimate capacity. This is a practical and effective tool for systems where there is significant ductility. For instance, a thin precast panel will have a lower ultimate resistance than a thicker panel and can be designed to perform well in ductility.

10.3 FENESTRATION

Windows, once the sole responsibility of the architect, become a structural issue once explosive effects are taken into consideration. Windows designed to mitigate the effects of explosions are first to be designed to resist conventional loads, and then to be checked for explosive load effects and balanced design.

Even if the building remains standing and no structural damage occurs, extensive injuries can occur due to nonstructural damages. Windows are typically the most vulnerable portion of any building. Although it may be impractical to design all the windows to resist a large-scale explosive attack, it is desirable to limit the amount of hazardous glass breakage to reduce the injuries. Typical glass windows break at low pressure and impulse levels, and the shards created by broken windows are responsible for many of the injuries incurred due to large-scale explosive attack.

Designing windows to provide protection against the effects of explosions can be effective in reducing the glass laceration injuries. Glass lacerations are caused when the shards of glass created when a window is broken are propelled into the building and penetrate the skin. This is perhaps the most common type of injury that results from large-scale explosion incidents and may occur hundreds or even thousands of feet from the source in urban areas, where the air blast attenuates more slowly as it propagates through the street "canyons."

Window protection should be evaluated on a case-by-case basis by a qualified protective design consultant, to develop a solution that meets established objectives. In the sections below, a number of generic recommendations are given for the design of the window systems to reduce injuries to building occupants.

To limit glass laceration injuries, there are several approaches that can be taken. One approach is to reduce the number and size of windows, which potentially will reduce the air-blast and glass shards entering the building, thus reducing the interior damage and injuries. Specific examples of how to incorporate these ideas into the design of a new building include:

- Limiting the number of windows on the lower floors where the pressures are higher due to an external explosive threat
- Using an internal atrium design with windows facing inward toward the atrium, rather than outward toward the facility perimeter
- Windows, which are close to the ceiling, above the heads of the occupants
- Angling the windows away from critical weapon locations to reduce the pressure levels

10.3.1 Glass

Glass is often the weakest part of a building, breaking at low pressures compared to other components such as the floors, walls, or columns. Past incidents have shown that glass breakage and associated injuries may extend many thousands of feet in large external explosions. High-velocity glass fragments have been shown to be a major contributor to injuries in such incidents. For incidents within downtown city areas, falling glass poses a major hazard to passersby and prolongs post-incident rescue and cleanup efforts by leaving tons of glass debris on the street.

As part of the damage-limiting approach, glass failure is not quantified in terms of whether breakage occurs or not, but rather by the hazard it causes to the occupants. The glass performance condition is defined based on empirical data from explosive tests performed in a cubical space with a 10-foot dimension. The performance condition ranges from 1, which corresponds to no glass breakage, to 5, which corresponds to hazardous flying debris at a distance of 10 feet from the window.

Design criteria established by the Interagency Security Council, which governs the antiterrorism efforts for most agencies, require performance condition 3b for buildings that are at moderate risk of attack. At this protection level, the window breaks, and fragments fly into the building but land within 10 feet of the window. The design goal for moderate-risk buildings is to achieve a performance level of less than 4 for 90% of the windows. Protection level 4 means that glass fragments will penetrate a witness panel located 10 feet behind the window for a height of two feet or less.

The preferred solution for new construction is to use laminated glass with structural silicone sealant (i.e., wet glazing) around the inside perimeter. The lamination holds the shards of glass together in explosive events, reducing its potential to cause laceration injuries. The structural sealant helps to hold the pane in the frame for higher loads. For insulated units, only the inner pane needs to be laminated because the inner pane will provide protection against the hazard of shards generated by the outer pane.

Annealed glass has a breaking strength that is about one-half that of heat-strengthened glass and about one-fourth that of tempered glass (Amstock 1997), thus reducing the loads transmitted to the supporting frame and walls. To reduce reaction loads, it is sometimes advantageous to consider annealed glass. The preferred interlayer thickness is 60 mil (0.06 in) unless otherwise specified by the criteria. At a minimum, 30 mil is to be used.

To make sure that the components supporting the glass are stronger than the glass itself, we specify a window breakage strength that is high compared to what is used in conventional design. The breakage strength in window design may be specified as a probability of the number of windows expected to break at that load. For instance, when designing for conventional loads, it is typical to use a breakage pressure corresponding to a probability of 8 breaks out of 1,000. Where there is a high probability of extensive glass breakage, as in an explosion incident, a pressure corresponding to 750 breaks out of 1,000 is used. Glass breakage strengths may be obtained from window manufacturers.

Smaller glass panes generally have higher capacities than larger panes. Consequently, smaller panes can cause significantly higher loads to be transmitted to the frames than larger panes. As a result, in blast-mitigating design, we avoid small panes. Also, since every size pane has a difference capacity, it is desirable to standardize the design as much as is practical.

There are several government-sponsored software products available for evaluating the response of window glass for use on federal projects, including HAZL (U.S. Army Engineer Research and Development Center 2001), WINGARD, and WINLAC. These codes are made available to government contractors who have government projects requiring this type of analysis. One approach that is publicly available is described in the criteria by the U.S. military (Unified Facilities Criteria Program 2007), using ASTM Standards (ASTM 2003, 2004a). For those who are designing unique systems for specific threats not covered in criteria documents, it is recommended that the system be tested (ASTM 2001).

Glass block is generally not recommended because of the heavy projectiles these walls may create due to failure at the mortar lines. However, blast-rated glass-block products are available in which the glass blocks are framed by a steel grate system or other method.

The way in which the glass is supported into the exterior envelope can have an effect on the severity of the glass hazard. A brief summary of various types of supports is given below.

Punched Windows: Punched windows (see Figure 10.3) often can be designed efficiently to be less vulnerable than other window types because the frame is attached directly into the wall, which is generally of more robust construction.

Strip Windows: Strip or ribbon windows (see Figure 10.4) generally consist of alternating bands of glazing and opaque material constructed using precast panels, poured concrete, or insulated metal panels. The opaque area conceals the floor structure and is referred to as a "spandrel." Ribbon windows have thin vertical metal framing separating the individual panes, and are supported by

Figure 10.3 Example of "Punched" Windows

the wall at the top and bottom. For this type of exterior facade, the spandrels are attached to each floor level instead. The system is very economical and a common facade type. To resist air-blast loads, steel angle kicker braces are often needed to laterally support the bottom of the spandrel panel.

Glass Curtain Wall: For this glazing type, a significant part of the building exterior is covered with windows supported by aluminum or possibly steel framing. Figure 10.5 shows three variations of curtain wall system.

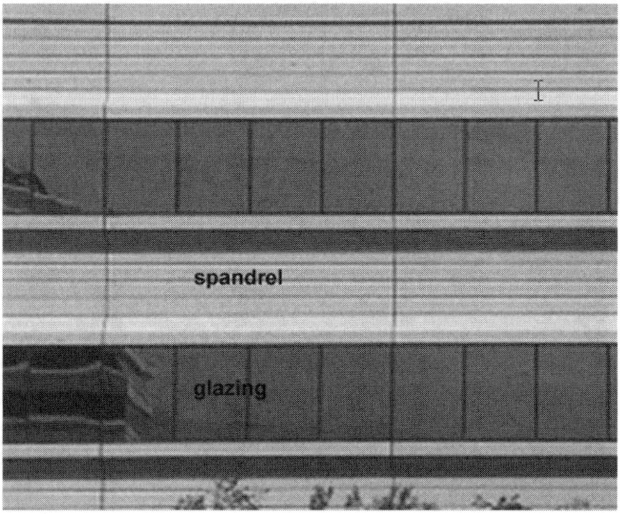

Figure 10.4 Example of Ribbon Windows, Showing Spandrel

(a)

(b)

(c)

Figure 10.5 Types of Metal and Glass Curtain Walls. (a) Curtain wall framing expressing main building structure. (b) Uniform curtain wall framing over the façade. (c) curtain wall framing draped over rectilinear building forms.

Figure 10.6 Example of Point and Cable Supported Curtain Wall

This support type has an increased risk of hazardous failure as compared with punched windows because strip window systems are supported by wall on two sides instead of four sides.. However, it has more flexibility, which allows it to deform significantly without failing.

Point and Cable Supported: In point and cable supported systems (see Figure 10.6), each window pane is connected at points near its corners. The bracket that supports the glass often has multiple arms, forming a spider, so that it can attach near the corner of several adjacent panes. The glass is drilled for a threaded connector to pass through, engaging the interlayer to secure the pane during an explosion. The panes are typically connected using a clear or translucent polymer material instead of the metal framing (referred to as "butt glazed").

Spandrels The term "spandrel" is used to describe the non-vision panels used above and/or beneath the window or at the floor levels (see also discussion regarding "strip windows" above). They are used to hide from view the structure or mechanical equipment behind. Sometimes a "shadow box" is used behind the spandrel, which is a metal pan designed to catch the glass fragments and keep the explosive loads from entering the building.

Operable versus Inoperable Although operable windows can be designed to meet modest explosion requirements, keep in mind that they will not keep air blast out of the building if they are open.

Inoperable window solutions have potential to be more reliable for protection during air-blast events, because occupants cannot void the windows' protective function by opening them. However, there are operable window solutions that are conceptually viable. For instance, if a window is designed to open outward about a horizontal hinge at the sill, the window will tend to slam shut in an exterior explosion. If this type of design is used, the governing design parameter may be the capacity of the hinges and/or hardware.

The controlling design component for sliding operable windows (and glazed doors) often is the supporting track. If possible, the track should be embedded in the supporting structure to provide added out-of-plane support to the system.

Skylights/Courtyards Skylights in roof systems have the advantage of not being subject to reflected pressures fromexterior explosions at ground level. Also, skylights are required by building code to be designed using laminated glass, which is preferred for explosion-mitigating design. However, they do create a falling fragment hazard. Therefore, these should be designed with a catch system beneath, or designed to remain in the frame for the design air-blast load. Ideally, skylights should be placed as far from the weapon as possible, to keep the pressures low.

For an atrium with an open courtyard in the center, the windows are protected by the building on all sides, and the glass will be subject to indirect air-blast pressures in the event of an explosion outside the building exterior. Because atria are protected on all sides, they may be designed for the reduced pressures from an exterior threat. These reductions in pressures can help significantly in reducing the design requirements imposed by explosion loads.

Doors Glazed doors and glass panes above a door are to be designed using the same methods as for windows. Glazed doors tend to use tempered glass for safety, in case of accidental breakage due to impact. As a result, they are high-strength and can complicate the design of the supporting system

10.3.2 Mullions/Transoms

The vertical frame members connecting adjoining windows are referred to as mullions. Horizontal members are termed transoms. These members may be designed in two ways. Either a static approach may be used, whereby the breaking strength of the window glass is applied to the mullion, or a dynamic load may be applied, using the peak pressure and impulse values. A static approach may be overly conservative because the designer using this approach assumes that the peak edge forces are sustained indefinitely, whereas in an explosion the forces are of very short duration, and the supporting structure has insufficient time to fully respond to the impulsive forces. Using this approach, the mullion can become very deep and heavy, driving up the weight and cost of the window system. Figure 10.7 shows heavy mullions in an older building.

Sometimes cables or steel bars or tubes attached to the supporting structure are placed behind the glass to prevent the laminated glass or glass with anti-shatter film from entering the interior (Crawfordk, Lan and Dunn 2006).

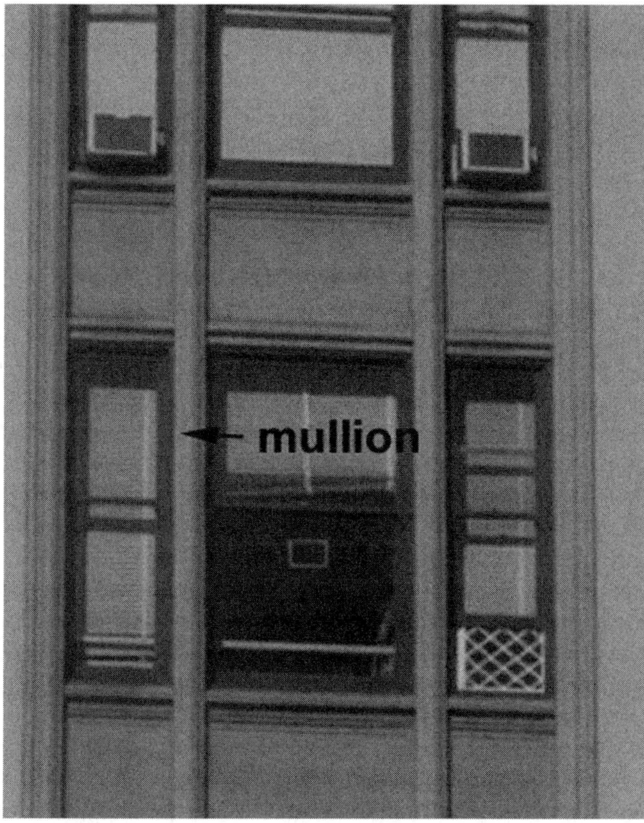

Figure 10.7 Example of Mullions in an Older Building

10.3.3 Frame and Anchorage

The window frames need to retain the glass so that the entire pane does not become a single large unit of flying debris.

To retain the glass in the frame, a minimum of a $^1/_4$-inch bead of structural sealant (e.g., silicone) is used around the inner perimeter of the window. For designs where a 60-mil polyvinyl butyral (PVB) interlayer is used, the silicone sealant should be designed to resist the shear forces caused by the membrane action forces using the ultimate tensile capacity of the PVB material. These membrane forces can significantly increase the reaction loads into the framing system, increasing the design requirements for the entire window system. The allowable tensile strength of the silicone sealant should be at least 20 psi. Also, the window bite (i.e., the depth of window captured by the frame) needs to be at least $^1/_2$-inch. For large windows, such as in the lobby area, the bite required for conventional loads may well exceed the $^1/_2$-inch minimum, and additional silicone sealant may be needed to avoid failure caused by the entire glass pane exiting the frame.

Frame and anchorage design is performed by applying the reaction forces from the window at breakage to the frame and the fasteners. In most conventionally designed buildings, the frames will be aluminum. In some applications, where the windows are designed to resist high pressures and for long spans, steel bar inserts, cable inserts, or built-up steel frames may be used.

For reinforced concrete construction designed to resist high-pressure loads, as is typical for embassy construction, anchorage of the steel window frames is provided by steel studs welded to a steel base plate. For this type of construction, the frame is typically constructed using a steel stop at the interior face and an angle with an exposed face at the exterior face. The frame is attached to the base plate using high-strength fasteners. Coordination between the wall and window subcontractors is required to ensure that the fastener locations are spaced so that they fit between the rebar in the wall.

For masonry walls, flat metal straps embedded in the mortar with adequate development length are recommended for anchoring the window into the wall.

10.3.4 Supporting Structure

It is inconsistent with balanced design and is potentially highly hazardous to have a wall system that is weaker than the windows it is supporting.

Anchoring window/wall systems into columns is generally discouraged because it increases the tributary area of lateral load that is transferred into these critical members, and may cause failure or instability.

Some window/wall designs will require additional lateral support. For instance, when the supporting walls are acting largely as cantilevers, such as for windows placed high on a wall, they might need to be supported with vertical braces spanning the clear floor height using for instance steel tubes to provide additional stiffness. For punched wall systems with narrow pilasters between them, vertical braces may also be needed. For lighter wall systems such as metal stud systems, suitable reinforcements such as back-to-back double studs framing the window are recommended.

The balanced design approach is particularly challenging in the design of ballistic-resistant and forced-entry-resistant windows, which consist of one or more inches of glass and polycarbonate. In this situation, the window may be framed directly to the structure using steel framing to transfer loads and not into the wall.

10.3.5 Other Penetrations

Similar to windows, other penetrations are designed to improve the safety of the occupants. Rather than explicitly designing to resist air-blast pressures, the focus is on reducing the hazard presented by the failure.

Doors Doors are handled differently in different criteria documents. Most criteria documents neglect the response of door systems. This may be for several reasons. Doors that are capable of resisting air-blast loads can be very expensive.

Also, doors are typically in transitory areas where people do not stay for very long. Some concepts for increasing the inherent protection offered by doors are as follows:

- Use double steel doors with internal cross braces.
- Orient doors to open outward so that they bear against the jamb during the positive pressure loading phase.
- Fill the jambs with concrete to increase their strength.
- Increase the number of fasteners used to connect the door into the wall system.

Revolving doors are required by code to use tempered or laminated glass, to prevent impact breakage during normal use. It is not typical to explicitly design these doors to resist the threat, but some precautions may be taken to reduce the hazard, such as increasing the bite or using silicone sealant between the glass and frame.

Louvers Louvers are another type of opening to consider. These components should be designed with connections that are able to resist the flexural capacity of the louver. A catch system, consisting of a well-anchored steel grate behind the louver, is another approach.

Air intakes that are at or close to the ground level should always have grates so that weapons cannot be lobbed into them. Also consider using a sloped grating (at an angle of at least 45 degrees from the horizontal) so that a potential weapon can roll off prior to detonation.

Blowout Panels Blowout panels are designed to disintegrate or break free at relatively low impulses, to limit the buildup of expanding hot gasses in an interior space. This type of wall fails easily and will act to vent the explosive forces outside the building, helping to protect vertically and horizontally adjacent spaces.

Historically, blowout panels also are used when designing for an accidental explosion inside a building. For deliberate explosions, the approach is somewhat different. Blowout panels typically are used when an internal space, such as a mailroom designed to resist a package bomb, is vulnerable to explosive attack. In these situations, mailrooms are placed adjacent to the exterior wall of the building, and the exterior wall is designed using lightweight construction using, for instance, metal studs. Alternatively, conventional construction, without consideration to explosion loads, may be used.

10.4 EXTERIOR WALLS

Exterior walls often are designed to resist reflection-magnified pressures from an external explosive threat located directly beyond the secured perimeter line. One important objective of the design is to detail these members to fail in a ductile mode such as flexure, rather than a brittle mode such as shear. The walls also need to be able to resist the ultimate loads transmitted by the windows and doors.

It may be cost-efficient to consider the reduction in pressure with height due to the increase in distance and the angle of incidence at the upper levels of a high-rise building. Even if pressure reductions are taken into account at the upper floors, minimum requirements such as balanced design and ductile response are to be met, to reduce the hazard to occupants in case the actual explosion is greater than the design threat.

Various types of wall construction are considered below.

10.4.1 Concrete Walls

Cast-in-Place Historically, the preferred material for explosion-mitigating construction is cast-in-place reinforced concrete. This is the material that is used for military bunkers, and the military has performed extensive research and testing on its performance. Reinforced concrete has a number of attributes that make it an attractive construction material because of its mass, ductility, and monolithic character.

Reinforced concrete is recommended for high-risk buildings that are vulnerable to large-scale attack, and require a high level of protection. It is also recommended for highly protected areas within buildings, such as primary egress paths or high-occupancy areas.

Note that for reinforced concrete to respond favorably to explosion loads, it is to be detailed in a ductile manner, such as is done in high seismic zones. Some attributes of ductile protective design (Federal Emergency Management Agency 2003) are as follows:

- Use symmetric reinforcement on both faces, so that load reversals may be accommodated.
- Span the wall from floor to floor, rather than from column to column, to reduce the chances of building collapse.
- Stagger splices away from high-stress areas, to maintain the monolithic character of the material.
- Space reinforcing bars no more than one wall thickness apart, but no less than one-half the wall thickness apart, to lessen the chances of localized brittle behavior.
- Use ductile special seismic detailing at plastic hinge locations, so that they are able to develop the full moment of the span.
- Use full tensile development lengths to develop the ultimate flexural capacity of the section.
- For progressive collapse prevention, consider the loss of an exterior wall that measures vertically one floor height and laterally one bay width.

- Use closed ties or spiral reinforcing along the entire length of beams and columns, including connections with a minimum bend angle of 135° and a spacing not exceeding d/2, to provide confinement.

Precast Panels For precast panels, consider a minimum thickness of five inches, exclusive of reveals, with two-way reinforcing bars to reduce the ultimate resistance, increase ductility, and reduce the chance of flying concrete fragments. This will also reduce the reaction loads transmitted into the connections. In high-pressure regions, using ribbed panels (see Figure 10.7) may be an effective way to resist the loads. These panels need to bear against the floor diaphragms.

The following are recommendations and considerations in designing precast elements for air-blast resistance.

Reinforcement: Two symmetric, continuous layers of two-way reinforcement are recommended to accommodate large deformations and rebound loads. For thin panels where it is difficult to place two layers of reinforcement, the use of two layers of heavy wire mesh, one layer of two-way reinforcement along the centerline, or staggered bars on either face may be considered.

If a single layer of reinforcement is used, it is critical to design the section so that the steel yields before the concrete fails in compression to obtain a ductile response. Enhanced protection may be provided by placing fiber-reinforced polymers, geotextile materials, Kevlar, or similar materials on the inside face to provide confinement, resistance to scabbing and spalling, and added tensile capacity. If reinforcing bars are used, a layer of wire mesh on the interior face may help to further restrain concrete fragments from entering the space. Closely spaced bars also help fragment restraint, but care must be taken not to increase the ultimate

Figure 10.8 Precast Concrete Ribbed Wall Panels

flexural capacity too much, so as to keep the reaction loads to a reasonable level. Note that centerline reinforcement will not work in the rebound direction due to the failure of the tension concrete during the positive phase. If the primary objective is to protect occupants and not passersby, then this may be acceptable. Above major egress points, however, where debris outside the building presents an obstacle to ingress and egress post-event, added protection is desirable.

Load-Bearing Systems: For progressive collapse resistance in load-bearing precast systems, panels need to be designed to span over failed areas by means of arching action, strengthened gravity connections, secondary support systems, or other means of providing an alternate load path.

Connections: Ductile connections are recommended. The connections and supporting structure need to be able to resist the loads transmitted by the panel loaded to its ultimate flexural capacity, so that the system is balanced for the possibility that the actual air-blast loading is higher than the design load. Reaction loads from the windows at ultimate capacity are to be included in the calculation of connection design loads. Using this approach, every panel with a different configuration will have a different set of design loads for the connections. Note that small panels will have higher reaction loads than larger panels, using this method, due to their increased stiffness. Standardizing the panel sizes greatly simplifies the connection load determination.

Also, the connections need to provide sufficient lateral restraint for the panels to accept large deformations. Depending on the design details of the connection used, lateral restraint design will require consideration of in-plane shear, buckling, flexure, and/or tension loads in the design. Punching shear through the panel also needs to be checked. The connection should provide a direct load path from the panel into the supporting structure to minimize P-Delta effects.

Floor-to-floor panels with continuous, gravity-type connections directly into the floor diaphragms are preferred. Multistory panels that are hung on the floor diaphragms may also be considered. Connections into exterior columns or spandrel beams are discouraged, to avoid the possibility of initiating structural collapse of the exterior bay. Redundant gravity connections are strongly recommended, to prevent falling debris if a single connection fails.

Connections should be checked for rebound loads. It is conservative to use the same load in rebound as for the inward pressure. More accurate values may be obtained through dynamic analysis or charts provided in military handbooks (Unified Facilities Criteria Program 2008).

Specifications: Specifications for precast elements should be in the form of a performance requirement, with the air-blast pressures and required response limits. The performance specifications give precast contractors more flexibility to provide the systems with which they are most familiar. This approach requires that the contractors have either in-house dynamic analysis capability or a relationship with a blast engineer who can work with them to customize the most cost-effective system.

For structural precast systems, the connections are the critical issue to be addressed. Connections need to permit the panel to reach its ultimate flexural capacity. Extensive use of transverse walls or an "egg crate" type of design is an effective way to achieve the needed lateral support. Special seismic detailing (American Concrete Institute 2008) is recommended for structural precast connections.

Tilt-Up Construction Tilt-up construction is acceptable, provided that the connections into the structure are able to withstand the ultimate capacity of the wall in both the inward and outward (i.e., rebound) direction.

Post-Tensioned Panels Pre-tensioned or post-tensioned construction without significant mild steel reinforcement provides little capacity for abnormal loading patterns and load reversals. If these systems are used, it is recommended that mild reinforcing be added, to provide the needed ductility

10.4.2 Masonry

Reinforced Masonry For concrete masonry unit (CMU) block walls, as a minimum use 8-inch block walls, fully grouted with vertical centered reinforcing bars placed in each void, and horizontal ladder type reinforcement at each layer. Connections into the structure are to resist the reactions associated with the wall loaded to its ultimate lateral capacity. For infill walls, avoid transferring loads into the columns if they are primary load-carrying elements.

The connection details may be very difficult to construct. It will be difficult to have all the blocks fit over the bars, near the top, and it will be difficult to provide the required lateral restraint at the top connection. A preferred system is to have a continuous exterior CMU wall that laterally bears against the floor system. For infill walls, use development lengths that develop the full capacity of the section. For increased protection, consider using 12-inch blocks with two layers of vertical reinforcement. Also, consider using CMU block units that encourage a homogenous response, such as an "I"-shaped unit with inner and outer faces that are connected with a small strut.

Unreinforced Masonry Brick load-bearing walls are unable to sustain tensile loads, so they must resist explosion loads mostly through mass to reduce the kinetic energy that must be absorbed. As a result, thicker solid walls on the order of approximately 18 inches are preferred and respond well at air-blast levels less than 10 psi or so. Unreinforced structural masonry is considered a very brittle material that may generate highly hazardous flying debris in the event of an explosion, and is to be avoided for new construction.

10.4.3 Steel

Metal Studs For metal stud systems, use metal studs back to back and mechanically attached. The two-stud system provides the benefit of lateral torsional resistance, and therefore can be more efficiently designed. To catch exterior cladding fragments, attach a wire mesh, steel sheet, or fiber-reinforced polymer to the interior side of the panel. The supports of the wall should be designed to resist the ultimate out-of-plane bending capacity load of the system.

If a single-stud system is used, deeper, thicker channels are preferred. As a minimum, consider 16-ga systems with depth of 6 inches, and laterally braced are preferred.

Special care is required at the connections to prevent failure prior to the stud reaching its ultimate capacity.

Enhanced protection may be provided by placing fiber-reinforced polymers, geotextile materials, Kevlar, or similar materials on the inside-face of the cladding to provide confinement, fragment restraint, and added tensile capacity.

Metal Panels Non-structural metal panels may be successfully used if they are braced behind with steel tubes or compact sections. Continuous welds are required around the perimeter braces to resist the membrane loads. At the end panels, the braces will need to be augmented to account for the lack of equalizing forces.

10.4.4 Other

Appurtenances Brick veneers, bris soleil, sunshades, and other nonstructural elements attached to the building exterior are to be avoided, to limit flying debris and improve emergency egress by ensuring that exits remain passable. If used, they should be designed using lightweight materials with connections designed to resist the ultimate capacity of the appurtenance.

Exposed Structural Systems For exposed exterior frame systems, there are two primary considerations. The first is to design the exterior columns to resist the direct effects of the specified threats. The second is to detail the exterior frame so that it has sufficient structural integrity to accept localized failure without initiating progressive collapse. As a rule of thumb to meet these goals, column spacing should be limited to 30 feet, and floor heights should be limited to not greater than 16 feet, wherever possible. It is recommended that the columns be designed to resist the explosive loads associated with a contact charge placed at the base, or that a column removal analysis be performed to verify that progressive collapse is not initiated.

Because columns do not have much surface area, air-blast loads on exposed standalone columns that are not supporting adjacent wall systems tend to be mitigated by "clearing-time effects" (Unified Facilities Criteria Program 2008). This refers to the fact that the pressure wave washes around these slender tall members, and consequently the entire duration of the pressure wave does not act upon them. Figure 10.9 shows an exposed column support on the corner of a high rise office building.

For columns subject to a vehicle weapon threat close by in the public street in an urban area, buckling and shear are the primary effects to be considered in

Figure 10.9 Example of column freely exposed to public access. The column supports the corner of a high-rise building.

analysis. This is a common scenario considered for office buildings in urban areas. Because exposed columns are often in public areas and are primary support members for buildings, the concern is that progressive collapse may be initiated.

A very large weapon close to a column can cause shattering of the concrete due to multiple tensile reflections within the concrete section, destroying its integrity. Circular columns shed load more rapidly than rectangular columns and can be beneficial.

Buckling is a concern if lateral support may be lost when a floor system that provides lateral support is damaged. This is a situation that arises frequently for office buildings that are close to public streets. In this case, exterior columns should be capable of spanning two or more stories without buckling.

Insetting the first line of columns a few feet into the building interior is a nonconservative approach to protecting the columns. Even if the columns are not publicly accessible, they are still vulnerable to air-blast effects, and their response still needs to be addressed.

Protect columns by using closely spaced closed ties or spiral reinforcing, to improve confinement and shear capacity. This also will improve the performance of lap splices in the event of loss of concrete cover, and greatly enhance column ductility. The potential cost-to-benefit ratio for providing closely spaced closed ties in exterior concrete columns is among the lowest and should be considered seriously. Closed ties are to be used in columns and spandrels for the entire span and across connections with a maximum spacing of d/2 and a minimum bend angle of 135°. Some other recommendations for reinforced concrete are given in wall subsection 10.4.1.

For exposed steel columns, splices should be placed as far above grade level as practical, and base plates should be recessed below the ground level.

For a contact or close-in package weapon, column breach is a major consideration. To mitigate this threat, some suggestions include:

- Do not use exposed columns that are fully or partially accessible from the building exterior. Arcade columns should be avoided (see Figure 10.10).
- Use an architectural covering that is at least 6 inches from the structural member. This will make it considerably more difficult to place a weapon directly against the structure. Because explosive pressures decay so rapidly, every inch of distance will help to protect the column.
- Use steel plates to surround the base of concrete columns where it is accessible. Plates need to extend several feet above the accessible location.
- Encase steel columns with concrete and add a plate, 3/8″–1/2″ thick around the perimeter for about 5 feet above the base, if necessary.

Load-bearing ductile reinforced concrete wall construction without columns can provide a considerable level of protection if adequate reinforcement is provided to achieve ductile behavior. This may be an appropriate solution for the parts of the building that are closest to the secured perimeter line (at scaled ranges less than 3).

Spandrel beams of limited depth generally do well when subject to air blast. In general, edge beams are very strongly encouraged at the perimeter of concrete slab construction, to afford frame action for redistribution of vertical loads and to enhance the shear connection of floors to columns. Confinement of concrete spandrels using closely spaced closed ties, such as those used for columns, is recommended. Transfer trusses and transfer girders are to be avoided because they increase susceptibility to progressive collapse.

Figure 10.10 Example of public walkway beneath second floor of high-rise building

10.5 ROOF SYSTEMS

The primary loading on the roof is the downward air-blast pressure. The exterior bay roof system on the side(s) facing an exterior threat is the most critical. Roof systems that are low and therefore closer to the explosion will be subject to higher pressures than high roof systems. The air-blast pressures on the interior bays are less intense and may require less hardening. Secondary loads include upward pressure due to the air blast penetrating through openings, and upward suction during the negative loading phase. The upward internal pressures may have an increased duration due to multiple reflections of the internal air-blast wave. It is conservative and recommended to consider the downward and upward loads separately.

Because roof systems are not exposed to the reflected wave from the ground level, they are subject to lower pressures than the walls facing the explosion. Because the angle between a sloped roof and the advancing shock sometimes is less acute than the angle between a flat roof and the advancing shock front, sloping roofs may be subject to somewhat higher pressures. They are usually of lighter construction than flat roof systems, making them particularly vulnerable to air-blast effects unless designed to respond in a ductile manner. For this reason, sloped systems sometimes require more robust designs.

10.5.1 Concrete

The preferred system is to use cast-in-place ductile reinforced concrete with beams in two directions. If this system is used, beams should have continuous, symmetrical top and bottom reinforcement with tension lap splices. Shear ties should develop the bending capacity of the beams and be closely placed along the entire span. All ties are to have a 135-degree bend minimum. Two-way slabs are preferred.

Precast and pre/post-tensioned systems, including hollow plank, are generally viewed as less desirable due to the lack of ductility. If they are used, a system that has continuous bond with the concrete is preferred, with anchors that are designed to be protected from direct air-blast effects. Also, additional mild reinforcement at top and bottom is recommended to ensure a ductile response. Connections need to be designed to resist both the direct and uplift forces.

Concrete flat slab/plate systems are also less desirable because of the potential of punching shear failure at the columns. Where flat slab/plate systems are employed, they should include features to enhance their punching shear resistance. Continuous bottom reinforcement should be provided through columns in two directions, to retain the slab in the event that punching shear failure occurs. Edge beams should be provided at the building exterior.

10.5.2 Steel

Lightweight systems, such as untopped steel deck, are considered to afford negligible resistance to air blast. These systems are prone to failure due to their low capacity for downward and uplift pressure.

10.5.3 Composite

Modest levels of protection are afforded by conventional steel beam construction with a steel deck and concrete fill slab. The performance of this system can be enhanced by using normal-weight concrete fill, increasing the gauge of deck and welded wire fabric reinforcement from that required for conventional loads, and making the connection between the slab and beams using additional shear connector studs. Tension membrane behavior along the edges should be considered in the design of the connections by using welding or fasteners along the edges. Since it is anticipated that the slab capacity will exceed that of the supporting beams, beam end connections and supporting columns should be capable of developing the ultimate flexural capacity of the beams, to avoid brittle failure. Beam to column connections should be capable of resisting upward as well as downward forces.

10.5.4 Penthouses/Gardens

Parapets, roof mechanical room enclosures, and tile roof systems are generally not a primary concern, since they are exterior to the building. Generally these members are designed to sustain heavy damage but not become flying debris. Although roofing aggregate may become a flying hazard, in the context of an explosion event, this hazard is not significant enough to warrant much concern.

Soil can be highly effective in reducing the impact of a major explosive attack. Bermed walls and buried rooftops have been found to be highly effective for military applications and can be effectively extended to conventional construction. This type of solution can also be effective in improving the energy efficiency of the building.

10.6 BELOW GRADE

For buildings that are very close to the secured perimeter, there is the possibility of the foundations becoming undermined by the cratering effects of an explosion. However, if this is an issue, then generally, it will be accompanied by heavy damages to the superstructure as well. If the crater is projected to reach the building, then the most cost-effective option may be to increase the building setback.

Ground shock effects are generally a secondary effect since most of the energy of a vehicle weapon is transmitted to the air rather than the soil. The weapon would have to be placed underground to have a significant effect on the structure. Currently, underground weapons are not considered by the governing federal criteria for civilian buildings.

There are significant benefits to placing secured areas below grade in terms of mitigating explosion effects from an exterior weapon. The massiveness and softness of the soil provide a protective layer that significantly reduces the impulse on the structural systems below grade.

For buildings with below-grade portions that are adjacent to the building, when creating a plaza level at ground level, keep in mind that the roof systems of these underground areas will need to be designed for the actual air-blast pressure levels, if these are occupied areas. If these are unsecured areas, such as a garage, consider letting the roof fail if adequate egress routes are available on other sides of the building, away from the failed plaza level.

Another consideration for belowground portions of the building is the design of the perimeter security barriers. The perimeter barriers often require deep foundations, which may interfere with the underground structure.

It is preferable to place underground garages adjacent to the main structure rather than directly underneath the building, to protect the structure against the effects of an internal weapon. The effects of an internal weapon are generally not a major concern for foundation walls. The soil on the other side of the wall provides a buffer that mitigates the response. One exception to this is a situation where the foundation is below the water table, where even a localized breach of the wall may cause extensive collateral damage.

Vaults for transformers placed beneath the ground close to public streets are of concern. If possible, place these away from public streets. If they are in a driveway, the vault lid needs to be designed to resist the downward pressure.

In high seismic regions, seismic isolators may be used at the base of a building (Figure 10.11) . In this case, the response of the building globally should be checked for the total air-blast loading acting on the side facing the explosion. Preliminary studies investigating this issue have shown that seismic response governs.

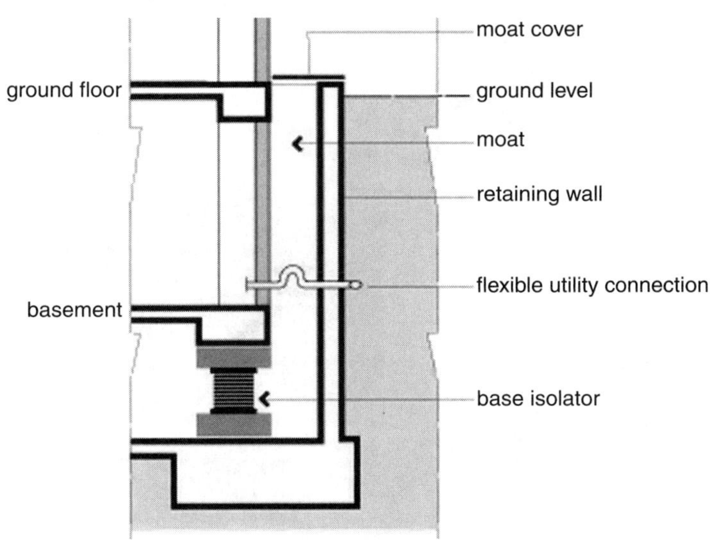

Figure 10.11 Diagram of seismic base isolation system. The building superstructure is detached from the foundation and supported by steel plat/rubber bearings.

10.7 REDUCTION OF BLAST PRESSURES

The placement of the building on the site can have a major impact on its vulnerability. Ideally, the building is placed as far from the property lines as possible. This applies not only to the sides that are adjacent to streets, but to the sides that are adjacent to adjoining properties as well, since we cannot be certain about how access will change for those neighboring properties during the life of the building. A common-practice example of this is the use of a large plaza area in front of the building, which often leaves little setback on the sides and rear of the building. This practice can diminish the vulnerability of the front of the building, but generally will increase the vulnerability of the other three sides.

The shape of the building can have a contributing effect on the overall damage to the exterior envelope. Reentrant corners and overhangs are likely to cause multiple reflections of the air blast, which may amplify the effect of the air blast (see Figure 10.12).

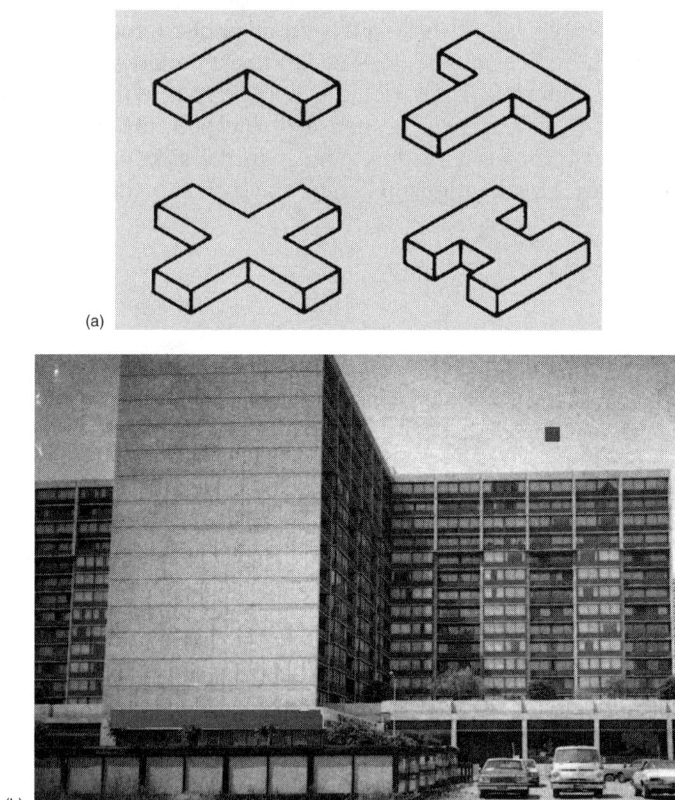

Figure 10.12 Reentrant corner plans: (a) Some plan types. (b) T-shaped building with two reentrant corners.

In general, convex rather than concave shapes are preferred for the exterior of the building (i.e., the shock front incidence angle on a convex surface increases more rapidly with lateral distance from a detonation location than on a planar surface, causing the reflected pressure on the surface of a circular building to decay more rapidly than on a flat building, See Figure 10.13). Similarly, the air-blast pressures decay with height, as the angle of incidence becomes more oblique. The sides of the building not facing toward the explosion do not experience reflected pressure and will typically perform with less damage.

(a)

(b)

Figure 10.13 Building shapes. (a) Convex, and (b) Concave.

Generally, simple geometries, with minimal ornamentation (which may become flying debris during an explosion), are recommended unless advanced structural analysis techniques are used. If ornamentation is used, it is recommended that it consist of a lightweight material, such as timber or plastic, which is less likely than brick, stone, or metal to become lethal projectiles.

REFERENCES

American Concrete Institute. 2008. *Building Code Requirements for Structural Concrete and Commentary* (ACI 318-08), Chapter 21. Farmington Hills, MI: American Concrete Institute.

Amstock, Joseph S. 1997. *Handbook of Glass in Construction*. New York: McGraw-Hill.

ASTM. 2001. *Standard Test Method for Structural Performance of Glass in Exterior Windows, Curtain Walls, and Doors Under the Influence of Uniform Static Loads by Destructive Methods* (E997-01). West Conshohocken, PA: ASTM International.

________. 2003. *Standard Practice for Specifying an Equivalent 3-Second Duration Design Load for Blast Resistant Glazing Fabricated with Laminated Glass* (F2248-03). West Conshohocken, PA: ASTM International.

________. 2004a. *Standard Practices for Determining Load Resistance of Glass in Buildings* (E1300-04e1). West Conshohocken, PA: ASTM International.

________. 2004b. *Standard Test Method for Glazing and Glazing Systems Subject to Airblast Loads* (F1642-04). West Conshohocken, PA: ASTM International.

Computers and Structures, Inc. 1990. *SAP2000/NonLinear User's Manual*. Berkeley, CA: Computers and Structures Inc.

Crawford, John, Shengrui Lan and Brian W. Dunn. 2006. Cable catcher systems for improving blast resistance of glazing facades. In *Proceedings of the 19th Military Aspects of Blast and Shock Symposium*, Alberta, Canada, October 2006. Burbank, CA: Karagozian & Case.

Federal Emergency Management Agency. 2003. *Primer for Design of Commercial Buildings to Mitigate Terrorist Attacks* (FEMA 427). Washington, DC: Risk Management Series of Publications, FEMA, Department of Homeland Security.

Hyde, David W. 1988. *Microcomputer Programs CONWEP and FUNPRO, Applications of TM 5-855-1*, Fundamentals of Protective Design for Conventional Weapons *(User's Guide)*. Vicksburg, MS: Army Engineer Waterways Experiment Station, Structures Lab.

Livermore Software Technology Corporation. 1999. *LS-DYNA User's Manual, Nonlinear Dynamic Analysis of Structures*, Version 950. Livermore, CA: Livermore Software Technology Corporation.

MathSoft. 2001. *Mathcad User's Guide*. Cambridge, MA: MathSoft Inc.

MathWorks. 2006. *MATLAB: The Language of Technical Computing*, Version 7. Natick, MA: The MathWorks, Inc.

Unified Facilities Criteria Program. 2007. *DoD Minimum Antiterrorism Force Protection Standards* (UFC 4-010-01) Washington, DC: U.S. Department of Defense, Unified Facilities Criteria Program.

_______. 2008. *Structures to Resist the Effects of Accidental Explosions* (UFC 3-340-02). Washington, DC: U.S. Department of Defense, Unified Facilities Criteria Program.

U.S. Army Engineer Research and Development Center (ERDC). 2001. *Window Fragment Hazard Level Analysis, the HAZL Model.* Vicksburg, MS: U.S. Army Engineer Research and Development Center.

11 Protection of Spaces

MeeLing Moy and Andrew Hart

Security can be integrated at the earliest stage of planning and continued throughout the design and construction phases for building structures requiring blast resistance, or other specialized security measures. Building occupant emergency plans may also be included in planning such that the design created can supplement the objectives of emergency operations. The goal is to increase the chances that emergency systems will remain operational during an explosion. The continued operation of particular building elements during emergencies is essential for building occupant life safety and evacuation. This chapter will focus on the following building components: areas isolating interior threats, stairwell enclosures, hardened plenums, and safe havens.

11.1 AREAS ISOLATING INTERIOR THREATS

Interior explosive threats may be carried or delivered into accessible areas within a building. These areas include, but are not limited to, loading docks, mailrooms, and lobbies. The magnitude of interior threats and the specific areas isolating them within a building are identified by a risk assessment. The Federal Emergency Management Agency (FEMA) defines a small explosive to be approximately 5 to 10 pounds of TNT equivalent, a large explosive to be approximately 50 to 100 pounds of TNT equivalent, and a mail bomb to usually be less than 10 pounds of TNT equivalent.

Protection of critical building components and systems within the identified interior high-risk areas may be achieved by installing electronic surveillance cameras, by implementing controlled access with entry and/or exit checkpoint screening, and by structural hardening.

Walls and slabs enclosing specific interior high-risk areas should be designed or hardened for the identified interior threats as determined by a risk assessment. FEMA recommends that blast effects should be vented through the walls isolating interior threats. For example, these walls may consist of blowout panels designed to provide security from blast loads applied on the outside due to an exterior threat, but to fail due to an interior threat, thus venting the blast loads outside. FEMA suggests that slabs isolating interior threats should be designed or hardened to consider downward and upward blast loads.

11.2 STAIRWELL ENCLOSURES

Exit and emergency egress stairwells are considered to be spaces that are accessible to the general public. They are designed for optimal performance with sufficient distribution and redundancy to meet current building code requirements, and building occupant needs. Stairwell planning and design are a fundamental part of a building occupant emergency plan. FEMA recommends that stairwells should maintain positive pressure for safe evacuation of building occupants, and for access by firefighters and rescue responders during an emergency. The installation of special filtering systems to provide a clean source of air may be required, to maintain positive pressure and minimize smoke and hazardous gases in stairwells. The National Institute of Standards and Technology (NIST) recommends that stairwell design capacities should also take into account the counterflow of rescue personnel during an emergency. Stairwell lighting should remain functional during an emergency, and stairwell exit signs should provide a clear and continuous route to the outside or to another area that is safe. In addition to the building code requirements, stairwells should be located in less vulnerable areas, and far from areas where blast events may occur in the building. Thus, stairwells should not discharge into high-risk areas.

Exit and emergency egress stairwell enclosures, which include walls and slabs, should be designed or hardened for the identified threats specified in a risk assessment. FEMA recommends that stairwell walls should be anchored to the floor slabs at each end, in order to adequately transfer the loads from a blast to the lateral resisting structural system of the building. FEMA suggests that stairwell slabs should take into account the uplift pressures from a blast by using an upward load equal to the dead load plus half the live load for the floor system. FEMA also recommends that doors to the stairwell enclosures should consist of steel, and be adequately anchored to the stairwell walls. FEMA also suggests that windows in the stairwell enclosures consist of laminated glass with a 60-mil polyvinyl butyral (PVB) interlayer. The laminated glass should be adhered within the window frames and mullions using a 1/2-inch bead of structural silicone. The window frames and mullions should be anchored to the stairwell walls in order to develop the full capacity of the window glazing.

11.3 HARDENED PLENUMS

Mechanical equipment and systems are considered to be primary nonstructural elements that are necessary for the life-safety procedures of a building. Their configurations and capacities are designed with adequate redundancy to meet current building code requirements and building occupant needs. They are identified as critical assets that should continue main operations during an emergency, to ensure the safe evacuation of building occupants. FEMA recommends that public access to critical mechanical equipment and systems should be restricted, in order to reduce the potential damage from a blast event. In addition to the building code requirements, primary and redundant mechanical equipment and systems should be situated in separate locations and far from high-risk areas, to reduce the possible hazard of both being damaged by a single blast event.

Structural elements supporting the plenum that surrounds the mechanical equipment should be designed or hardened to resist reflected blast pressures and impulses, such that debris and fragments are not propelled into the protected spaces. Mechanical vents and exposed plenums are designed or hardened to the same level of protection as required for exterior walls. Designs should take into account the potential infill of blast overpressures through mechanical openings. Structural elements and mechanical equipment mountings and attachments should be designed to resist these internal pressures.

11.4 SAFE HAVENS

A safe haven can be defined as a secure area designed to protect occupants from various hazards. Such shelters may be located outside or within a structure. Safe havens are fully enclosed structures designed to resist the effects of blast loads, and the impact of fragments and debris due to an explosive event. This section will focus on safe havens as documented by FEMA.

11.4.1 FEMA Documents

The FEMA document that deals with the protection of safe havens is primarily FEMA 453, *Safe Rooms and Shelters, Protecting People Against Terrorist Attacks*, dated May 2006. Within the document two types of shelters are defined:

- Standalone shelter
- Internal shelter

The difference between the two types of shelters is that a standalone shelter is a defined as a building that is not attached to or within any other building; it is considered as a separate building. This separate building is to be constructed to withstand a range of natural or man-made hazards and may be situated away from any debris or potential fragment hazards. The building is to be structurally separated from any other building, and thus not be affected by any weakening from an adjacent building collapsing. Since this is a separate building, it does not need to be integrated into the building design, but it is costly compared with an internal shelter.

An internal shelter is a specially designed room within or attached to a building. The room is designed to be structurally independent of the building and is able to withstand a range of natural or man-made hazards. Since the shelter is partially shielded by the building, it may not get the full blast pressure. The location of the room within the building has to be easily accessible by all building occupants, and when designing for blast mitigation, no resistance from any of the surrounding areas should be included. Having it within the building is less costly than building a standalone shelter, as it can be part of a renovation or a later addition to the building design.

11.4.2 Multi-Hazard Threats

FEMA 453 does not quantify a specific threat such as a terrorist threat from a vehicle bomb; rather, the document gives general guidelines on different types of building construction and reasonable mitigating measures for providing a secure shelter.

Prior Warning In some instances, prior to an event, a warning is given. All types of warnings given about any type of threat should be treated in a similar manner. The standard procedure for a prior warning is laid out in FEMA 453: Once the warning has been received, the contacted person calmly asks for the location of the weapon (e.g., the building where it is housed), the type of weapon, and any other pertinent information that he or she can obtain. Once obtained, the contacted person must immediately call his/her local enforcement agencies—such as local and federal law agencies—and emergency services, and pass on all information from the perpetrator. All warnings should be treated as reliable, regardless of their apparent validity.

Sequential Events after/during an Event Every type of shelter, regardless of its size and number of occupants, should have an Emergency Operations Center, or EOC. This center or group of people serves as an information management center where decisions are made, such as those made during the attacks on the twin towers on September 11, 2001. Each EOC should be equipped with communication equipment, reference materials such as medical books, log books to note down what happened and when, and any other tools necessary to respond quickly and appropriately to an emergency, such as a defibrillator and a first aid kit. Within this center, the Emergency Management Group (EMG) makes decisions based upon the information provided by the Incident Commander (IC) and other personnel. The relationship between the EMG and the emergency services or Emergency Operations Group (EOG) is shown in Figure 11.1. Note that both the EMG and the EOG have directors, each in constant contact with each other throughout the duration of the event, coordinating the responses to the event as it unfolds.

The use of an IC is vital, as this person is responsible for the front-line management of the response to the incident, tactical planning and execution, determining if help is required, and relaying requests for assistance, if required,

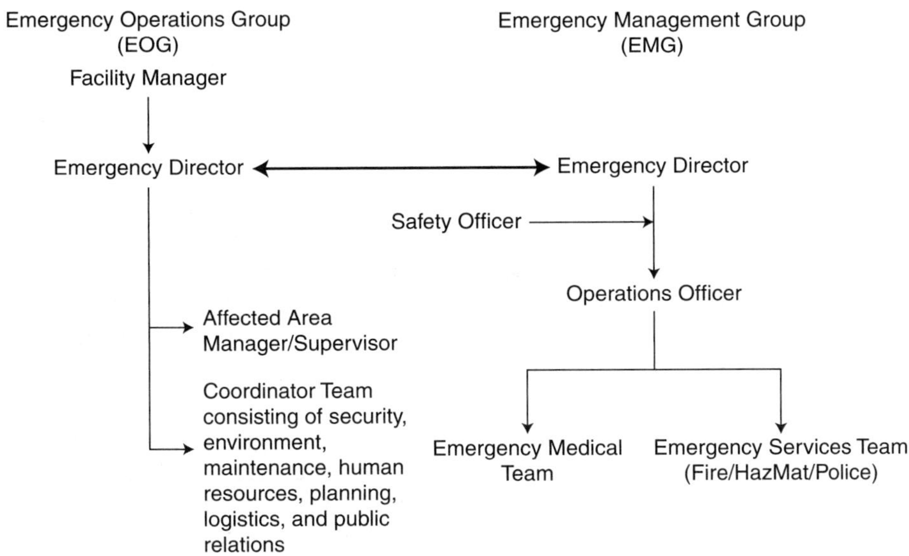

Figure 11.1 Relationship between the EMG and the EOG

through the EOC. It is recommended that the IC be a member of management with the ability to make critical time decisions and to act on them.

Subsequent Fires Because of the lack of fuel, fires from an explosion event do not usually occur unless the weapon is an incendiary-type device or the event has set off a chain of events that have given rise to a fire. In the event of an incendiary bomb, whose sole purpose is to cause a fire, a fuel source close by is required to sustain the fire. In the case of a chain reaction, the fire usually results not directly from the explosion itself but from a different source of ignition, such as a hot fragment igniting a burst gas pipe. Again, these fires, unless they have an external fuel supply, are usually small and can be easily extinguished by either cutting off the fuel supply or suffocating the flame.

11.4.3 Design Requirements for Protective Shelters

As in any design, protective shelters need to follow the basic construction codes: American Concrete Institute (ACI) 318 for concrete (American Concrete Institute 2008), American Institute for Steel Construction (AISC) 360-05 for steel construction (American Institute for Steel Construction 2005).

It is advisable, before beginning construction of a protective shelter, that a preliminary threat analysis be performed, by the designer, potential shelter owner, or other responsible party, utilizing the Building Vulnerability Assessment Checklist included in FEMA 426, *Reference Manual to Mitigate Potential*

Terrorist Attacks Against Buildings (Federal Emergency Management Agency 2003).

When the preliminary threat analysis has been performed, an independent design professional should perform a more thorough assessment of the shelter to either confirm the preliminary assessment or modify it accordingly if any deficiencies are found within the design/assessment.

If an existing building is to be used as a shelter, an assessment utilizing the FEMA 426 Checklist will help to determine the building vulnerabilities and thus aid in its modifications. It should be noted that FEMA 426 provides guidance to the building construction industry in how to reduce physical damage to buildings. The purpose of the manual is to show various approaches that can be taken to minimize the effects of terrorist threats.

Resistance to Progressive Collapse Progressive collapse occurs when a localized failure occurs and the adjoining members are overloaded. This causes failure of the members, and a cascade or "pancaking" failure effect occurs. This failure may result in a partial or total failure of the structure. The initial localized failure could be the result of a small parcel bomb in direct contact with a critical structural element, or the source could be a vehicle-borne explosive device located a relatively short distance away. A larger explosive device located a longer distance away is not likely to cause a single member to fail, but will probably affect a number of structural elements instead, which is not deemed failure by progressive collapse. In addition, according to FEMA 453, *Safe Rooms And Shelters: Protecting People Against Terrorist Attacks* (Federal Emergency Management Agency 2006), progressive collapse is not an issue for buildings with three stories or fewer.

The use of transfer girders and nonductile, nonredundant construction is prohibited in the building construction, as it may give rise to structural systems that are not tolerant of localized damage. The columns that support transfer girders—and the transfer girders themselves—may be critical to the stability of a large floor area. For interior shelters, the building surrounding the shelter must be sufficiently hardened so that progressive collapse is not an issue. For these buildings, it is required that the structural integrity of the building must still be intact if a member fails.

Some of the characteristics that a building containing a shelter must have to resist progressive collapse are:

- *Mass*: The larger the inertial resistance, the more resilient it is to failure.
- *Shear Capacity*: Connections and members need to be designed to prevent failure.
- *Capacity for Load Reversals*: Some members may undergo several large deformations before damping down. These load reversals can be in the form of uplift pressures, which defy the normal convention of gravity load design. Structural elements subject to blast effects, therefore, should be designed for load reversal. For design purposes, it is usualto provide equal areas of

reinforcing steel on both sides of reinforced concrete members, to provide the necessary resistance. The connections also need to be designed for load reversals.

- *Redundancy*: The use of an alternative load path in the vertical load-carrying system. This allows the gravitational loads to redistribute in the event of a structural member failure.
- *Ties*: The use of an integrated system of ties can serve to redistribute the loads. The ties need to be located perpendicular to the principal line of structural framing.
- *Ductility*: The ability of a structural member or connection to have a large deformation while still maintaining strength. Allowing the member to inelastically deform (e.g., plastically deform) causes large amounts of blast energy to dissipate; however, special detailing is required.

Using the above guidelines for designing buildings with shelters, or shelters themselves, robustness is greatly improved and progressive collapse may be avoided. For existing structures that are being modified to accommodate a shelter, it may be possible to allow only one floor adjacent to the failed column to fail, but if the structural members are retrofitted to develop catenary behavior, the adjoining bays must be upgraded to resist the lateral forces. This may result in more extensive retrofitting than is feasible or desirable. When such cases occur, isolate this region and risk the collapse of the adjoining bays by upgrading the vulnerable localized columns. When upgrading these columns, a balanced design approach must be obtained. It should be noted that upgrading an existing building for an alternative load path system is potentially counterproductive.

For buildings that can provide a continuous load path that will support all vertical and lateral loads, all structural components and fasteners used in the connection system must be able to develop the full capacity of the member. In order for this to take place, the capacity of each component must be balanced with the capacity of the other members and their connections. As all applied loads must be transferred eventually to the foundation/ground, the load path must be continuous from the top structural member to the ground.

Location within the Building One of the most important factors when dealing with a shelter is its location, including determining the best location for saving lives. The location of the shelter will depend on how many occupants the shelter has to hold. In addition to the effects of blast, the location should also consider effects of chemical, biological, and radiological threats. The shelter should be located such that all persons who are to take refuge can reach it within a minimal travel time from where they are located. A way to ease travel time is to clearly mark the route to the shelter.

Standalone shelters should be located as far as deemed possible from the surrounding buildings, to avoid both progressive collapse and/or impacts from the collapse of nearby buildings. Internal shelters require that the main building be designed for progressive collapse.

Access to Egress Shelter locations should be such that all designated persons may take refuge within the shelter with a minimum of travel time. Travel routes to the shelter should be clearly marked, allowing easy access. Exit routes from the shelter should direct the persons away from the threat, and hazards signs should be located using Crime Prevention through Environmental Design (CPTED) principles. These principles make use of natural access control, natural surveillance, and territoriality, which can also help to deter the aggressor.

Access to the building must be as "natural" as possible, in that the siting of the entrances, exits, fences, lighting, and landscaping tends to guide people who are entering or exiting the building. This is called "natural access control," and when it is coupled with limited access, the opportunity for a terrorist attack is greatly reduced. The placement of external features such as bushes, trees, and fountains must also be considered so that visibility is maximized. This is termed "natural surveillance," and it can reduce the opportunity for terrorists to attack, by revealing their intentions; in addition, it gives a feeling of safety, as the public can be easily seen. Finally, the use of fences, bollards, signage, and the like, termed "territoriality," clearly defines the property lines between what is public and private. This can be a gateway into a community or neighborhood.

Fire Rating There are only two fire ratings, H and A (Steel Construction Institute 1990). The H fire rating is solely for hydrocarbon fires such as those caused by a petroleum product. The A rating is for normal fires such as those that have wood or paper as their source of fuel. A number usually follows the letter, such as 60 or 120, and the value represents the time that structural integrity can be maintained during a fire—e.g., how long a member or area can be kept below a designated temperature, usually 40°C, before the temperature rises, causing that member to plasticize and fail. In H-rated fires, the maximum temperature is designated as 1100°C within 10 minutes, and for A, the maximum temperature is 950°C, over 60 minutes (Steel Construction Institute 1990). See Figure 11.2. It is usual for commercial buildings to be designated as A-rated, as the chances of a hydrocarbon fire within one of these buildings is remote.

Currently, the fire protection of buildings, based on FEMA, is based on a four-level hierarchy that comprises alarm and detection, suppression, compartmentation, and passive protection.

- Fire alarms and detectors are usually used to activate the system and notify the building occupants and emergency services through verbal and visible systems.
- Suppression systems, usually in the form of water sprinkler systems, are primarily designed to control small to medium fires and to prevent the fire spreading beyond the water zone, which is typically 1,500 square feet.
- Use of compartmentation will mitigate the spread of more severe fires. These typically require fire-rated partitions between the rooms and can cover an area of 12,000 square feet.

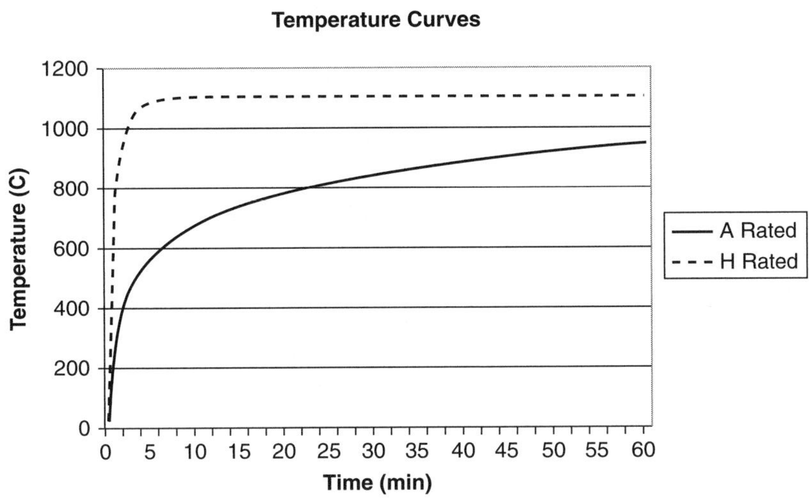

Figure 11.2 Temperature profile of H- and A- Rated Fires

- Passive protection is the use of a protective covering on a steel structure. This protection system will ensure that the temperature of the steel system does not reach an undesirable level such that the steel becomes plasticized and cannot withstand the vertical loading applied to it without buckling. There are numerous different types of ratings for these types of products. The purpose of this system is to prevent structural failure of the building before the rated time of the product has elapsed, by which time the building can be safely evacuated.

REFERENCES

American Concrete Institute. 2008. *Building Code Requirements for Structural Concrete and Commentary* (ACI 318). Farmington Hills, MI: American Concrete Institute.

American Institute of Steel Construction. 2005. *Specification for Structural Steel Construction* (AISC 360-05). Chicago, IL: American Institute of Steel Construction.

Federal Emergency Management Agency. 2003. *Reference Manual to Mitigate Potential Terrorist Attacks against Buildings* (FEMA 426). Washington, DC: Federal Emergency Management Agency, Department of Homeland Security.

________. 2006. *Safe Rooms and Shelters: Protecting People Against Terrorist Attacks* (FEMA 453). Washington, DC: Federal Emergency Management Agency, Department of Homeland Security.

Steel Construction Institute. 1990. *Interim Guidance Notes for the Design and Protection of Topside Structures against Explosion and Fire* (SCI P-112). Silwood Park, Ascot, Berkshire, UK: Steel Construction Institute.

12 Defended Perimeter

Joseph L. Smith and Charles C. Ellison

The four main functions of any comprehensive physical security program are to **deter**, **detect**, **delay**, and **respond** to a threat. A defended perimeter serves an important role in achieving these aims. A well-planned perimeter can **deter** an aggressor by increasing the perceived difficulty of attack, can support the implementation and function of **detection** sensors, and can **delay** an aggressor, providing the security force time to **respond** to an attack.

12.1 GOALS

The perimeter barrier system will generally be the outermost zone in a tiered system of protections that includes the building facade and internal screening zones such as visitor lobbies, mailrooms, and loading docks. These zones, as shown in Figure 12.1, can be viewed as concentric layers of protection.

In terms of the primary topic of this book, the goal for each tier should be to reduce the size of the likely explosive threat to a level that can be mitigated by reasonable levels of blast hardening. As the first line of defense, the perimeter is the foundation on which all other blast-mitigation options are selected. As such, the defended perimeter must address the largest threats to the facility. Failure of the perimeter to perform as expected can have severe consequences for the adequacy of subsequent blast-mitigation methods.

Heavy barriers are not the only option when creating a defended perimeter. Perimeter protection may include effective site planning and landscape design that, if deployed properly, can substantially decrease the risk to occupants and mitigate damage to a facility. The effective use of landscaping and terrain can also minimize the design requirements and cost of other man-made barriers employed to stop vehicles or other threats.

12.2 STANDOFF

The standoff distance is generally considered to be the range between the center of the explosive device and the nearest structural element requiring protection

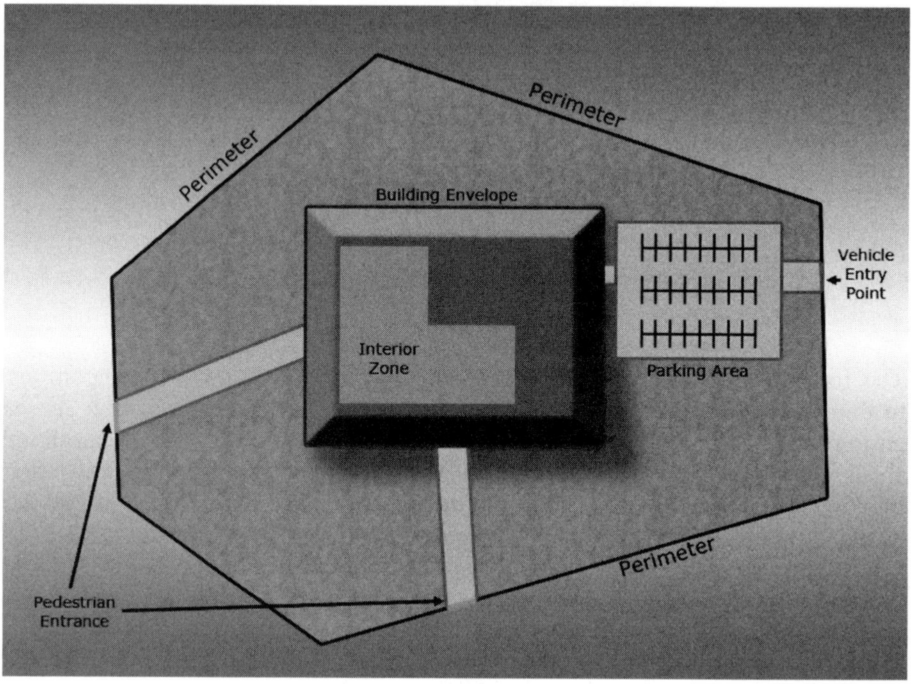

Figure 12.1 Zones of Protection

(i.e., the target). The term "setback" is used when discussing the distance between a facility and the defended site perimeter. These terms are often used interchangeably; however, standoff and setback do not necessarily have to be equal.

Since blast loads decay rapidly with distance, setback can be an effective means of protecting a facility from an explosive attack. The first few feet provide the most benefit. Figure 12.2 illustrates peak pressure versus standoff curves for a range of explosive devices [Based on Equations in UFC 3-340-02, Structures to Resist the Effects of Accidental Explosions (Unified Facilities Criteria Program 2005b)].

As shown in Figure 12.2, at very small standoff distances (i.e., less than 5 ft) just an extra foot of additional standoff can reduce the blast load on a structural member by an order of magnitude (i.e., a factor of 10) or more. It may only take a few pounds of explosive in direct contact with a structural column or wall to cause devastating structural collapse. However, that same column or wall may resist several hundred pounds of explosives if only a few feet of standoff are provided. The benefits of a few additional feet of standoff can also be significant at much greater distances. For example, the pressure loads generated by a 500-lb TNT bomb at 112 ft are over 20% less than the loads generated at 100 ft.

Even though the benefit per foot of standoff decreases with distance, providing standoff is often more cost-effective than structural hardening for significant distances from the facility. To illustrate this point, the approximate relationship

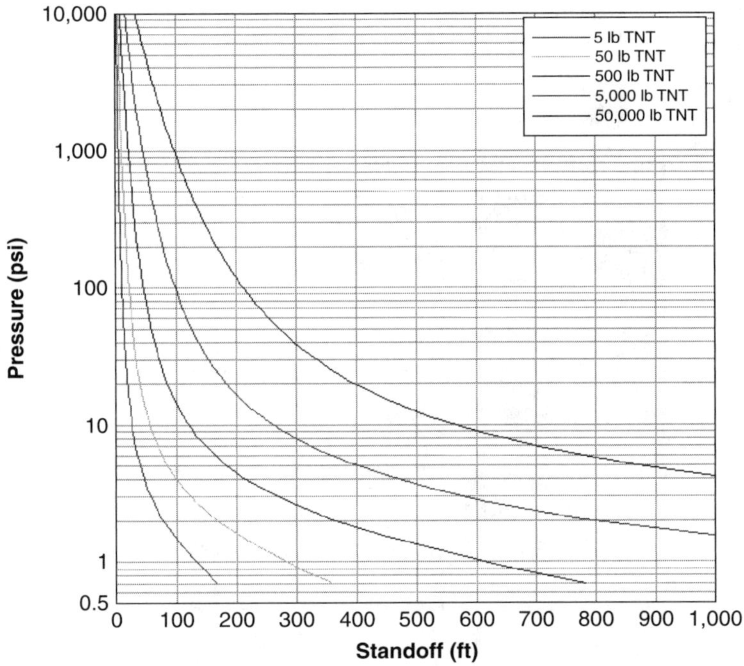

Figure 12.2 Blast Pressure Loading versus Standoff

between defended standoff and protection offered by conventional construction is shown in Figure 12.3 (Smith et al. 2005).

Figure 12.3 illustrates the level of protection offered by a wide range of conventionally constructed buildings for a given setback. The left portions of the bars indicate that no significant protection from blast effects is readily attainable at these distances with conventional construction. The next area is an indication of a low level of protection. At these distances, conventionally constructed buildings will typically sustain moderate to heavy damage. Occupants in exposed structures may suffer temporary hearing loss and injury from the force of the blast wave and building debris fragmentation. Other assets may receive damage from these effects. The third area is an indication of a medium level of protection. At these distances, conventionally constructed buildings will generally sustain light to moderate damage. Occupants of exposed structures may suffer minor injuries from secondary effects such as building debris. The rightmost area indicates a high level of protection. At these distances, conventionally constructed buildings will generally sustain only minor damage.

12.2.1 Balancing Hardening with Standoff

A common question that designers face is "Where do I put my barriers?" Typical security personnel will declare that barriers must be as far away from the protected asset as possible. Typical architects will respond that barriers intrude

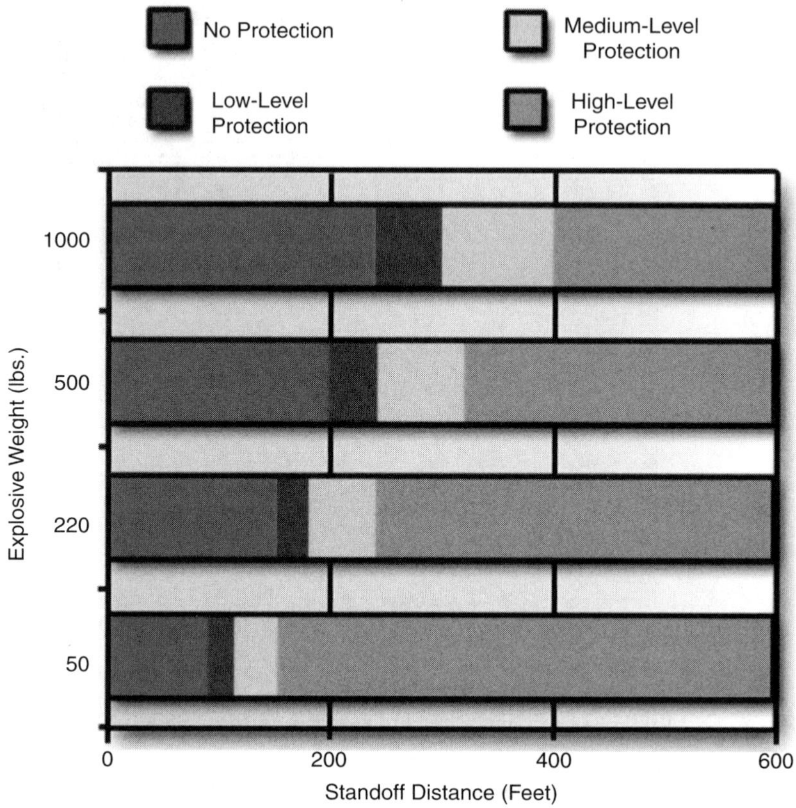

Figure 12.3 Effects of Standoff Distance for a Typical Conventionally Constructed Facility (explosive weight is in lbs of TNT equivalent)

on the appearance and function of the project and must be minimized. Typical owners will say they want to be safe but, at the same time, not inconvenienced. Typical community planning boards will want free, unrestrained access to the site for community use. In most practical cases, especially for government facilities, there may be design criteria that prescribe a specific minimum standoff for the defined threats. In other cases, there may be no such clear requirement. In the end, politics or other nontechnical factors may also play a deciding role in the selection and placement of barrier systems.

So what's a designer to do? It is important to consider all aspects of the design in determining where barriers are placed, the type of barriers to be used, and the protection level that such barriers will provide. The site features may play a role in determining the placement of the protected asset or building within a protected site. From a protection engineer's perspective, balancing setback requirements with structural hardening and hazard-mitigating features should be of primary concern. Designing to resist explosive effects increases the requirements of structural components; thicker walls, additional reinforcement, and blast-resistant glazing and frames may be needed. An increase in explosive weight or a decrease in standoff generally increases structural requirements. For an efficient design, one must balance the effect of available standoff with the effect of incorporating blast-resistant design and/or hazard-mitigation measures.

The types of perimeter protection should also be balanced with the types of attacks considered in the risk assessment. A crash-rated barrier line is not needed if the attackers are unwilling to risk drawing attention to themselves. However, heavy barriers may be the only option to stop a truly determined attacker. Potential attacks can generally be reduced to three types:

- A true covert attack in which the attacker desires to completely escape notice and repercussion. Attackers of this type will avoid anything that draws attention to themselves, such as high speeds and illegal parking. Eliminating public parking near the building and presenting a notable guard presence will deter this type of attacker and provide standoff.
- An attack in which the aggressor is dedicated to the attack, but intends to escape. This attacker will break traffic laws, but wants to get out of the area before a police response can be mounted. Designated parking or even illegal parking will not deter this attacker since he or she can flee the area before a response is mounted. However, heavy-rated barrier systems are not necessarily required, since this type of attacker will be deterred by the risk of a potentially fatal crash or a significant delay in escaping.
- A suicide attack by an attacker who is willing to be captured or killed in the attempt. If vehicle barriers are not in place, this attacker will drive onto the site and through the front door of the building. Stopping this threat will require significant and consistent hardening of the perimeter against vehicular attacks. Developing appropriate limits on achievable barrier penetration is crucial. Sally ports and other redundant barriers may be necessary at all vehicular entrances.

12.2.2 Balancing Costs

The effects of standoff provided by barrier systems on various structural and nonstructural components are conceptually illustrated in Figure 12.4 (Smith et al. 2005). This figure generally illustrates, at no specific scale, the typical relationships between standoff and the cost of protection. A number of the various components of incremental security cost are shown, including structural and nonstructural component contributors. The relative magnitude and scale of these relationships will vary from project to project.

For example, the cost associated with hardening the mailroom, loading dock, and lobby to meet the design requirements may not vary with the available standoff to a vehicle-delivered bomb along the site perimeter. The cost associated with progressive collapse considerations may also be constant with standoff, since it is often treated as threat-independent by security design criteria. However, there may be a point at shorter standoffs where the structural framing design is further impacted by the blast loading on the frame (above the progressive collapse requirements), resulting in larger framing members and additional cost. Some criteria may place limits on the design blast pressure and impulse for certain secondary building components such as doors and windows, resulting in the relatively flat curves seen in the figure. However, some criteria do not use such restrictions and are likely to experience a significant cost increase at low standoffs. The sum of the varying costs of hardening for the various components results in the "cost of hardening" curve indicated on Figure 12.4.

One cost component that increases with increasing standoff is that for site area and perimeter protection. For example, to provide increased standoff, the distance to the defended perimeter must increase, thereby increasing the area of

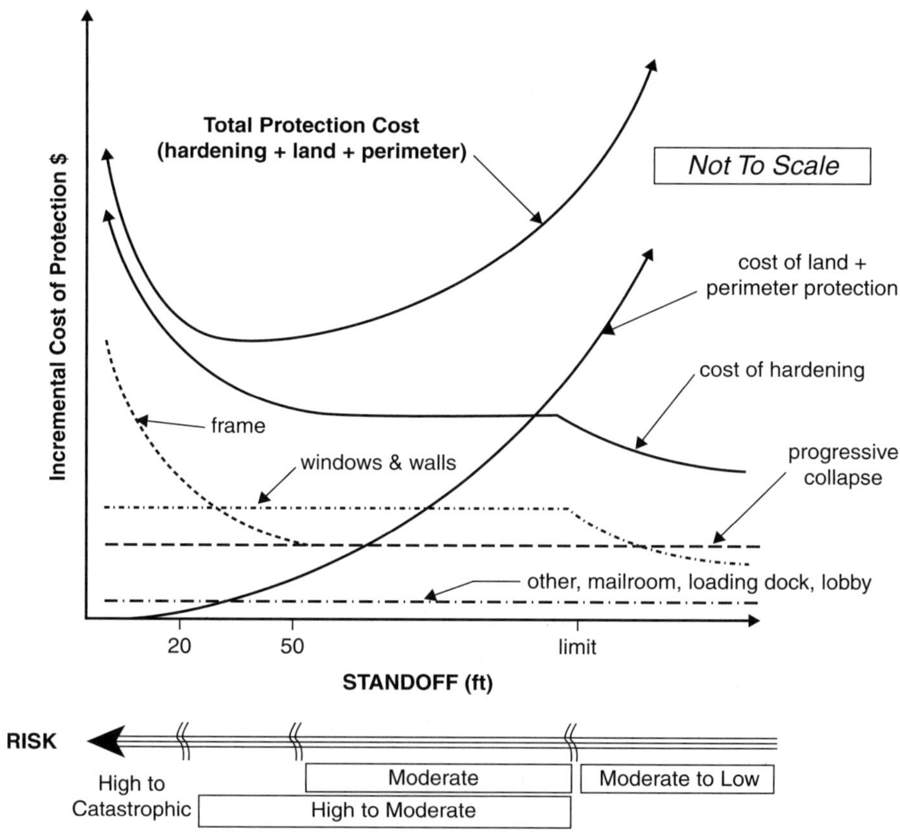

Figure 12.4 Impact of Standoff Distance on Total Protection Costs (Compliments of Applied Research Associates, Inc.)

the site and the length of the perimeter that must be protected. Adding the cost of hardening and the cost of land and perimeter protection results in the curve indicated as "Total Protection Cost." In the particular example shown in Figure 12.4, a standoff distance in the range of 30 to 50 feet would provide the desired protection at minimum total cost. A project-specific analysis that examines these variables can help in determining the optimal standoff for the proposed site.

12.2.3 Site Planning

Good site planning is the first step in ensuring that the desired standoff distances are achieved at a given site. An effective site plan will minimize or eliminate places where explosive devices can be concealed, will ensure that all activity around the facility is subject to casual observation by the occupants or actively monitored by a security force, and will discourage or prevent pedestrian or vehicular traffic in vulnerable areas.

There are an almost endless number of possibilities when it comes to implementing security features on a new site, and there is a great deal of room for creativity. As shown in Figure 12.5, numerous types of barriers can be integrated into the site plan. The implementation of landscaping and architectural features can also have an important effect on the vulnerability of a facility. The positioning of roadways, parking, and sidewalks relative to the building can discourage or encourage traffic. These features can also affect the likelihood of detecting an attack or the time between detecting an attack and responding to it. In the case of vehicular attack, roadway planning and landscaping features can limit

Figure 12.5 Integrated Site Plan

the impact speed and approach angle of an attacking vehicle, resulting in lighter barrier requirements. However, it is important that security be considered early in the design process, when civil, architectural, and structural changes can be made with minimal impact on ongoing designs in other nonsecurity disciplines.

Controlling the maximum impact velocity of an attacking vehicle is one very important part of site planning, since the velocity term is squared in the relationship for kinetic energy and has a very large effect on the size and cost of vehicular barriers. For example, doubling the impact speed increases the impact energy for a given vehicle impact by a factor of four, thus greatly increasing the design requirements and potential cost of a barrier system.

Using basic engineering and physics, there are formulas and methods available to estimate the maximum attainable impact speeds that a vehicle may obtain. Performing such a study early in the planning process can significantly reduce the costs associated with vehicle barriers.

Attainable Vehicle Speed on a Straight, Level Path The most basic vehicle speed calculation is based on a straight and level path between the starting point and the vehicle barrier. The final speed for such an approach can be calculated using the following basic equation (Unified Facilities Criteria Program 2005b):

$$v_f^2 = v_0^2 + 2as$$

where: v_f = final vehicle velocity
v_0 = initial vehicle velocity
a = constant rate of acceleration
s = distance traveled

Initial vehicle velocity and distance traveled are easy concepts to grasp and develop input values for. However, rate of acceleration can be difficult to determine. First, acceleration is very dependent on the weight and power output of the vehicle. Average accelerations between zero and 60 mph can range from 25 feet per second squared for a modern sports car to 5.8 ft per second squared for a 2.5-ton commercial truck. Second, acceleration is not constant. Neglecting drag and rolling friction, a vehicle producing a constant 200 hp at the rear wheels and weighing 4000 lb will accelerate from 20 mph to 30 mph in 0.61 sec (average acceleration = 24 ft/sec^2). However, the same vehicle will require 1.34 seconds to accelerate from 50 mph to 60 mph (average acceleration = 11 ft/sec^2). As can be seen, a 30-mph increase in speed reduced the rate of acceleration by one half.

These numbers were obtained by solving the following equations through time:

$$v_i = v_{(i-1)} + a_{(i-1)}(dt)$$
$$a_i = P/(mv_i)$$

where: $P =$ power
$m =$ mass
$v_i =$ current velocity
$v_{(i\text{-}1)} =$ velocity at last time increment
$a_i =$ current acceleration
$a_{(i\text{-}1)} =$ acceleration at last time increment
$dt =$ time increment

Slopes A downhill slope will decrease the required acceleration distance, and an uphill slope will increase it. A correction factor can be obtained using the following equation:

$$s'/s = 1/[1 + (g/a)\sin(\text{theta})]$$

where: $s' =$ acceleration distance needed to attain final speed on a sloped path
s = acceleration distance needed to attain final speed on a horizontal path
a = acceleration
g = gravitational constant = 32.2 feet per second squared
theta = angle of slope (0 is level, positive is downhill slope)
(Unified Facilities Criteria Program 2005b)

Maintainable Vehicle Speed on a Curved Path Centrifugal force will result in loss of control of a vehicle driving on a curved path if it exceeds the available friction force on the tires. By equating centripetal force to friction, the following equation can be derived for a vehicle traveling on a flat surface:

$$v_s = sqrt(fgr)$$

where: $v_s =$ skid velocity
$f =$ friction coefficient
$g =$ gravitational constant = 32.2 feet per second squared
$r =$ radius of curvature
(Unified Facilities Criteria Program 2005b)

The value of the friction coefficient, f, is highly variable. It depends on the tire and its condition, the material and condition of the drive path, and any other factors such as oil or water on the drive surface. The value of f should generally fall between 0 and 1. For common vehicles under normal conditions, $f = 0.6$ is usually used. Values greater than 1 are possible for a dedicated race vehicle on specially prepared tracks, but a value of $f = 1$ is generally safe when road conditions are unknown, or a more conservative value is desired.

Curve Banking Banking can increase the achievable speed of a vehicle in a curve and is commonly used in roadway design. The following equation adds an additional term to the curved path equation for calculating maintainable vehicle speed on a banked curve, to cover that possibility (Hickerson 1967):

$$v_s = sqrt[(f + \tan(\text{alpha}))gr]$$

where: v_s = skid velocity
f = friction coefficient
g = gravitational constant = 32.2 feet per second squared
r = radius of curvature
alpha = angle of embankment, positive for outer edge of curve higher than inner edge

12.3 VEHICLE CONTROL BARRIERS

Since larger explosive weapons are difficult to carry and conceal on an individual, vehicles play a critical role in many explosive attacks, and vehicular control should be considered when protecting a facility from large explosive threats. A successful vehicle control system will prevent an unscreened vehicle from reaching a point where it can be used to damage or destroy the facility.

A well-planned site with good detection equipment and a properly trained guard force can be effective in keeping what is generally referred to as a stationary vehicle-delivered threat away from a facility. That is, a vehicle driven to the target, parked, and left by an aggressor intent on escape. However, barriers designed to defeat moving vehicular threats will be required to stop a more determined aggressor willing to be captured or killed in a successful attack.

Barriers designed to defeat a moving vehicular threat (that is a threat actively driven into the site or facility) must be designed to arrest the energy of the attacking vehicle. For most standards, kinetic energy is used as a basis for establishing vehicle barrier performance requirements. The gross mass of a vehicle (vehicle mass plus the mass of explosives or any other cargo) and its maximum attainable speed at the point of impact define the kinetic energy that must be absorbed by the barrier. Kinetic energy can be expressed as:

$$KE = 0.5\, mv^2$$

where: KE = kinetic energy
m = gross mass of vehicle
v = velocity of vehicle

12.3.1 Crash Testing

Both the U.S. Department of State (DoS) and the U.S. Department of Defense (DoD) rate barriers based on full-scale crash tests conducted by independent test laboratories or government-approved facilities. The testing standard traditionally used for these tests is designated SD-STD-02.1, *Specification for Vehicle*

Crash Test of Perimeter Barriers and Gates (U.S. Department of State 1999). The DoS test standard rates the anti-ram barriers within three categories: K4, K8, and K12. The K rating corresponds to the kinetic energy of the impacting vehicle at the moment of impact. A K12 rating corresponds to impact energy of 1,200,000 ft-lb. These tests are conducted using a flatbed truck with a designated mass of 15,000 lb. The standard originally provided an additional L rating that designated the level of penetration achieved by the vehicle. However, in March 2003, with the release of Revision A, acceptable penetration was limited to 1 m, eliminating the need for an L rating. All previous barriers where the penetration exceeded 1 m were designated unacceptable by the revised standard, and the L rating was dropped from use. While significant penetration is not allowed by the DoS standards, penetration may be permitted in other applications especially where large setback is available.

ASTM has also developed a test standard, with input from both the U.S. Department of State and the U.S. Department of Defense, that is beginning to grow in acceptance in the barrier industry. This standard, *Standard Test Method for Vehicle Crash Testing of Perimeter Barriers* (ASTM F2656-07) is likely to grow in use and is gradually being accepted by both the U.S. Department of State and the U.S. Department of Defense.

12.3.2 Crash Modeling

There is no one-to-one correlation between the kinetic energy and a static design load. The impact forces vary significantly through time and are a function of the crushing of the vehicle as well as the strength and stiffness of the barrier. One method for calculating barrier response is to equate the energy in the vehicle (KE) with the work done on the barrier and the vehicle:

$$
\begin{aligned}
KE &= W_b + W_v \\
W_b &= \int F(x) \cdot dx_b \\
W_v &= \int F(x) \cdot dx_v
\end{aligned}
$$

where: KE = kinetic energy
W_b = work done on the barrier
W_v = work done on the vehicle
F = force generated between the barrier and the vehicle
x_b = displacement of the barrier
x_v = displacement of the vehicle

Calculating work on the barrier and the vehicle depends on obtaining good force and displacement histories. This is difficult to accomplish with simple models, given the complex interactions that occur between the barrier and the deforming vehicle during impact. A conservative simplified analysis can be performed using a single-degree-of-freedom (SDOF) model of the barrier, if the work done on the vehicle is set to zero. However, an SDOF design will generally

Figure 12.6 Comparison of Vehicle Impact Test to Finite Element Analysis (Compliments of Applied Research Associates, Inc.)

be very conservative. Testing has shown that the real force and displacement histories that occur during impact are quite complex, and the energy absorbed by the vehicle can be significant.

Given the need to make assumptions that ensure conservatism, barrier system designs resulting from simplified analysis can be significantly overdesigned. Therefore, barrier systems are generally certified through full-scale crash testing or sophisticated analysis.

Properly performed nonlinear finite element analysis (FEA) techniques have shown good correlation to test data and can be an effective means of developing new barrier systems. FEA models can replicate many of the complex interactions that control the force and displacement histories in actual crashes. Examples of an actual vehicle impact compared to a successful pretest finite element analysis are shown in Figures 12.6 and 12.7.

Finite element analysis is a powerful tool, but it should generally be supplemented with crash testing to verify the analysis results. When such testing is not possible, a qualified firm or consultant who is independent of the manufacturer should be contracted to perform or review such analysis. FEA is most effective

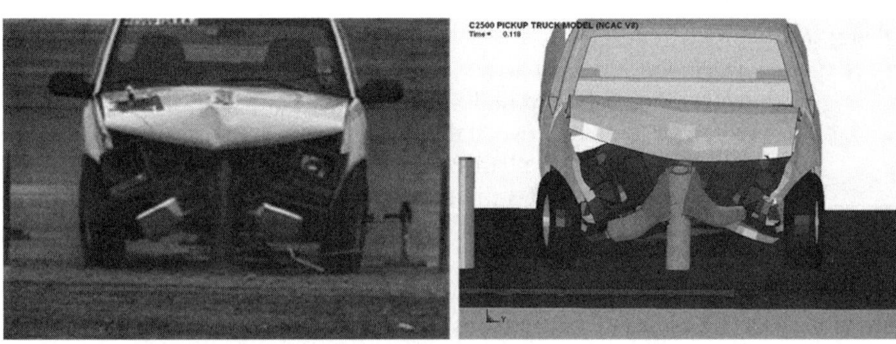

Figure 12.7 Front View of Impact versus Analysis Result (Compliments of Applied Research Associates, Inc.)

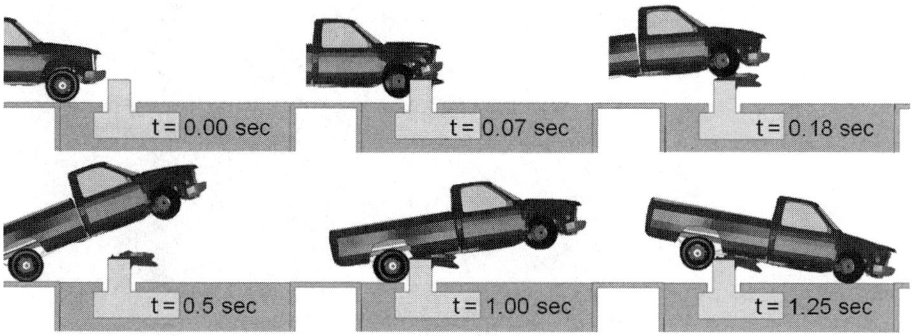

Figure 12.8 Low Wall (Compliments of Applied Research Associates, Inc.)

when the analyst has access to crash test data for similar systems and vehicles in order to validate the FEA model performance. Therefore, the more the proposed system differs from known test parameters, the more doubt and scrutiny should be given to the results. Ideally, the analyst will be able to provide FEA results that demonstrably replicate tests for similar barrier systems.

12.3.3 Walls

Walls and retaining walls, in particular, are some of the most robust crash barrier systems available. Commercial wall systems are available that have been designed and tested to stop a 60,000-lb vehicle traveling at 50 mph (KE = 5,000,000 ft-lb). Reinforced concrete is a common construction material for crash-rated walls. The wall thickness is generally between 12 and 24 inches. Crash-rated walls have traditionally risen 36 inches above grade, but recent testing and analysis have supported the use of 30-inch-tall barriers for stopping typical cars and some trucks. Figure 12.8 illustrates what can happen when the wall height is set too low.

12.3.4 Bollards

Bollards are one of the most common vehicle barrier systems in use. Bollards are usually cylindrical posts spaced close enough to prevent vehicle passage and are typically constructed of steel pipe anchored in a reinforced concrete foundation. A few bollard systems are designed with foundations less than 1 foot deep, but most have deeper foundations. Four feet is typical. Bollard spacing is also frequently set at 4 ft, but bollards can in many cases be effective at spacing up to 5 ft. Spacings beyond 5 ft are generally avoided, since small cars or trucks are capable of squeezing between bollards spaced farther than 5 ft apart. A typical passive bollard design is shown in Figure 12.9.

The bollard shown is capable of a K4 rating. K12-rated bollard designs are also available.

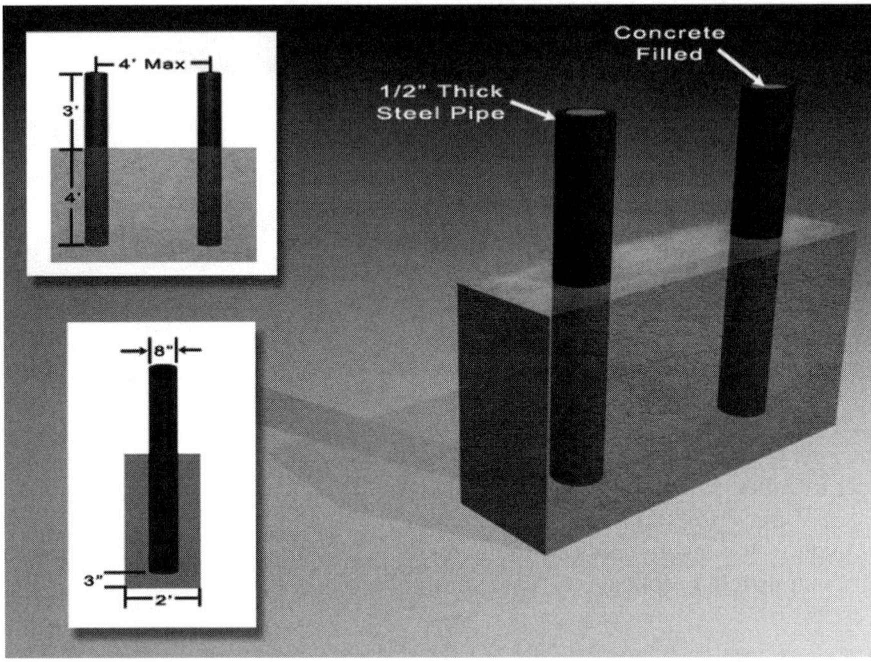

Figure 12.9 Eight-Inch-Diameter Fixed Bollard System

Active bollards can also be obtained. The most basic can be manually detached from the foundation and carried away (Figure 12.10).

The more advanced systems can be retracted into the foundation, using hydraulic or electric motors. Figure 12.11 illustrates one such retractable hydraulically operated bollard system.

12.3.5 Active Wedge

Wedges are a common type of active barrier and are capable of achieving K12 ratings. Wedges are typically installed to be flush with the road surface when lowered. When the barrier is raised to stop traffic, steel plates pivot up from the road surface to form a wedge shape. Wedges are typically operated electrically or hydraulically. Many variations are available. Figure 12.12 illustrates one common configuration of wedge barrier.

12.3.6 Beam Barriers

Beam barriers generally swing, pivot, or slide out of the path of the vehicle to permit traffic flow. Beam barriers generally do not achieve impact ratings as high as those of other active wedge barriers, but they generally cost significantly less and are easier to operate. The photo in Figure 12.13 shows a beam barrier.

Figure 12.10 Removable Bollard

Figure 12.11 Operable (Pop-Up) Bollards

Figure 12.12 Wedge Barrier

Figure 12.13 Beam Barrier

12.3.7 Cable-Based Systems

Cable-based systems generally consist of horizontally spanning steel cables anchored to the ground through anchor posts or other means and positioned to capture an attacking vehicle. When a cable system is impacted by a vehicle, the intermediate anchor posts respond progressively away from the point of impact until the vehicle is arrested. Cable-based barrier systems generally allow more penetration than other more rigid barrier systems. However, they also generally require lighter or more widely spaced foundations. Another advantage of cable-based systems is their ability to absorb the energy of the impact in a fashion that limits vehicle damage and injury to the vehicle occupants. The anchored cables can also be integrated into various fence systems to prevent intrusion of people as well as vehicles. See Figure 12.14.

Cable barriers are a good choice for large perimeters where a few feet of penetration is acceptable in exchange for reduced installation costs. Cable-based systems are most effective for protecting long stretches of perimeter, since they

Figure 12.14 Cable Fence Barrier

Figure 12.15 Concrete Planter

perform best when splice points are widely spaced. Fences with cables are common at airports throughout the United States. These systems are generally vulnerable to impacts near corners and ends, where the impact forces cannot be transferred to adjacent anchor posts through tensile-membrane action. Special detailing or supplemental barriers may be required near corners to ensure a consistent level of protection.

Active cable-based systems also exist and typically consist of a cable net that spans the road to prevent traffic flow. The cables can be relaxed to allow vehicles to pass over the cable net or tensioned to activate the barrier.

12.3.8 Planter and Surface Barriers

Surface barriers typically consist of large containers filled with soil, water, sand, or some other readily available material. One of the more common and aesthetically pleasing barriers in this category is concrete planters (see Figure 12.15).

True surface barriers (ones that do not include a foundation) initially absorb the energy of impact through momentum transfer and must have significant mass to perform well. Lighter surface barrier systems may allow significant penetration before stopping a vehicle. Figure 12.16 shows the aftermath of a real-world collision with a relatively lightweight planter system. In this case, the planter was driven nearly to the building before coming to a full stop. Ground or concrete anchors are available for such systems and will significantly reduce the penetration distance.

12.3.9 Berms, Ditches, and Other Landscaping Features

Berms, ditches, and hillside cuts, as well as trees, boulders, and other large objects can be effectively used to stop vehicles from penetrating a restricted

Figure 12.16 Concrete Planter Penetration

boundary. The authors are not aware of detailed analysis or testing that verifies the kinetic energy resistance of this category of barrier, but these systems can provide significant vehicular control. Figure 12.17 diagrams several commonly used barriers in this category.

Triangular ditches, trapezoidal ditches, and hillside cuts are easy to construct and are effective against a wide range of vehicle types. The designs presented in Figure 12.18 are based on typical vehicles with respect to tire size and ground clearance. Dedicated off-road vehicles may defeat these systems, and careful consideration should be given to likely threats. It is also important to consider the speed and trajectory of an approaching vehicle when implementing any of these systems. For example, it may be possible for a vehicle to jump a ditch or hillside cut, given an ideal approach. The width of the ditch may need to be increased where a high-speed approach is possible.

Boulders, trees, and other vegetation can also be used to stop vehicles. One such application is shown in Figure 12.18.

12.4 PEDESTRIAN CONTROL BARRIERS

Pedestrian control is designed to direct and moderate the flow of people around or through a facility. Good pedestrian control will deter or prevent visitors or uncleared personnel from entering areas of the facility vulnerable to attack. It is important to remember that a properly placed man-portable explosive device in close contact with a protected asset or structure can have a devastating effect.

Good site planning is the first step in pedestrian control and may be the only pedestrian control system implemented outside the building envelope for many facilities. An effective site layout will eliminate places where explosive

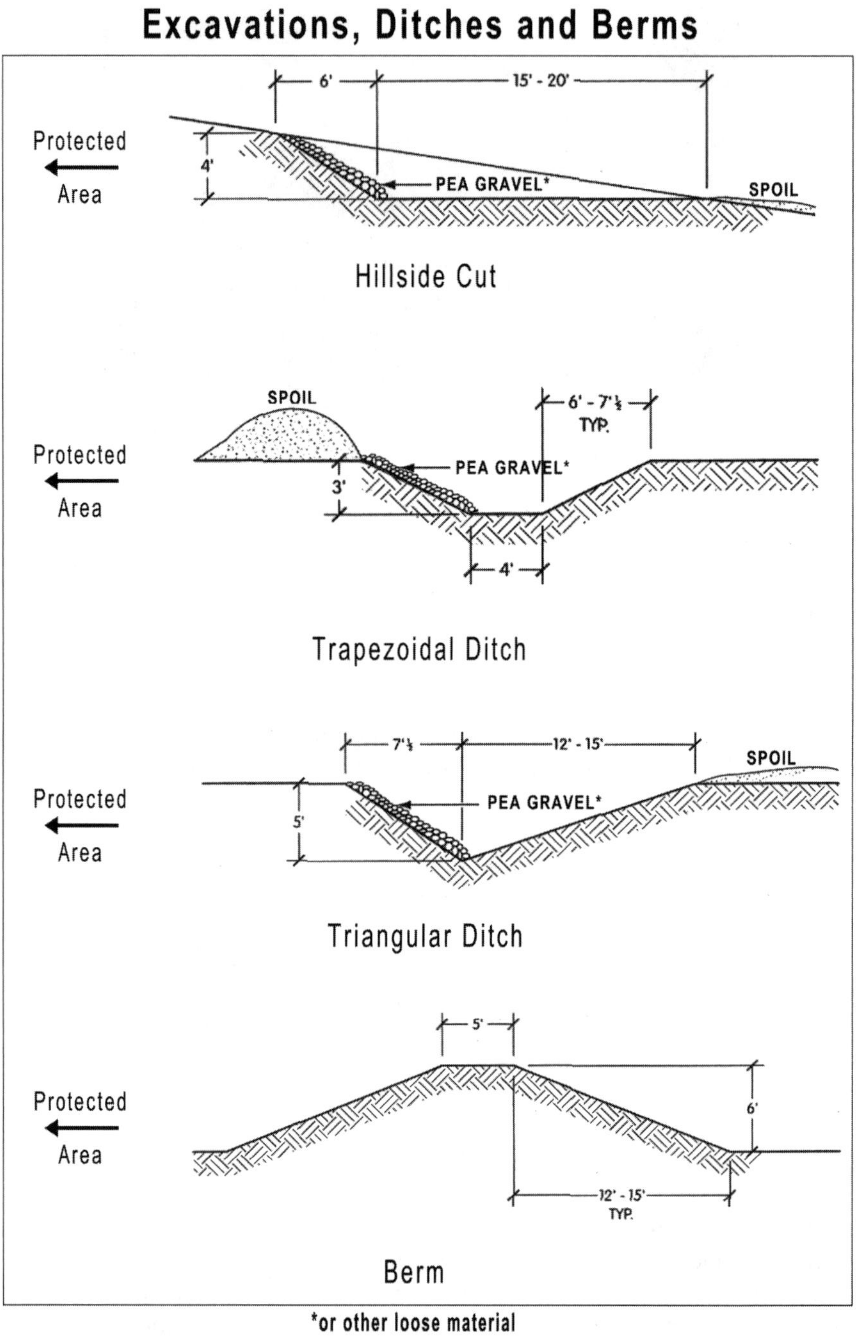

Figure 12.17 Excavation Barriers

Figure 12.18 Trees and Boulders Used as Barriers

devices can be concealed, will ensure that all activity around the facility is subject to casual observation by the occupants, and will discourage pedestrian traffic in vulnerable areas without drawing attention to those areas. Sidewalks, landscaping, lighting, and signage can all contribute to achieving these goals.

True physical pedestrian barriers are prudent for facilities where there is a significant risk of pedestrian attacks. Fencing or high walls are the most common type of pedestrian control barrier for the building perimeter. Fencing and walls come in many sizes and architectural styles, and each style carries a different delay factor based on height and structure. However, for the purpose of security, fencing or walls should generally be a minimum of eight (8) feet high and be durable enough to impede or withstand attacks from at least simple hand tools such as bolt cutters. In some cases, fences and walls can be built to also act as vehicle barriers. An example of this is shown in Figure 12.19. This fence application includes heavy cables secured to massive, buried anchors that serve to effectively stop a vehicle-borne threat.

12.5 BLAST WALLS AND BERMS

Obtaining adequate standoff to reduce the blast loads to an acceptable level is not always possible. One way to reduce loading is the implementation of blast walls. Properly implemented blast walls and revetments can mitigate the peak air-blast pressure and impulse from an explosion.

When placed across the path of the air-blast shock wave, the barrier may reflect a portion of the incident pressure back toward the explosion, or cause the blast to diffract over the barrier. The diffracted pressure will be significantly

Figure 12.19 Anti-Climb Fence with Integrated Catch Cables

reduced for some distance behind the barrier before the shock reforms to its original intensity. The area of effectiveness behind the barrier is controlled by the dimensions of the blast wall or revetment, the size of the explosive device, and the relative locations of the bomb, the wall, and the building.

Properly designed blast walls can also capture vehicle and bomb fragments before they reach the building. This can be an important consideration, especially when dealing with military-grade weapons designed to generate hazardous fragments.

Proper hardening of the wall is an important consideration in implementing a blast wall. Testing has shown that walls located close to the explosive device can break up and become dangerous fragments during an explosion. In many cases, the energy imparted to the facility by wall debris can negate the benefits provided by the reduction in air-blast loads. Using wall types that are inherently resistant to generating hazardous fragments is an option. Lightweight containers filled with water and sand are examples of systems with low secondary fragmentation risks.

The phenomena that occur behind a blast wall and that generate load reductions are very complex. A site-specific analysis is required to properly locate the wall and calculate the benefits. An improperly implemented blast wall will provide little benefit, and it is possible for a blast wall to actually increase the blast loads on a structure. A documented procedure for analyzing the benefits of blast walls is presented in UFC 4-020-03, *Security Engineering, Final Design* (Unified Facilities Criteria Program 2005a).

The primary sources of information about the benefits of blast walls and berms are experiments conducted on blast walls and a study on revetments. These studies only covered a limited range of devices and barrier heights, but the following trends have been observed. The wall must be close to the building or the explosive charge to achieve useful load reductions. For walls placed close to the charge, peak pressures are reduced dramatically directly behind the wall but gradually approach the free-field blast at large distances. There are some combinations of large distances between the bomb and the wall and short distances between the wall and the building that may cause an increase in the reflected impulse due to multiple reflections of the blast wave between the wall and the structure.

The shape of the wall or barrier and its height have a significant effect on the load reduction. Walls and revetments with faces oriented perpendicular to the shock wave (and the ground surface) should be a goal, since this arrangement provides by far the most benefit. Mounded or sloped barricades (i.e., typical berm construction) provide little reduction in blast loads.

The current blast wall data are based on situations where the wall height was selected by taking the cube root of the charge weight in pounds and multiplying by values between 0.8 ft and 1.3 ft. For example, the cube root of 500 lb is 7.9. Therefore, 6-ft to 10.5-ft walls fall within the range of the data. Trends in the data show that the wall is most effective if the bomb is within 1 wall height of the wall and within approximately 10 wall heights of the building. Ranges from the wall to the building greater than 20 wall heights generally result in negligible reductions.

REFERENCES

Hickerson, Thomas F. 1967. *Route Location and Design* (5th ed.) New York: McGraw Hill.

Smith, Joseph L. *et al.* 2005. *Protective Design and Security Implementation Guidelines*. Prepared by Applied Research Associates, Inc. for the U.S. General Services Administration.

Unified Facilities Criteria Program. 2005a. *Security Engineering, Final Design* (UFC 4-020-03). Washington, DC: U.S. Department of Defense, Unified Facilities Criteria Program.

________. 2005b. *Selection and Application of Vehicle Barriers* (UFC 4-022-02). Washington, DC: U.S. Department of Defense, Unified Facilities Criteria Program.

________. 2008. *Structures to Resist the Effects of Accidental Explosions* (UFC 3-340-02). Washington, DC: U.S. Department of Defense, Unified Facilities Criteria Program.

U.S. Department of State. 1999. *Specification for Vehicle Crash Test of Perimeter Barriers and Gates* (SD-STD-02.1). Washington, DC: U.S. Department of State.

13 Blast-Resistant Design of Building Systems

Scott Campbell and James Ruggieri

13.1 BACKGROUND

Public awareness of the need to protect structures against blast effects has risen sharply since September 11, 2001. A number of public agencies are engaged in research and development of blast-resistant technology, and its application to domestic and foreign structures that are potential targets of terrorist activity. The federal agency responsible for developing technology in this area is the Defense Special Weapons Agency/Technical Support Working Group (DSWA/TSWG), which receives funding from Congress through Department of Defense (DoD) appropriations. This funding is used to support physical testing of candidate blast-resistant design concepts, computational modeling, and understanding of phenomenology. A range of government agencies and private consultants participates in the program, whose results are available to the General Services Administration (GSA), the DOD, and other U.S. and foreign government agencies. The technology derived from this and similar previous programs is currently being implemented for the protection of U.S. Secret Service headquarters and FBI laboratory in the Washington, D.C.. area, the federal courthouse in Brooklyn, NY, and the passenger terminal at Midway Airport, Chicago, to name only a few locations. The U.S. State Department plans to request between \$1 and \$3 billion to raise security at U.S. embassies to the standards set by the Department in the 1980s.

The events of 9/11, as well as similar follow-on events, have introduced new performance requirements for all building systems. New performance requirements for such structures have introduced a need to revisit and modify the traditional assumptions governing building design. New systems, including atmosphere segregation systems (i.e., positive pressure, collective protection, and hazardous agent monitoring systems, etc.) as well as changes to the environment, including consideration to blast and shock, introduce constraints not previously considered in the design development of building mission and life-safety systems.

Although existing commercial consensus standards, such as NFPA 70A (National Fire Protection Association 2005), have historically performed well for general "nontarget" buildings and facilities, they do not address the special considerations now facing "target" buildings and facilities serving critical infrastructure needs. Therefore, new practical and readily expedient methods need to be identified that could better fulfill this new environment.

13.2 INTRODUCTION

The detonation of explosives near or in a building affects not only the building facade and structure, but also the building contents. In seismic events, the improvement in earthquake-resistant structural design has led to a significant decrease in structural damage. However, the result is that nonstructural damage now accounts for a majority of the costs associated with earthquakes. Similarly, increasing the blast resistance of building structures will reduce the associated damage, but may increase the relative importance of nonstructural component damage. Significantly, a building may be structurally undamaged, but rendered useless due to damage to HVAC, electrical, plumbing, and other systems. Therefore, a balanced design approach is suggested wherein the building systems are designed to withstand either the appropriate blast loads or resulting structural movements.

Very little information is available regarding the design of building systems to resist the effects of explosions. This lack of design guidance has resulted in the use of better-developed seismic guidelines, such as the *Uniform Building Code* (UBC) (International Code Council 1997) or *International Building Code* (IBC) (International Code Council 2006), the use of extremely conservative designs, or the neglect of the protection of equipment and systems. Traditionally, equipment has been designed for blast by being placed in shelters or internal to the building and checking acceleration resistance against expected levels. This remains the best practice, although particularly rugged equipment—for example pumps—may be left exposed with anchorage designed for the applied loads.

Third-party testing by nationally accredited conformity assessment organizations, such as Underwriters Laboratories (UL), is often used to verify certain performance characteristics of a system element. Traditional testing often includes verification of desired material properties such as strength, ductility, chemistry, and composition; and random sampling procedures are used to audit performance of the subject components for use, in comparison to design requirements and/or regulation. Examples of this include equipment operability, controllability and speed, current-carrying capacity, insulation resistance, power required, power developed, vibration, and many others. In most cases, the requirements invoked for electrical equipment and devices used in building systems only address electrical shock safety (e.g., electrocution risk) and risk of fire. However, such criteria presume a benign environment, not subject to risk of blast. The standards and building code organizations readily admit that such documents only identify a consensus for a common denominator that speaks to the minimum acceptable performance or criteria, assuming a safe and relatively comfortable environment. Such documents do not address blast. Although there is one U.S. national standards development activity presently underway to provide enhanced criteria to harden building electrical systems to the effects of blast—ASCE/AEI *Recommended Hardening Techniques for Control, Communication and Power (C2P) Systems of Critical Facilities*—there is nothing available in the nonmilitary community at this time that can be used for design guidance.

Explosions generate loads on building systems through three basic mechanisms. Direct blast pressures can impact systems exterior to the building from explosions outside the building envelope and from internal explosions. Indirect pressures are generated when a blast external to the space where the equipment is located leaks into the room through openings. Finally, ground shock and air shock generate loads on building systems through motions developed in the structure. All three types of loads are considered in this chapter. Note also that fragments can pose a severe threat to building systems and equipment. However, calculation of fragmentation and the resulting damage to nonstructural components is beyond the scope of this section (see Chapter 8).

The general design approach for building systems is to determine the load levels, both blast pressures and base accelerations, for the chosen location. The equipment and anchorage capacities are then checked against the design load. If the equipment's allowable load is exceeded, then either it must be moved or protected. Each of these steps is detailed in subsequent sections of this chapter.

13.3 DESIGN CONSIDERATIONS

The key philosophy for providing continuity of services for mechanical and electrical systems in buildings is SHR: Separation, Hardening, and Redundancy. Segregation of critical and vital circuits from normal services provides design economy, since hardening-type construction methods can be focused on only those specialized areas. Judicious use of steel conduits, pipes, and ducts associated with life-safety systems provides increased likelihood of survivability for these systems, and a greater likelihood that emergency systems will remain operational post-event to assist rescuers in the evacuation of the building, and/or provide continuity of critical business services.

Although each design situation is unique, it is possible to provide some general guidance regarding blast-resistant design of equipment and systems. This section will discuss choosing appropriate design goals, specific points of concern for the different loads that can be imposed, and some guidelines regarding select types of equipment and systems. However, good engineering practice and judgment are the best approach for these situations, given the amount of uncertainty present in this field.

13.3.1 Level of Protection

A critical step in the design of building systems for blast load is to determine the goal of hardening. In general, there are two levels of performance normally considered in system blast-resistant design:

1. *Life Safety*: Also called position retention. This design level is intended to keep the equipment in place, with no large pieces or components coming loose and potentially injuring personnel. The equipment or system is not expected to remain operational after the event.
2. *Continued Operation*: As the name implies, the equipment is expected to remain operational after the event. This might entail resetting the equipment, but no major repairs should be required to maintain basic operational capabilities.

Continued operation of equipment subjected to blast or shock loads is often difficult to predict. Seemingly minor failures can lead to a shutdown and loss of critical systems. In general, only inherently rugged equipment can be considered for continued operation if directly exposed to blast pressures. Less rugged equipment may be designed to remain operational for reduced pressures due to protection by shelters or interior location. If the continued operation of a system is required, it is essential that all aspects of the loading condition be considered, and testing of the equipment is likely to be necessary.

The life-safety level of design is easier to achieve, as no evaluation of the working parts of the equipment is required. It is only necessary to analyze the equipment structure and anchorage. Note, however, that the analysis includes all external components, including the sheet metal shell of the equipment, if any, to ensure that they remain attached to the equipment. Large internal components that have the ability to break out of the shell must also be considered. Another response quantity that must be considered for suspended systems is deflections. A system may be capable of resisting the pressure or acceleration load on its own, but large deflections may cause an impact with other equipment and lead to an anchorage or equipment failure.

13.3.2 Blast Pressures

The best way to protect equipment and systems from the effects of blast is to locate them in areas protected from direct air blast. If the building envelope is properly designed to resist blast effects, location of equipment in mechanical rooms without openings to the exterior of the building will preclude pressure loading from external events. However, equipment that is located inside protective structures will often be subjected to blast pressure that leaks in through openings in the walls and/or roof. Likewise, properly securing hardened mechanical rooms, and thus limiting access to authorized personnel, will significantly reduce the likelihood of internal blast loads on the equipment. However, concentrating mechanical equipment in a single location will render the overall system more vulnerable to unauthorized access if proper precautions are not maintained.

Removal of equipment from vulnerable areas is not always possible, and is likely not feasible for all distributed systems. In these situations, it is necessary to require design to higher than desired levels of pressure loading. This may entail providing barrier walls, if not complete shelters, or testing equipment to the anticipated design load. Particularly critical distributed systems may be enclosed in hardened runways, or redundant systems located elsewhere in the building may be provided. Obviously, these additional precautions can be extremely expensive, and care should be taken early in the design process to eliminate or reduce to a minimum the number and types of systems that require special design.

13.3.3 Shock Induced by the Structure

The impact of blast-generated air and ground shock on the structure will produce motions that are potentially damaging to the building systems. The resulting structural accelerations can be estimated using the methods outlined later in this chapter. After the structural motions are calculated, the resistance of the equipment to shock loads can be checked. Typical values of equipment shock resistance are given in Table 13.1. One major caveat applies to the use of these values. In many cases they are based on testing that is decades old. In the interim many changes have occurred in the design of some equipment. The tabulated values may no longer be applicable, and caution should be used in applying them to modern equipment. Significant testing is often required for equipment used in the military and the nuclear industry, and manufacturers may be able to provide capacities for specific pieces of equipment. The nuclear industry Seismic Qualification Utility Group (SQUG) has a large database of equipment response during earthquakes, but these data are not generally available to the public.

A common practice is to specify design loads for the equipment in terms of an equivalent earthquake load. This is particularly true where it is desirable that the actual load source remain confidential. It is perfectly acceptable to give only an equivalent "g" acceleration value. The acceleration level to be specified can be determined using the methods detailed in Chapter 7 or a more detailed calculation that considers the stiffness and mass distribution in the structure. However, there are several caveats that must be observed.

First, unless a dynamic analysis of the building is performed and the floor accelerations specified as time histories, the frequency content of the input motion to the equipment is lost. Second, it is not correct to specify a design to, for example, 0.75g using the 2006 International Building Code (IBC). The IBC method is used to determine earthquake loads for anchorage design and equipment qualification. A base ground acceleration is used that is then modified to account for location within the building, natural frequency of the equipment, ability of the equipment to undergo nonlinear deformations without failure, and the importance of the equipment.

Table 13.1 Selected Equipment/System Fragility Level Guidelines (ASHRAE 2007)

	Accelerations (g)	
Equipment/System Type	No Damage Probable	Minor Damage Probable
Pumps up to 100 hp	4.5	8
Control Panels	3.5	8
Fans up to 100 hp	4	9
Welded steel piping (18 in. dia max, Sch 30 and up)	4	8
Air-handling units	4.5	10
Duct, steel	4.5	9
Motor control centers	2.5	N/A
Uninterruptible power supplies	3	N/A
Cable trays (up to 36 in.)	4.5	8
Chillers		
Centrifugal	4	N/A
Absorption	4.5	N/A
Screw	3.5	N/A
Air-cooled	3.0	N/A
Boilers	3.5	N/A

There are several contributing factors as to why it is not good practice to apply this methodology to in-structure shock loads. First, earthquake loads on equipment depend not only on the size of the earthquake, but also on the vertical location within the building. If the floor accelerations have been determined, and that is what is being specified, it must be made clear that the factor for vertical location (an increase in load of as much as three times) should be disregarded in the calculation. It is also not clear if the other factors are applicable. However, the most critical point is that many parts of the country are exempt from providing seismic restraint to equipment regardless of the specified acceleration level. For these reasons it is best to provide an acceleration level, either ground or floor level; require use of the building code load equation as appropriate; and explicitly state that no exemptions are permitted.

Shock Isolation: Isolation of equipment to reduce the effects of shock loading has been used successfully on many projects. Isolation systems work on the principle that if the loading and isolation system's natural frequencies vary greatly, the isolated equipment will undergo motions much less severe than the base structure. The amount the motion is reduced depends upon the ratio of the forcing frequency to the natural frequency of the isolation system (frequency ratio), as illustrated in Figure 13.1. Thus, to achieve a high level of reduction, the isolation system should have a much lower natural frequency than the input

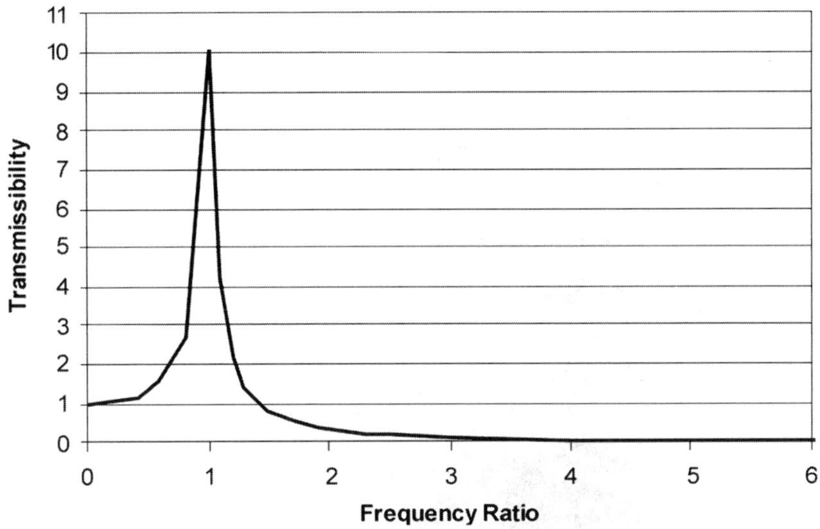

Figure 13.1 Transmissibility of Motion for Isolated Systems with 5% Damping

motion. However, this leads to a soft isolation system (low stiffness) and potentially high static deformations and problems with coupling between displacement and rotation deformation modes. A balanced system using a properly selected isolator, additional weight (inertia mass), and possibly supplemental damping can help to alleviate the problems associated with soft isolation systems.

There are several types of isolators that are commonly used, including steel springs, wire coils, air-springs, elastomeric mounts, and damping devices (Figure 13.2). Most of the isolators can be used in either direct-mounted configurations, where the inertia mass and equipment are placed on top of the isolators, or in a pendulum configuration where the load is hung from the isolators. In either case, the systems are designed to carry the gravity load and accommodate the deformations expected during a shock load. The equipment thus remains relatively motionless while the supporting floor undergoes large motions, with the isolation system deforming while maintaining vertical load capacity. Shock isolation systems are available from a number of manufacturers who provide design services along with the isolation systems. The project engineer must determine the level of isolation required and the input motions, while the manufacturer will choose the correct isolator to achieve the desired results.

13.3.4 Equipment/System Anchorage

The anchorage of equipment to the structure must be checked for all loading cases and desired levels of performance. The procedure is relatively simple: Analytically apply the load to a mathematical model of the equipment, calculate

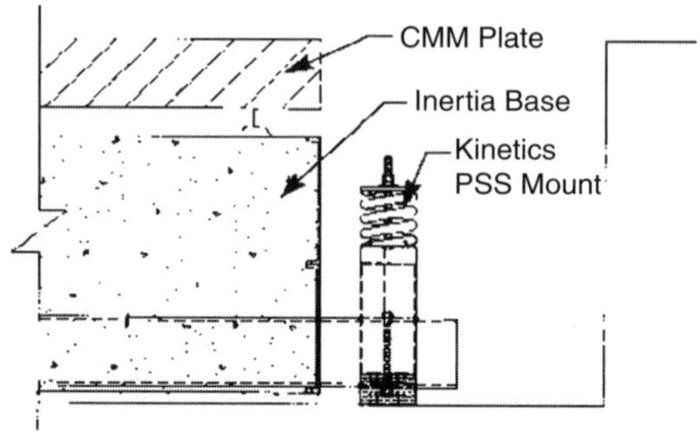

(a)

(b)

Figure 13.2 Examples of Spring Isolators for Shock Isolation: (a) Spring isolator in a pendulum configuration (Kinetics Noise Control, Inc.); (b) Wire rope isolators (VMC Group, Inc.); (c) Air-spring isolator (Kinetics Noise Control, Inc.); (d) Elastomeric isolators (Kinetics Noise Control, Inc.)

the forces at the support location, and design the anchorage to accommodate the maximum forces. The equipment is usually assumed to be rigid for the purpose of calculating anchorage forces, since the dynamic effects are not considered. If desired, a more detailed model that considers the mass and stiffness of the equipment may be used. The load is applied at the center of the exposed surface for pressure loads, and at the center of mass for accelerations. It may also be necessary to account for nonsymmetric loading and/or mass by varying the load application point. In addition, the load should be considered to act in all

(c)

(d)

Figure 13.2 *Continued...*

possible directions. Note that this is not simply all directions, as in seismic design, since the explosive location may be restricted and therefore limit the load direction. All seismic design guidelines should be observed regarding limitations on the type of anchors. For example, power-actuated fasteners should not be used unless specifically qualified, and undercut anchors are required for equipment over 10 hp that is not vibration-isolated. In addition, the structural component to which the equipment is mounted must also be capable of carrying the anchorage force. Details on the anchorage calculations are provided in the design examples.

13.3.5 Placement of Critical Systems Equipment and Control Stations

Unsecured areas such as the lobby, loading dock, mailroom, garage, retail, and other public access areas must be separated from the secured areas of the building, particularly areas of a building housing switchgear, computer server data centers, centralized command centers, and machinery plant control and monitoring stations. Ideally, unsecured areas are placed exterior to the main building or along the edges of the building. Occupied areas or areas providing emergency or critical functions should not be placed immediately adjacent to the lobby, but should be separated by a substantive buffer area such as a storage area or corridor. From an operational security standpoint, it is important to restrict and control access to air-intake louvers, mechanical and electrical rooms, telecommunications spaces, and rooftops by means of such measures as visitor screening, limited elevator stops, closed-circuit television (CCTV), detection, and card access-control systems.

13.3.6 Staffing and Building Operations

How building staffing (i.e., managers, maintenance crew, boiler-men, security personnel, attendants, etc.) is determined can make a substantial difference in minimizing loss of life and minimizing injury, as well as improving the ability of a facility to sustain continuity of critical services in the event of a blast. Determining staffing is integral to the design-development process of building systems. For instance, automation and systems monitoring rely on certain assumptions regarding staffing. Increased use of centralized equipment/systems health-determining methods, such as computerized condition-based monitoring (vibration, temperature, differential pressures, etc.) significantly reduces the need for periodic roving manned inspections.

An excellent model for effective staffing can be taken from the military. U.S. warships are designed to absorb a certain number of strikes to minimize loss of life, ensure continued reliable maneuvering and control (survival of steering gear, propulsion systems, etc.), and ensure reliable operation of both defensive and offensive systems. Such vessels rely on an architecture of self-sufficiency using highly trained crews who are assigned damage control functions in addition to their normal duties (e.g., cook and fireman). As with warships in battle, buildings in the face of a civilian mass casualty incident need be self-sufficient, since it is unlikely that first responders will be able to respond promptly, if at all.

Many elements of the military model can be easily adopted by the civilian community, and much can be gained by investing in certain added training and emergency duty assignments for critical personnel. Although using better-trained personnel may increase operational costs, strategic optimization of variable resources will offset cost increases. For instance, since such personnel are better trained in building systems, they could also perform certain maintenance and repair functions routinely outsourced to specialists—serving not only to save money but likely to reduce downtime of building systems.

The availability and familiarity of selected record drawings to such designated personnel (i.e., electrical one-line, list of feeders and mains, equipment location drawings, etc.) and the periodic drilling of personnel promote familiarity of systems while also identifying weaknesses in capability.

13.3.7 Construction of Hardened Spaces

Spaces dedicated to housing critical equipment should be constructed to withstand the effects of blast and fire in unsecured adjacent spaces. Historically, the preferred material for explosion-mitigating construction has been cast-in-place reinforced concrete. Reinforced concrete has a number of attributes that make it the practical construction material of choice. The use of the new tough polymer spray-on coatings on exposed concrete surfaces, as commonly used to protect the beds of commercial pickup trucks (e.g., Line-X®, Rhino®.) checks the release of airborne debris in the event of blast. Outfit and furnishings located in secure spaces should be selected to minimize fire load, while electrical equipment should be mounted on resilient foundations to minimize the effects of shock. Depending upon the function and equipment located in the secure space, consideration should be given to automated conflagration control systems, such as CO_2 deluge systems, the use of fire-retardant cable and cable systems, and the use of intumescent coatings. Positive pressure systems and/or gas-tight construction should be considered in critical spaces to control air flow and thwart and control undesired migration of contaminants.

13.3.8 HVAC and Plumbing Systems

Equipment that is deemed essential should be placed in secure locations. That is obviously not possible for all system components, as they must be distributed throughout the building. Because of that, design of HVAC/plumbing systems usually focuses on position retention (adequate anchorage). Systems that service areas that must remain functional after an event—for example a safe haven or control center—should be served by redundant systems or have the equipment and delivery systems placed in secure areas and hardened runs. The high cost associated with these countermeasures necessitates a thorough evaluation of the expected threat and the consequences of system failure.

Air-intake locations should be located as high up in the building as practical, to limit access and reduce the blast pressure level. Generally, air intake should be located in secure areas. Systems that serve public access areas such as mail-receiving rooms, loading docks, lobbies, freight elevators/lobbies should be isolated and provided with dedicated air-handling systems capable of 100% exhaust mode. Air-intake locations and fan rooms should be coordinated with the security surveillance and alarm system so that surveillance is provided and unauthorized access can be responded to by the security force.

Building HVAC systems are typically controlled by a building automation system, which allows for quick response to shut down or selectively isolate air-conditioning systems. These systems should be coordinated with the smoke-control and fire-alarm systems to allow for automated, proper response to fires.

Access to building mechanical areas (e.g., mechanical rooms, roofs, elevator equipment access) should be restricted and monitored. These areas provide access to equipment and systems (e.g., HVAC, elevator, building exhaust, and communication and control) that could be used or manipulated to assist in smoke evacuation after a blast or during a chemical, biological, or radiological (CBR) attack. Additional protection may be provided by including these areas in those monitored by electronic security and by eliminating elevator stops at the levels that house this equipment. For rooftop mechanical equipment, ways of restricting (or at least monitoring) access to the roof that do not violate fire codes should be considered.

Particular care should be taken to ensure that HVAC equipment and attached ducts/pipes can accommodate any expected deformations. This generally means using flexible connectors between duct/pipe and equipment, an example of which is shown in Figure 13.3. In addition, distributed systems often cross areas with

Figure 13.3 Example of Flexible Connectors Near Equipment Attachment Points (Courtesy of Kinetics Noise Control, Inc.)

large differential motions, including stories with different floor motions and building joints. Attachments at these locations should also provide adequate deformation capacity to prevent damage to the duct or piping. Possible configurations and options for allowing displacements while retaining functionality at a pipe penetration through a floor are shown in Figure 13.4.

Locations where building systems intersect with the building envelope can require additional considerations. Air intakes can be fit with blast valves (Figure 13.5) that allow for air flow under normal conditions but close when excess pressure is detected. These systems can be either passive or active. Penetration of piping or conduit through the wall can allow blast pressures

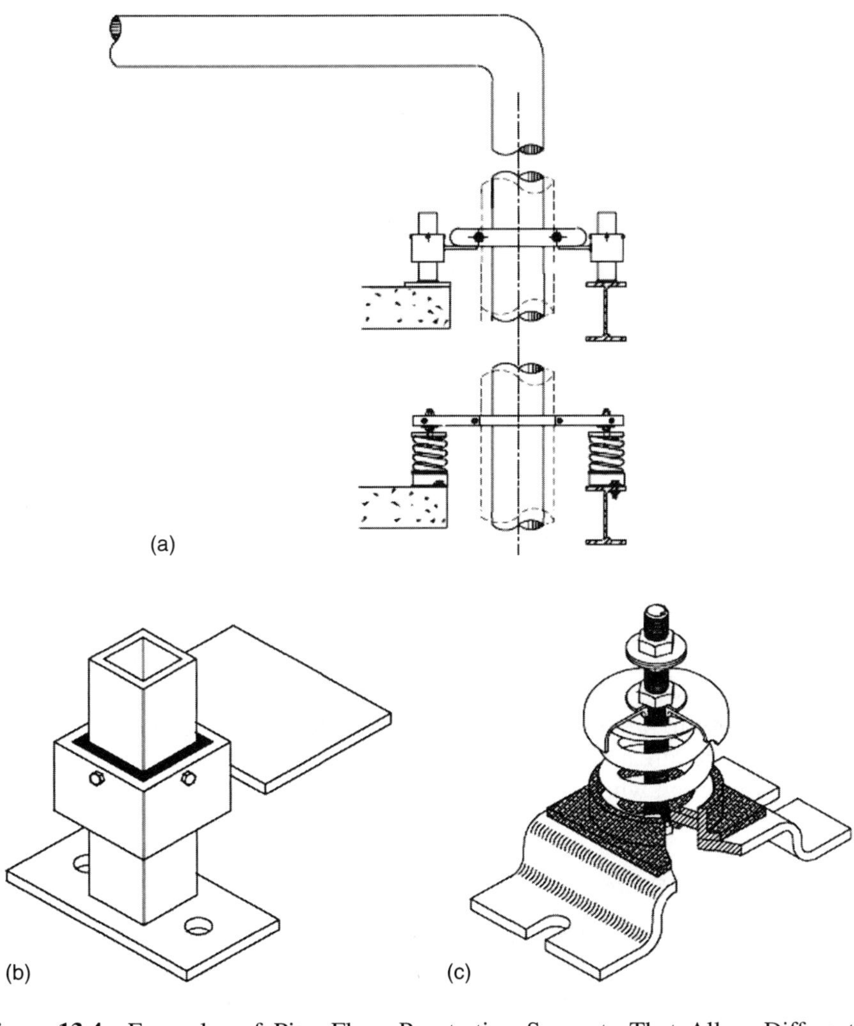

(a) (b) (c)

Figure 13.4 Examples of Pipe Floor Penetration Supports That Allow Differential Movement: (a) Pipe riser configurations; (b) Riser guide; (c) Riser isolator (Courtesy of Kinetics Noise Control, Inc.)

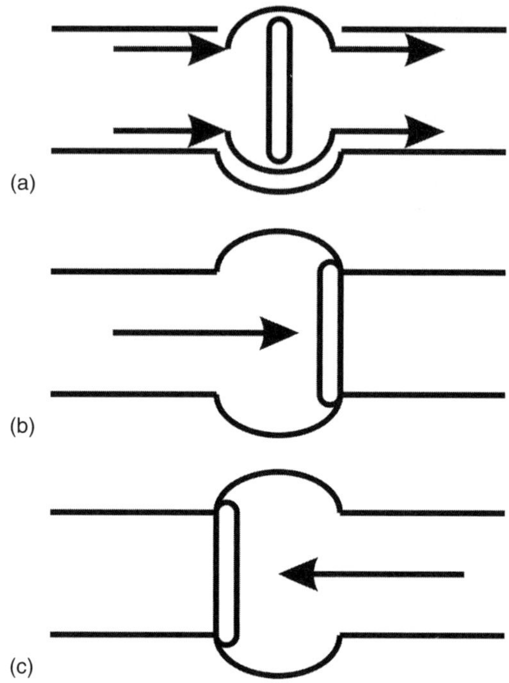

Figure 13.5 Conceptual Operation of a Blast Valve: (a) Normal flow; (b) Excessive pressure from outside; (c) Excessive pressure from inside

to enter the building and/or damage the mechanical system. Various available schemes, including sleeved openings, special accumulators (Figure 13.6) that allow for pressure buildup and gradual release, and flexible connections can protect both the piping and the building interior.

Fire protection systems may be considered a subset of piping systems and designed using the same general principles. However, while the failure of water piping systems is unlikely to jeopardize the building occupants, leaking or inoperability of fire protection piping (and pipes containing hazardous material) is potentially life threatening. For this reason it is recommended that additional caution be exercised when designing these systems.

13.3.9 Electrical Systems

Electric power supply, command, control, and communications systems are essential to all building systems and a proper threat-assessment matrix evaluation is central to the reliability and survivability of such services. A proper threat-assessment evaluation consists of identifying critical system components, dependencies, vulnerabilities, and redundancies, and mapping such items in a

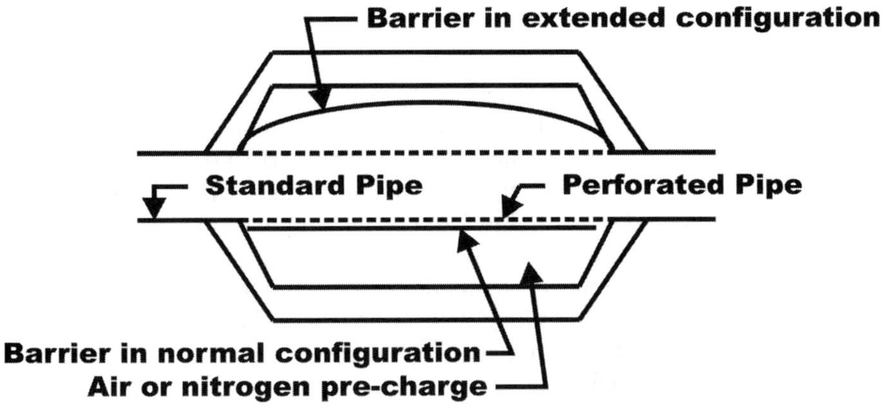

Figure 13.6 Conceptual Design of Accumulator (After U.S. Department of the Army, 1986)

documentable "matrix." Deliverables resulting from such an effort frequently result in a Failure Mode and Effects Analysis (FMEA).

For electric systems, certain methods or design techniques are frequently used in guarding against unconventional threats, including:

- Vertical and horizontal zonal definition of equipment and feeds to provide redundant paths and feeds
- Self-healing power supply and distribution network designs comprising multiple, segregated power feeds to minimize human-element intervention
- Command, control, and communication (C^3) services combining both distributed and centralized control techniques to further minimize the need of human resources
- Tactical location of critical equipment and personnel
- Co-location of critical equipment in hardened/controlled-access spaces
- Application and integration of automation systems to monitor and orchestrate power, control, and communication systems healing actions
- Appropriate personnel matrix and personnel training to increase the likelihood of survivability/mission continuity, while reducing variable resource costs

Utility power transformer(s) and switchgear should be located interior to the building if possible, and in secure spaces. For larger buildings, multiple transformers, separated from each other, enhance reliability should a blast damage one transformer or its associated equipment. In circumstances where a redundant utility transformer arrangement is not provided, consideration should be given to locally stocking a replacement transformer and associated equipment when replacement lead times may be unacceptably long.

13.3.10 Lighting Systems

Lighting system design for blast protection is similar to electrical system design. The same principles of separation, zoning, emergency power supply, and so forth, all apply to lighting. One recent development that should be considered in lighting design is the use of light-emitting diode (LED) technology. More frequently recognized in the electronic equipment arena, where LEDs are used for equipment pilot and status lights, LED technology has now evolved to offer very practical and cost-effective design alternatives in applications where incandescent and fluorescent lamps presently dominate. With respect to susceptibility to blast damage, LEDs stand alone as a clear leader. In addition, LED lighting reduces the environmental impact of the lighting system and offers reduced maintenance and increased life.

LED lighting methods do not depend on moving parts or parts that are motion-sensitive, as do incandescent and fluorescent lamp technologies (filaments, standoffs, glass, etc); that is, LEDs indeed live up to their designation as "solid-state," and their elemental constituents, such as device leads, are solidly embedded into inert and tough structural polymers. Their substantively smaller sizes and physical profiles minimize concerns related to weight and moment and susceptibility to blast damage, while also eliminating the need for support devices/equipment (e.g., ballasts) and the weight/room required for these other support devices.

13.3.11 Other Systems/Considerations

Emergency Power System An emergency generator should be available to provide an alternate source of power if utility power becomes unavailable to support critical life-safety systems such as alarm systems, egress lighting fixtures, exit signs, emergency communications systems, smoke-control equipment, and fire pumps. In cases where continuity of critical business missions is necessary, the emergency generator system can also support these missions. Practically, diesel-fueled emergency generator systems are preferred over other conventional fuels (i.e., gasoline, propane, etc.) because of fuel availability, lower flash point of diesel fuel, and greater power supply capacity from a relatively small footprint. Economically, maintenance costs are substantially lower for a diesel-driven emergency generator system when contrasted to other forms of emergency power, including fuel cells and battery-based systems.

Emergency generators typically require large louvers to allow for ventilation and cooling of the generator. Care should be taken to locate the generator so that these louvers are not vulnerable to attack or are fit with blast valves. A remote radiator system could be used to reduce the louver size and thus reduce air-intake requirements.

Ideally, emergency power-distribution feeders are contained in hardened enclosures, or encased in concrete, and configured in redundant routing paths to minimize vulnerability and thus enhance reliability. Emergency distribution panels and automatic transfer switches should be located in hardened spaces isolated from the normal power system. Emergency lighting fixtures and exit signs along the egress path could be provided with integral battery packs to provide close-transition services (uninterrupted silhouette illumination) between the time the utility system fails and the emergency generator system assumes loading.

Emergency Generator System and Fuel Storage A nonexplosive fuel source, such as diesel fuel, should be considered for emergency generators and diesel-driven fire pumps. The preferred philosophy is to locate all support equipment, including the diesel fuel day tank, in the same space as the emergency generator, as a means of minimizing system dependencies. Consideration should be given to locating the emergency generator day tank in a fire-hardened space adjacent to the emergency generator space. Day tanks should include sufficient fuel to power the generator for a minimum of eight hours.

All control equipment, including starting provisions (i.e., air start, battery start, hydraulic start, etc.), emergency generator switchgear, cooling water tank, local controls, and so forth, should co-locate with the emergency generator. Diesel storage tanks that augment the day tanks should be located away from the building in secure, fire-rated, hardened structures. Fuel piping within the building should be located in hardened enclosures, and redundant piping systems could be provided to enhance the reliability of the fuel distribution system.

Fire Control Center A centralized Fire Control Center (FCC) should be provided to monitor alarms and life-safety components, operate smoke control systems and fire doors, communicate with occupants, and control the firefighting/evacuation process. Consider providing redundant Fire Control Centers remotely located from each other, to allow system operation and control from alternate locations. The FCC should be located near the point of firefighter access to the building. If the control center is adjacent to the lobby, separate it from the lobby using a corridor or other substantive buffer area. Provide hardened construction for the Fire Control Center. Further, the FCC can be co-located with other critical automated building service functions, such as with the machinery plant control and monitoring systems. The proliferation of computer-based monitoring and control systems permits much flexibility in this regard, particularly for collecting multiple, monitored systems into a single space through building intranet cabling. If configured, a simple laptop computer can be used to access and control monitored systems from many possible locations.

Emergency Elevator System Elevators are not used as a means of egress from a building in the event of a life-safety emergency event, as conventional elevators are not suitably protected from the penetration of smoke into the elevator shaft. An unwitting passenger could be endangered if an elevator door opened onto a smoke-filled lobby. Firefighters may elect to manually use an elevator for firefighting or rescue operation. A dedicated elevator, within its own hardened, smoke-proof enclosure, could enhance the firefighting and rescue operation after a blast/fire event. The dedicated elevator should be supplied from the emergency generator, fed by conduit/wire that is protected in hardened enclosures. This shaft/lobby assembly should be sealed and positively pressurized to prevent the penetration of smoke into the protected area.

Smoke and Fire Detection and Alarm System A combination of early-warning smoke detectors, sprinkler-flow switches, manual pull stations, and audible and visual alarms provides quick response and notification of an event. The activation of any device will automatically start the sequence of operation of smoke control, egress, and communication systems to allow occupants to quickly go to a safe area. System designs should include redundancy such as looped infrastructure wiring and distributed intelligence such that the severing of the loop will not disable the system. Install a network of fire-alarm systems consisting of distributed intelligent fire-alarm panels, such that each panel can function independently, process alarms and initiate sequences within its respective zone, while at the same time communicating with other panels.

Smoke-Control Systems Appropriate smoke-control systems maintain smoke-free paths of egress for building occupants through a series of fans, ductwork, and fire-smoke dampers. Stair pressurization systems maintain a clear path of egress for occupants to reach safe areas or to evacuate the building. Smoke-control fans should be located higher in a building, rather than at lower floors, to limit exposure/access to external vents. Vestibules at stairways with separate pressurization provide an additional layer of smoke control.

General Announcing Communication Systems A general announcing system facilitates the orderly control of occupants and evacuation of the danger area or the entire building. The system is typically zoned by floor, by stairwell, and by elevator bank for selective communication to building occupants. These systems may be integrated with other building announcing and communication systems. Emergency communication can be enhanced by providing extra emergency phones separate from the telephone system, connected directly to a constantly supervised central station, and an in-building repeater system for police, fire, and EMS (emergency medical services) radios.

13.4 LOADING CALCULATION

Nonstructural components can be subjected to several different load types due to blast. This section details the procedures used to calculate the blast loads, including direct loading, leakage through openings, and propagation through ducts. In addition, calculation of the building motions for design of systems is addressed, as is the load due to accidental blastlike loads such as short circuits in electrical equipment.

13.4.1 Blast Pressure

Building systems can be subjected to blast pressure from three sources: direct external blasts, direct internal blasts, and indirect pressures. Equipment located external to the building shell will be exposed to direct blast pressures. These pressures are characterized by a short rise time, high peak pressure, and relatively fast drop of pressure. In contrast, pressures resulting from internal blasts are confined by the walls and consequently have longer duration. The calculation of blast pressures due to both external and internal blasts is covered elsewhere in this handbook (see Chapter 7).

Leakage through Openings Equipment housed in shelters or in rooms with a direct connection to the building exterior will experience blast pressures from leakage through openings. These loads may be applied directly to the equipment with connections to the outside, such as air handlers with external intake/exhaust, or indirectly as the pressure fills the room. The calculation of interior pressure loads due to leakage is complex and usually involves many assumptions.

One procedure, presented below, incorporates the prominent assumption that the openings are too small to allow a shock front to develop, and that the pressure inside the room increases uniformly. The method calculates the average pressure in the room. The pressure near the opening will be somewhat larger than the average, but the approximation of uniform pressure within the space is adequate for design.

The interior pressure loading history is calculated in a step-by-step manner. The procedure is as follows:

1. Calculate the external pressure-time history.
2. Divide the duration (t_o) of the external load into equal intervals of length δt equal to approximately $t_o/10$ or $t_o/20$.
3. For each time interval, calculate the change in internal pressure over the time step, using

$$\delta P_i = C_L \left(A_o / V_o\right) \delta t$$

   where C_L is the leakage pressure coefficient (Figure 13.7) (Departments of the Army, the Navy and the Air Force 1990), taken as negative when P is negative; A_o is the area of the opening; and V_o is the room volume. The internal pressure at the end of the time interval is then calculated from the pressure at the beginning of the step as

$$P_i = P - \delta P_i$$

4. Repeat until the internal pressure drops to the initial ambient pressure.

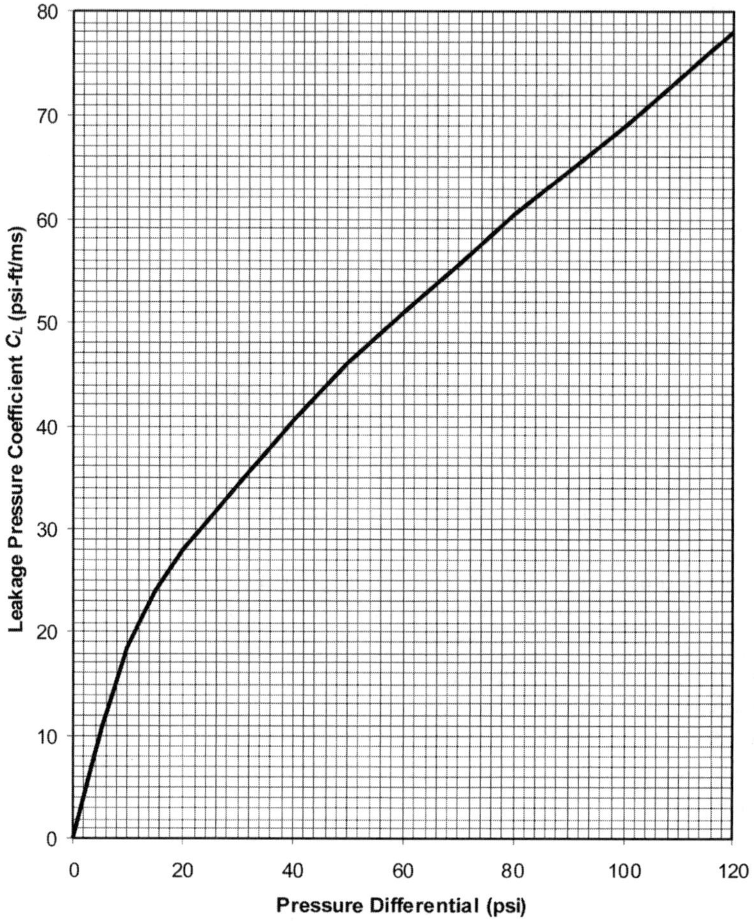

Figure 13.7 Leakage Pressure Coefficient (After Departments of the Army, the Navy and the Air Force, 1990)

Example: Calculate the internal pressure-time history for an equipment penthouse with a small opening in one wall.

Given:

$$A_o = 4 \text{ ft}^2 \ (2' \times 2' \text{ opening})$$
$$V_o = 1000 \text{ ft}^3 \ (10' \times 10' \times 10' \text{ penthouse})$$

Assume peak external pressure at time 0 equal to 3 psi with $t_o = 50$ ms.
Solution:

$$\text{Use } \delta t = t_o/10 = 50/10 = 5 \text{ ms}$$

Perform calculations in a tabular format. Results are shown in Figure 13.8.

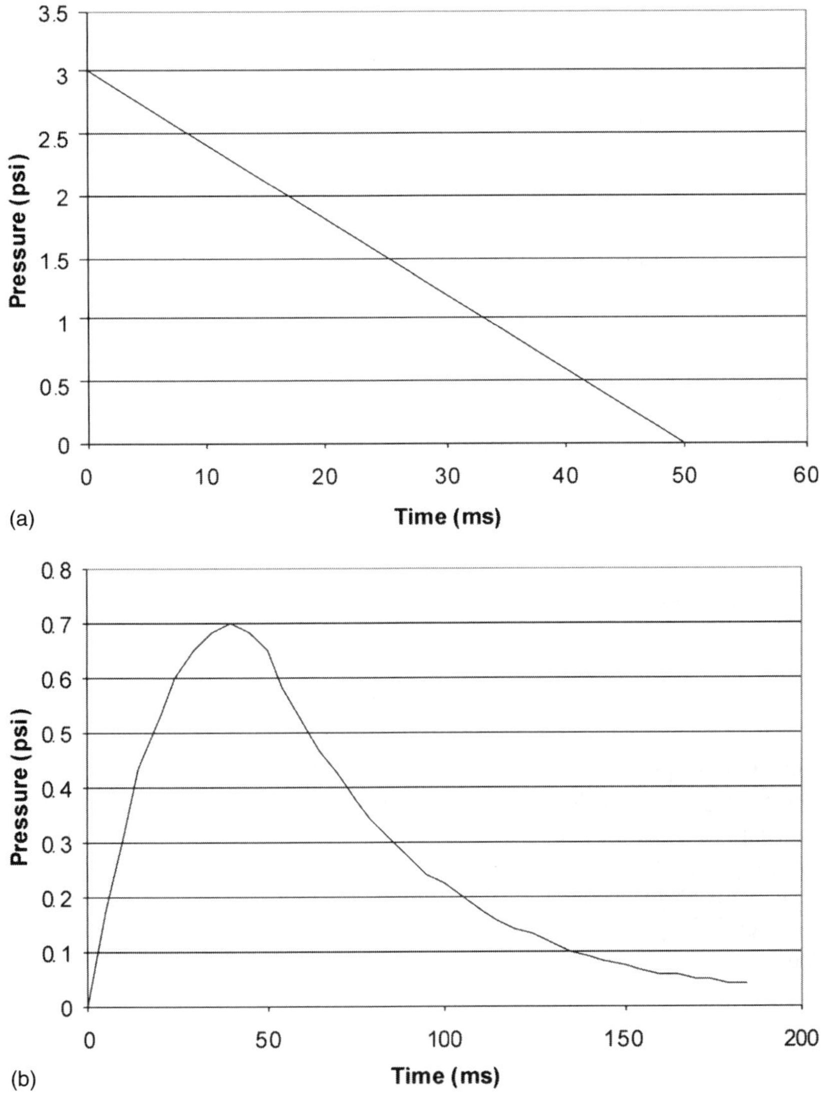

Figure 13.8 External and Internal Pressure-Time Histories for Leakage Example: (a) External pressure-time history; (b) Internal pressure-time history

Propagation of Pressure through Ducts Blast pressure will propagate through ducts into the interior of a building. The pressure entering the duct, and the change in pressure at different branching configurations, can be calculated using Figure 13.9. These figures are valid for peak overpressures less than 50 psi, which is appropriate for most civilian structures. In addition, 90° bends in the duct reduce the pressure by approximately 6% each. Thus, after each

T (ms)	P (psi)	P_i (psi)	P-P_i (psi)	C_L	δP_i
0	3.00	0	3.00	8.5	0.17
5	2.70	0.17	2.53	7.17	0.14
10	2.40	0.31	2.09	5.91	0.12
15	2.10	0.43	1.67	4.73	0.09
20	1.80	0.53	1.27	3.61	0.07
25	1.50	0.60	0.90	2.55	0.05
30	1.20	0.65	0.55	1.56	0.03
35	0.90	0.68	0.22	0.62	0.01
40	0.60	0.69	−0.09	−0.26	−0.005
45	0.30	0.69	−0.39	−1.10	−0.02
50	0	0.67	−0.67	−1.89	−0.03
55	0	0.63	−0.63	−1.78	−0.04
60	0	0.59	−0.59	1.68	−0.03
65	0	0.56	−0.56	−1.58	−0.03
70	0	0.53	−0.53	−1.49	−0.03
...					
185	0	~0	—	—	—

bend, the peak pressure, P_T, can be calculated from the pressure before the bend, P_o, as

$$P_T = 0.94P_o$$

There is also pressure loss along the length of smooth ducts. However, the calculation is not suggested for the typical ranges and charge weights assumed in this chapter, because it is only slightly conservative to assume no loss along straight portions of ducts. This assumption will be overly conservative for either long duct runs (over 50 feet) or short ranges. Further details can be found in TM 5-855-1 (U.S. Department of the Army 1986) if more detailed calculations are desired.

13.4.2 In-Structure Shock

In-structure shock refers to the structural motions that result from a blast. Although equipment located within the structure may not be subjected to direct or indirect blast pressures, it will experience the resulting structural accelerations. As a result, building systems not exposed to pressure loads must still be designed to resist the in-structure shock.

Structural motions due to blast arise from three sources: air-induced ground shock, direct-induced ground shock, and air shock. Air-induced ground shock is the result of the pressure wave acting on the ground surface and producing deflections and hence motion on the ground. Direct-induced ground motions are due to direct transfer of energy to the ground, typically due to buried or surface detonations. Air shock occurs when the blast pressure impacts the building,

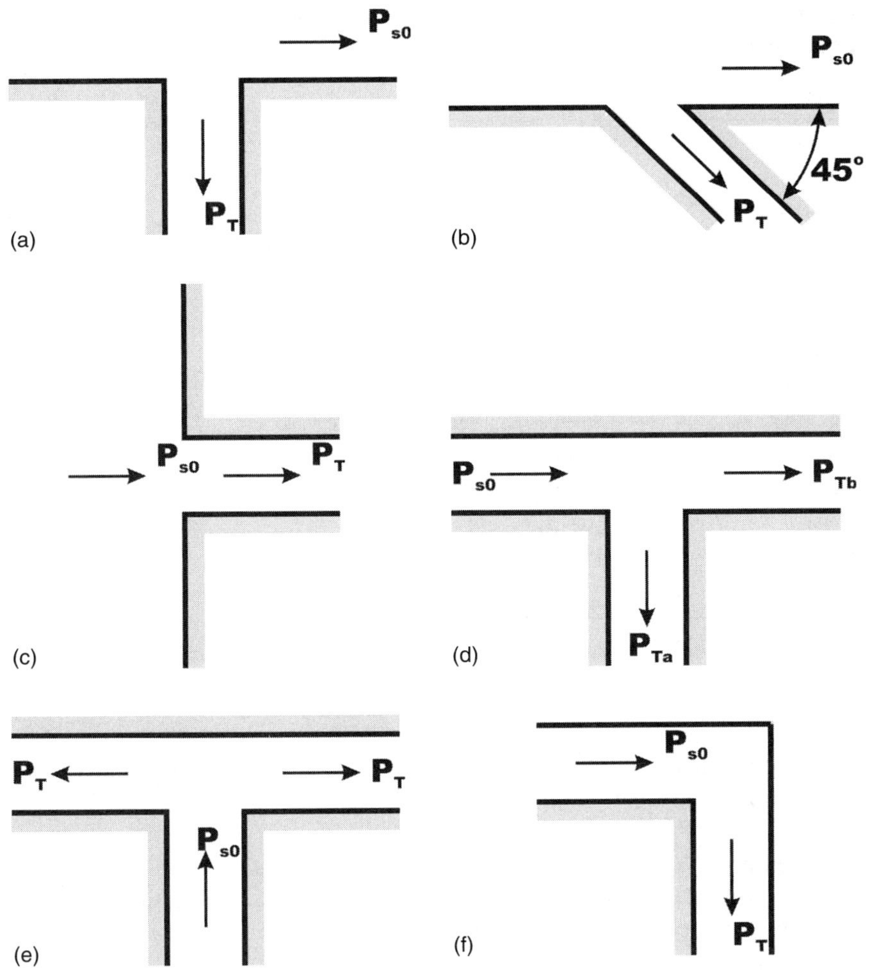

Figure 13.9 Transmitted Overpressures in Ducts (After U.S. Department of the Army, 1986)

imparting structural motions. The effect of all three sources must be considered when determining the design motions for building systems.

Air-Induced Ground Shock The peak vertical and horizontal accelerations due to air-induced ground shock are calculated as

$$A_{Vai} = 100P_{so}/(\rho C_p g)$$
$$A_{Hai} = A_V \tan\left[\sin^{-1}\left(C_p/(12000U)\right)\right]$$

where A_{Vai} and A_{Hai} are the air-induced vertical and horizontal peak ground accelerations (g), P_{so} (psi) and U (ft/ms) are the peak positive incident pressure and

Table 13.2 Typical Values of ρ and C_p (After Departments of the Army, the Navy and the Air Force, 1990)

Soil Type	ρ (lb-sec^2/in^4)	Soil Type	C_p (in/sec)
Loose, dry sand	1.42E-04	Loose and dry soils	7200–39,600
Loose, saturated sand	1.79E-04	Clay and wet soils	30,000–75,600
Dense, dry sand	1.65E-04	Coarse and compact soils	36,000–102,000
Dense, saturated sand	2.02E-04	Sandstone and cemented soils	36,000–168,000
Dry clay	1.12E-04	Shale and marl	72,000–210,000
Saturated clay	1.65E-04	Limestone-chalk	84,000–252,000
Dry, sandy silt	1.57E-04	Metamorphic rocks	120,00–252,000
Saturated, sandy silt	1.95E-04	Volcanic rocks	120,000–270,000
Basalt	2.56E-04	Sound plutonic rocks	156,000–300,000
Granite	2.47E-04	Jointed granite	9600–180,000
Limestone	2.25E-04	Weathered rock	24,000–120,000
Sandstone	2.10E-04		
Shale	2.17E-04		
Concrete	2.25E-04		

shock front velocity, ρ (lb-sec^2/in^4) is the soil mass density, C_p (in/sec) is the soil seismic wave velocity, and g is the acceleration due to gravity (32.2 ft/sec^2). Typical values of ρ and C_p are given in Table 13.2. If $C_p/12000U$ is greater than 1, take $A_{Hai} = A_{Vai}$.

Direct-Induced Ground Shock Direct-induced ground motions depend upon the charge weight, distance from the structure, and the soil conditions. The peak vertical ground acceleration from direct-induced ground shock can be calculated from

$$A_{Vdi} = 10{,}000 \Big/ \left(W^{1/3} Z_G^2\right)$$

where W is the TNT equivalent charge weight (lb), and Z_G is the scaled ground distance (ft). The peak horizontal acceleration depends upon the soil conditions and is

$$A_{Hdi} = \begin{matrix} 0.5A_V & \text{dry soil} \\ 1.0A_V & \text{wet soil or rock} \end{matrix}$$

The arrival time of the direct-induced ground shock at the structure can be calculated from

$$t_{AG} = 12{,}000 R_G / C_P$$

where R_G is the ground distance from the explosion (ft).

Air Shock Air-shock motions can be calculated in two ways. A simplified method is presented in this chapter that assumes a rigid structure and is based on sliding of the foundation. For this method, only the horizontal motions are considered, as rocking and vertical motions due to air shock are assumed to be negligible. Optionally, a more sophisticated analysis can be performed that accounts for the time history of the load and the flexibility of the structural elements. Although this type of analysis is beyond the scope of this handbook, some general guidelines are presented. For all but the smallest blast loads, nonlinear behavior of the structural elements, at least locally near the blast location, is expected and must be included in the analysis model. The load should be based on the design charge weight and standoff and the time history of pressure applied to the exposed surfaces of the structure. Care must be taken to ensure that the stiffness, strength, and mass distribution are modeled as accurately as possible. The general level of detail used in a dynamic, nonlinear analysis for earthquake loads is appropriate. One exception to this is that the out-of-plane behavior of walls is critical in determining the structure response to blast loading, and the simple linear models used for this behavior mode in earthquake analyses are not adequate. (See Chapters 6, 7, and 9 for additional information.)

The simplified method starts by assuming that the structure has a single degree of freedom and that the only resistance to motion is the foundation stiffness. The resulting equation is

$$F(t) - F_R(t) = M A_{Has}(t)$$

where F is the applied (blast) load, F_R is the resistance of the structure foundation, M is the structure mass, and A_{Has} is the resulting horizontal acceleration. For a shallow footing, the resistance function can be assumed to depend only upon friction between the foundation and soil, and is calculated as

$$F_R = \mu N$$

where μ is the coefficient of friction (Table 13.3), and N is the normal force acting on the footings. N is a combination of the structure weight and the vertical load applied to the roof through the blast pressure. Thus N, and subsequently F_R, varies with time.

Ideally, the equation of motion would be solved at each time step, and the maximum acceleration extracted from the results. However, since the load is only slightly dependent upon the motion, movement in the direction of the load reduces the effective pressure, and the resistance is assumed to be independent of motion, a practical alternative is available. The peak pressure and minimum value of N are assumed, and the resulting acceleration is used as the peak value. This results in a peak air-shock horizontal structural acceleration of

$$A_{Has} = \frac{P_{so} A_S - \mu W_s}{M}$$

where A_S is the exposed area of the structure, and W_s is the structure weight.

Table 13.3 Coefficient of Friction between Concrete and Soil (After Departments of the Army, the Navy and the Air Force, 1990)

Soil Type	μ
Clean sound rock	0.70
Clean gravel, gravel-sand mixture, coarse sand	0.55–0.60
Clean fine to medium or silty medium to coarse sand, silty or clayey gravel	0.45–0.55
Clean fine sand, silty or clayey fine to medium sand	0.35–0.45
Fine sandy silt, nonplastic silt	0.30–0.35
Very stiff and hard residual or preconsolidated clay	0.40–0.50
Medium stiff and stiff clay and silty clay	0.30–0.35

Load Combination The air-induced ground shock and air shock may be assumed to arrive at the structure simultaneously, since both are the result of the blast pressure. The ground-induced shock wave will arrive at a later time, since the wave speed through soil is lower than through air. Since only the peak motions are considered, whether or not the motions directly combine depends not only upon the arrival times, but also upon the actual peak motion times induced within the structure. The actual time histories of the in-structure motions are extremely difficult to calculate due to the lack of adequate data on the time history of the input loads for ground shock and the uncertainty in structural response.

Whether or not the direct-induced ground shock adds to the air-induced ground shock and air shock depends upon the arrival time, t_{AG}. If t_{AG} is greater than the combined arrival time and duration of the blast load ($t_a + t_0$), then the air-induced motions can be assumed to dissipate before the direct-induced shock arrives, and the peak accelerations should not be added. However, if t_{AG} is less than or equal to $t_a + t_0$, accelerations from all three sources should be combined.

The equations for all three sources of shock were derived assuming a rigid structure. Based on this assumption, the motion at all levels of the structure will be identical. However, although structures tend to be very stiff vertically, this is not true for the horizontal motions. The ground motion will induce horizontal deformations within the structure, resulting in differing levels of acceleration throughout the building height. This variation of in-structure motion with height has been codified for earthquakes, with accelerations increasing through the height due to the response of the building. This is not necessarily true for blast loads, as the short duration does not allow for structural response in the same manner as earthquake loads. For example, an analysis of a typical three-story steel building subjected to a 10g ground shock load of duration 50 ms, and accounting for nonlinear structural response, results in absolute accelerations of 10g at the base, 2g at the second story, and 1g at the roof. Relative displacements between floors are less than 1 inch. Of course, this is only an example, and the response in other buildings may be vastly different, but it does illustrate the effect of the short load duration on building response and the difference with response to earthquake loads. Thus, great care must be taken when considering the effects of location within the building, and it is conservative to assume that the ground accelerations are present at all levels.

The simplified method for calculating air-shock motions does not allow for differentiation based on location within the structure. Thus, the vertical accelerations in the structure, assuming simultaneous application of the direct- and air-induced ground shock, are calculated as

$$A_V = A_{Vai} + A_{Vdi}$$

and the horizontal ground accelerations are

$$A_H = A_{Hai} + A_{Hdi} + A_{Has}$$

Example Given: Loose, dry sand

$$\begin{aligned} \rho &= 1.42\text{E} - 04 \\ C_p &= 7200 - 39{,}600 \\ \mu &= 0.45 \\ U &= 1.4 \end{aligned}$$

$$1000 - \text{lb TNT surface burst @ } 100'$$

$$\begin{aligned} W &= 1000 \text{ lb} \\ P_{so} &= 5.5 \text{ psi} \\ Z_G &= 10 \end{aligned}$$

Building: 100′ wide × 40′ tall, assuming typical density gives

$$\begin{aligned} W_s &= 3{,}000{,}000 \text{ lb} \\ M &= 7764 \text{ lb} - \text{s}^2/\text{in} \end{aligned}$$

Solution:

$$\begin{aligned} A_{Vai} &= 100P_{so}\big/(\rho C_p g) = 100(5.5)/[(.000142)(23{,}400)(32.2)] = 5.14g \\ A_{Hai} &= A_V \tan\left[\sin^{-1}\left(C_p/(12000U)\right)\right] = 5.14g \\ A_{Vdi} &= 10{,}000\Big/\left(W^{(1/3)} Z_G^2\right) = 10{,}000\Big/\left((1000)^{(1/3)}\,(10)^2\right) = 10g \\ A_{Hdi} &= 0.5A_{Vdi} = 5g \\ A_{Has} &= \frac{P_{so}A_S - \mu W_s}{M} = \frac{(5.5)(1200)(480) - (0.45)(3{,}000{,}000)}{7764} = 0.61g \end{aligned}$$

The total in-structure motions are

$$\begin{aligned} A_V &= A_{Vai} + A_{Vdi} = 5.14 + 10 = 15.14g \\ A_H &= A_{Hai} + A_{Hdi} + A_{Has} = 5.14 + 5 + 0.61 = 10.75g \end{aligned}$$

Note that these accelerations are much higher than typically seen in seismic events. For earthquakes, the largest accelerations considered in the lower 48 states for nonstructural component design in the 2006 *International Building Code* (International Code Council 2006), not including the effect of component flexibility, are approximately 5g horizontally and 0.6g vertically.

Anchorage Force Calculation The anchorage forces are calculated by applying the shock load at the equipment center of mass. The applied force is equal to the shock acceleration multiplied by the equipment mass. The force should be applied in all directions, unless the possible blast locations limit the shock direction. Usual practice is to also consider eccentricity of the center of mass, as well as the actual attachment point locations.

Seismic design principles require that the applied load be further modified based on the flexibility of the equipment and the amount of nonlinear behavior that can be tolerated before damage. The flexibility component is intended to magnify the forces if the equipment will move significantly relative to the floor (increasing the effective force). For two pieces of equipment of identical mass, the more flexible equipment will undergo larger deformations and experience larger accelerations at its center of mass. Another factor that can modify calculated forces recognizes that most equipment can undergo some nonlinear behavior without sustaining significant damage. Since the analysis used to evaluate the equipment and anchorage assumes linear behavior, a lower effective force will be transmitted during the actual event. The amount of the reduction that is used depends on the ability of the equipment to accept nonlinear deformations without damage. However, no formal guidance exists regarding the applicability of these factors for blast loading, and great care should be taken if they are to be applied for blast analysis.

Example: Motor Control Center Equipment data (Figure 13.10):

$$\begin{aligned} W &= 2800 \text{ lb} \\ A &= 80 \text{ in} \\ B &= 20 \text{ in} \\ H &= 92 \text{ in} \end{aligned}$$

Load Data: Use values from previous examples.

$$\begin{aligned} A_V &= 15.14g \\ A_H &= 10.75g \\ P_{so} &= 5.5 \text{ psi} \end{aligned}$$

Shock calculation:

- Apply the loads A_V and A_H at the center of mass (assume at center in plan and midheight). Account for 5% eccentricity in any direction.

(a)

(b)

Figure 13.10 Example Calculation Equipment Layout: (a) Motor control center; (b) Anchorage layout

- Calculate the anchorage forces assuming that static loads equal to $A_V W$ and $A_H W$ are applied simultaneously. The horizontal load can be applied in any direction with the maximum anchorage forces governing.

The applied loads are:

$$\begin{aligned}\text{Vertical} &= 15.14g * 2800\text{ lb} = 42{,}392\text{ lb (uplift or compression)}\\ \text{Horizontal} &= 10.75g * 2800\text{ lb} = 30{,}100\text{ lb}\end{aligned}$$

The anchorage forces with the horizontal load applied in the short direction with zero eccentricity are:

$$\begin{aligned}\text{Uplift} &= (42{,}392 - 2800)/4 + (30{,}100)(46\text{ in})/[2(20)] = 44,513\text{lb}\\ \text{Compression} &= (42{,}392 + 2800)/4 + (30{,}100)(46\text{ in})/[2(20)] = 45,913\text{lb}\\ \text{Horizontal} &= 30{,}100/4 = 7525\text{lb}\end{aligned}$$

and the resulting maximum anchorage forces considering eccentricity are:

$$\begin{aligned}\text{Vertical} &= 47{,}632\text{ lb uplift}\\ \text{Horizontal} &= 8455\text{ lb}\end{aligned}$$

Note that these are considerably higher than 1/4 of the applied loads due to the location of the load application and the assumed eccentricities. In particular, the vertical forces are greatly increased due to the overturning moment imposed by the horizontal force.

These anchorage load values exceed the capacity of $1^1/_4{}''$ anchors by approximately 25%. Adequate anchorage would require multiple anchors at each support location. This compares to one 1/2″ anchor required at each support location in an area of moderate seismic activity (UBC, Zone 3) (International Code Council 1997).

Blast Calculation: Three basic approaches can be used to apply blast loads to equipment. First, the peak (or average) blast pressure can be applied as a static force. This is an extremely conservative approach for most equipment, as it ignores the short duration of the applied load. The second approach is to consider the blast load to be impulsive in nature and to use a simplified method to determine the maximum response. A drawback to this approach is that the dynamic characteristics, mass and stiffness, of the equipment must be known, which is usually not the case. A third alternative is to explicitly calculate the time history response of the equipment to the blast load, but this approach is rarely justified due to the—at best—approximate nature of the input parameters and the requirement to know the equipment mass and stiffness distribution.

The first two approaches will be considered in this example.

Method 1 – Static Load Application

- Apply the pressure as a static load distributed evenly across the exposed face.
- Calculate the anchorage forces for the given loads. The horizontal load can be applied in any direction, but for symmetric equipment, applying the load along each principal axis is usually sufficient.

The equivalent applied forces are:

$$\begin{aligned} \text{Strong axis} &= 92 \text{ in} * 20 \text{ in} * 5.5 \text{ psi} = 10{,}120 \text{ lb} \\ \text{Weak axis} &= 92 \text{ in} * 80 \text{ in} * 5.5 \text{ psi} = 40{,}480 \text{ lb} \end{aligned}$$

The resulting maximum anchorage forces are:

$$\begin{aligned} \text{Strong axis loading vertical} &= 5119 \text{ lb (uplift)} \\ \text{Strong axis loading horizontal} &= 2530 \text{ lb} \\ \text{Weak axis loading vertical} &= 185{,}508 \text{ lb (uplift)} \\ \text{Weak axis loading horizontal} &= 10{,}120 \text{ lb} \end{aligned}$$

The anchorage forces generated by these loads are clearly beyond the capacity of a cost-efficient anchorage system. Several simplifying assumptions were made in this calculation. In addition to the static application of the load, no load was applied to the top, side, or rear surfaces. Loads applied on the top and rear face will reduce the anchorage forces.

The exposed front face of the motor control center must be able to resist the 5.5 psi load. The ability of most sheet metal to sustain such a load is doubtful at best, with maximum duct pressures approaching 1 psi. Thus, although this example demonstrates the procedure for calculating the anchorage forces to a load of 5.5 psi, the equipment would have to be protected through barriers or relocation to reduce the load to a level that could be resisted.

Method 2 – Impulsive Load

A typical natural frequency for lateral motion of a motor control center might be 0.5 sec. For this example, that is a lateral stiffness of roughly 1000 lb/in. From Figure 13.11 the dynamic amplification factor for a ratio of

$$\frac{t_0}{T} = \frac{0.050}{0.50} = 0.1$$

is $D = 0.35$. This leads to a maximum total horizontal reaction force of

$$H_{total} = \begin{matrix} 0.35(5.5)(92)(20) \\ 0.35(5.5)(92)(80) \end{matrix} = \begin{matrix} 3542 \text{ lbs} \\ 14{,}168 \text{ lbs} \end{matrix}$$

for the strong and weak directions, respectively. This calculation yields horizontal support loads of 885 and 3542 lbs, respectively. Similarly, the vertical forces

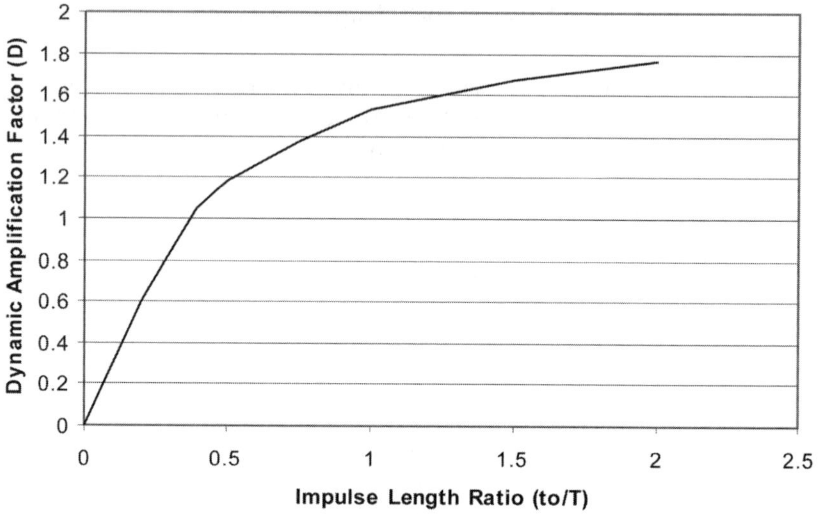

Figure 13.11 Dynamic Amplification Factor for Triangular Impulse Loads

due to blast are D times the static values. This results in

$$V = \frac{2246 \text{ lbs}}{65{,}382 \text{ lbs}}$$

The vertical loads (uplift) are still very large, but are likely able to be resisted with proper anchor design, certainly if direct attachment to steel is possible.

Stiffer equipment (smaller T) will have larger dynamic amplification factors, while more flexible equipment will see greater reductions in force.

13.5 SUMMARY

Presently, there is little guidance and few standards for blast-resistant design of building systems for civilian structures. Although there are several efforts currently under way, familiarity and selective adoption of methods used in other arenas, such as used by the DoD in hardening combatant-type systems, can be economically and practically exported to the civilian building theater. Likewise, adoption of new and available technologies, such as LED lighting, can provide reasonable stopgap measures.

This chapter has attempted to highlight the main areas of concern, provide basic guidance as to the issues to be addressed, and demonstrate computational techniques where appropriate. However, given the lack of recognized, standardized design procedure, it is imperative to recognize that, as in all engineering, there is no substitute for common sense and engineering judgment. The principles of design remain unchanged, despite the lack of exact knowledge regarding the loading and capacity for resistance.

REFERENCES

American Society of Heating, Refrigerating and Air-Conditioning Engineers. 2007. *ASHRAE Handbook: Applications*. Atlanta, GA: American Society of Heating, Refrigerating and Air-Conditioning Engineers.

Departments of the Army, the Navy and the Air Force. 1990. *Structures to Resist the Effects of Accidental Explosions*, rev. 1 (Department of the Army TM 5-1300, Department of the Navy NAVFAC P-397, Department of the Air Force AFM 88-22). Washington, DC: Departments of the Army, the Navy and the Air Force.

International Code Council. 1997. *Uniform Building Code*. Washington, DC: International Code Council.

________. 2006. *International Building Code*. Washington, DC: International Code Council.

National Fire Protection Association. 2005. *National Electrical Code Requirements for One- and Two-Family Dwellings* (NFPA 70A). Quincy, MA: National Fire Protection Association.

U.S. Department of the Army. 1986. *Fundamentals of Protective Design for Conventional Weapons* (TM 5-855-1). Washington, DC: U.S. Department of the Army.

IV Blast-Resistant Detailing

14 Blast-Resistant Design Concepts and Member Detailing: Concrete

Steven Smith and W. Gene Corley

The next four chapters provide design and detailing guidelines for members constructed from reinforced concrete, steel, and masonry, as well as members strengthened using fiber-reinforced polymers (FRP).

The general approach for all of these materials systems is to use two independent approaches to providing the toughness that is critical to blast resistance. The approaches are informally associated with strength and ductility (which, taken together, provide toughness). Strength is scaled through the response limits for each level of protection (LOP). For example, the response limit targets for LOP IV are elastic response. Thus, members designed to this LOP will be larger and stronger than those designed to lower LOP (which allow greater displacement). However, recognizing the universal need for ductility, the detailing recommendations are often constant for all LOP. For the lower LOP, this provides the ductility necessary to achieve the expected response limits. For higher LOP, this detailing provides a higher margin against the uncertainties inherent to blast-resistant design.

14.1 GENERAL

14.1.1 Scope

Detailing recommendations for reinforced concrete structures are provided in this chapter. The nature of high-rise structures in a dense urban environment introduces significant complexity, including near-field blast loads and intricate structural members such as transfer girders. For this reason, the methods presented in this section, while generic in nature, are best applied to low-rise structures with moderate standoff distance (say, $Z > 3.0$).

The seismic provisions in ACI 318-08, *Building Code Requirements for Structural Concrete* (American Concrete Institute 2008), provide the basis for the detailing, although many references are also made to UFC 3-340-02, *Structures to Resist the Effects of Accidental Explosions* (Unified Facilities Criteria 2008). Chapter 21 of ACI 318-08 contains requirements for design and construction of structures subjected to earthquake motions. The requirements are scaled to the severity of earthquake motions and are identified as ordinary, intermediate, and special moment frames. Special moment frame (SMF) requirements provide the highest level of performance and are the primary resource for the ACI provisions referenced in this chapter.

The general intent of seismic detailing is to provide greater ductility and post-yield strength than conventional detailing. This is consistent with the blast-resistant design intent of improving toughness. The Portland Cement Association's *Blast Resistant Design Guide for Reinforced Concrete Structures* (Portland Cement Association 2009) contains additional discussion of structural aspects of seismic and blast phenomena, as do Hayes (Hayes et al. 2005) and Bilow (Bilow and Kamara 2007).

14.2 FAILURE MODES

14.2.1 Flexural

Flexure is in many instances intended to be the primary structural mode for members resisting blast loads. This is due to two related concepts: the relatively predictable nature of, and the ductility associated with, flexural response.

Columns subjected directly to lateral air-blast pressures are often the single most critical design consideration for blast-resistant design. Columns directly exposed to blast pressures will respond primarily in flexure. In fact, UFC 3-340-02 recommends disregarding the column axial load and analyzing it solely as a beam. Neglecting the compressive preload is generally a conservative approach to sizing columns, as long as a separate stability check is performed that includes the axial load.

Columns may develop a three-hinged mechanism with plastic hinge zones located at end supports and at an intermediate point along the length of the column. The location of the intermediate plastic hinge can be determined by analysis and depends on the location of the blast source as well as the height and properties of the column. However, this analysis can be complex and may require assumptions that render results of little use. For most situations, the best approach is either to conservatively design the column for the potential of a hinge anywhere along the entire length, or to make general assumptions of intermediate hinge location based on the anticipated blast loading. If the column is in the far field, the blast pressure will be essentially uniform, and the intermediate hinge will form near the center of the column (assuming no other structural details that would alter the symmetry of the column response). If the column is in the near field, and subjected to blast from a package bomb, the blast pressures will be concentrated at the base of the column, and move the intermediate hinge downward.

Seismic design philosophy as applied in FEMA 350, *Recommended Seismic Design Criteria for New Steel Moment-Frame Buildings* (Federal Emergency Management Agency 2000) employs a target hinge rotation of 0.03 rad (approximately 2°). This hinge rotation can also be associated with reinforced concrete design (Bilow and Kamara 2007). Potential hinge rotations under blast loading, however, span a wide range. as described elsewhere in this handbook. The target seismic deformation magnitude, and overall response intent, would be best associated LOP II–III. The extreme deformations, structural damage, and allowance of tension membrane associated with LOP I are beyond the intent of seismic design.

14.2.2 Diagonal Tension

Diagonal tension shear is a brittle failure mode and must be avoided. Care must be taken to provide sufficient shear capacity to allow for the full flexural development of the section. The phi-factors provided in ACI 318-08 are intended to promote preferential failure modes; a greater strength reduction is applied to shear than flexure. However, the structural analyst has the task of obtaining the correct member reactions, particularly when conducting single-degree-of-freedom (SDOF) analysis and mapping the results back to the physical system.

14.2.3 Direct Shear

The primary focus of blast analysis is typically the dynamic flexural response of members. However, earlier during the blast event, direct shear forces are transmitted to the element supports. The magnitude of the direct shear force may be significantly greater than the reaction due to dynamic flexural response, though the resistance to direct shear is also typically greater that diagonal shear resistance.

Direct shear can be an even more brittle mode of response than diagonal tension. Direct shear is the phenomenon of failure at a shear plane, typically at the face of a support. Direct shear is resisted by shear friction and inclined reinforcement. Specific detailing guidelines for beams and columns are provided later in this chapter.

14.2.4 Membrane

Slabs and beams that develop extreme deformations from large, directly applied blast loading may be designed to transition from a flexural response to a tension membrane response. In addition to careful detailing of the slab or beam, this mode of response requires careful consideration of the large in-plane forces that will be transmitted to the surrounding structural elements.

14.2.5 Stability

Stability is a critical design factor for columns. Designing the column to support some or even the entire design load after loss of intermediate lateral support is an indirect means of providing substantial buckling resistance. This equates to a scenario of the loss of bracing from floor framing, and essentially requires designing the column for an unbraced length equal to two story heights.

This is not a blast-resistant design method per se, although it clearly relates to the way the building will resist progressive collapse. However, this is such an important design consideration that it should be carefully considered, even if formal progressive collapse design is not utilized. And, this is an essential design concept whenever a strong-column, weak-beam approach is used.

14.3 DETAILING

14.3.1 General

General procedures that will promote toughness include:

1. Use of concrete with a minimum strength greater than 3,000 psi. This will help to ensure against brittle shear failure.
2. Concrete with strength greater than 10,000 psi should be used with caution due to its potentially brittle failure mode in flexure.
3. All concrete should be normal weight. Existing test data and analysis of the response of reinforced concrete to blast and seismic loading are primarily based on normal-weight concrete. Thus, for normal-weight concrete, concepts such as strain-rate effects, fragmentation, and pressure-impulse (P-I) calibration for deformation limits are best understood.
4. The ACI 318-08 design compressive strain in concrete of 0.003 should also be applied to blast-resistant design, to avoid crushing of concrete and an associated brittle failure in flexure.
5. Reinforcing bars are recommended to be ASTM A706 (ASTM 2006). Where ASTM A615 (ASTM 2008) bars are used, limitations as specified in ACI 318-08, Section 21.2.5, should be applied. The primary concerns for reinforcing steel are the ratio of yield to ultimate strength and the actual yield versus design yield strength. A large ratio of yield to ultimate strength is necessary to provide sufficient ductility and energy absorption. Tests indicate that reinforcing that has actual tensile strength exceeding the actual yield strength by at least 25% provides substantial yield region. Additionally, the actual yield strength should not exceed the design yield strength by more than 18,000 psi. The intent of this limitation is to prevent the development of undesirable failure modes (such as shear) prior to the formation of plastic hinges.
6. Reinforcing bar sizes should be smaller than No. 11.

7. Straight reinforcing bars should be used to avoid reduced ductility at bends.
8. *Continuity with In-plane Members*: Every opportunity should be taken to provide a load path from framing members to plane elements. For example, beams should be tied to floor slabs, and columns to bearing walls. Particularly in situations when blast forces would be directed against the weak axis of the member (often the case for beams), this approach can provide the required blast resistance without having to substantially increase the dimensions of the member.
9. *Balanced Design*: Where appropriate, members should be designed to resist the full capacity of supported elements (as defined by a governing failure mechanism), rather than the design loads transmitted by those elements. For example, it may be desirable for a girder supporting a floor slab to be designed to resist in-plane forces developed by catenary action in the slab, even though these forces might be significantly greater than forces developed from the design loads alone with the slab responding in a flexural mode.
10. *Weak-Beam, Strong-Column*: Beam failure is generally preferable to column failure, since the beam failure typically only affects the structure in close proximity, whereas the column failure could lead to progressive collapse.
11. ACI 318-08 includes provisions for spiral reinforcement (referred to as "laced" in UFC 3-340-02). Spiral reinforcement provides confinement and torsion and shear capacity that are superior to stirrups and rectangular hoop reinforcement. References to selected ACI 318-08 guidelines for spiral reinforcement are included below. *UFC 3-340-02* includes more detailed information on detailing members with laced reinforcement. In general, the higher performance of spiral reinforcement is most necessary for very close-in blasts ($Z < 1.0$), which are beyond the scope of this guide. But any situation requiring a high level of structural performance merits consideration of spiral reinforcement.

14.3.2 Splices

Splicing can be achieved by lapping bars (No. 11 bars and smaller), mechanical connectors, or welding. It is good practice to locate splices in noncritical regions, where low stress states are expected under blast conditions. In particular, splices should be avoided in locations where plastic hinges are likely to form and at member intersections. The percentage of bars spliced in a single location should be limited to 25 percent if possible. Splices should also be separated along the axial length of a member by at least twice the member depth (Figure 14.1).

Lap splices should satisfy ACI 318-08 Class B requirements, which provide for a 30% increase in development length over Class A lap splices. Lap splices should be enclosed by lateral reinforcement conforming to Sections 21.6.4.2, 21.6.4.3, and 21.6.4.5 of ACI 318-08, to provide adequate confinement of the

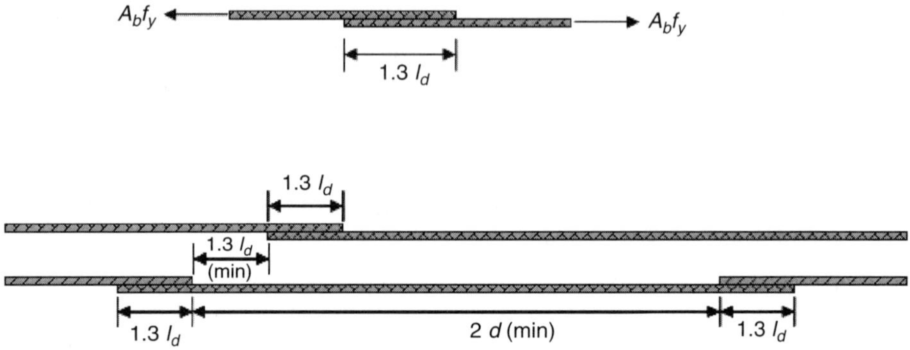

Figure 14.1 Schematic of ACI 318-08 Lap Splice Spacing and Grouping Recommendations.

concrete surrounding the splices, and to protect the splices from the effects of spalling. Noncontact lap splices are not recommended. According to UFC 3-340-02, Section 4-21, splices of adjacent bars should be staggered by at least the required lap length, since overlap of splices of adjacent bars is undesirable. Section 21.5.2.3 of ACI 318-08 advises that lap splices in flexural members should not be used within joints (a) within a distance of twice the member depth from the face of a joint, or (b) at regions of flexural yielding. Since hinge locations can be difficult to anticipate for blast loading, welded and mechanical joints are preferred for blast-resistant design.

If mechanical splices are used, they should satisfy ACI 318-08 Type 2 requirements, which call for at least 125% of the specified yield strength of the bar. UFC 3-340-02 necessitates testing to exhibit the adequacy of mechanical splices under dynamic conditions prior to acceptance for use in hardened structures. UFC 3-340-02 requires that mechanical splices develop both equivalent tensile capacity *and* ductility of a continuous bar; however, no mechanical splices have been shown to satisfy this ductility requirement. Thus, where ductility is of paramount importance, mechanical splices should be avoided.

UFC 3-340-02 suggests that welded splices should be avoided, since welding may result in a reduction of the ultimate strength and ductility of the reinforcing steel. However, if welding is necessary, strict adherence to the requirements of ANSI/AWS D1.4, *Structural Welding Code—Reinforcing Steel* (American Welding Society 2005), should permit proper development of the reinforcing steel strength. Use of testing to demonstrate the performance of welded reinforcing steel is also good practice.

14.3.3 Columns

Columns subjected directly to lateral air-blast pressures are often the most critical single design consideration. These columns experience large shear forces, out-of-plane deformations that can lead to instability, and reflected pressures that may have the potential to shatter concrete.

The flexural response of the columns may develop a three-hinged mechanism with plastic hinge zones located at end supports and at an intermediate point along the length of the column. The location of the intermediate plastic hinge can be determined by analysis and depends on the location of the blast source as well as the height and properties of the column. This analysis can be challenging, and may require assumptions that render results of little use. For most situations, the best approach is either to conservatively design the column for the potential of a hinge anywhere along the entire length, or to make general assumptions of intermediate hinge location based on the anticipated blast loading. If the column is in the far field, the blast pressure will be essentially uniform, and the intermediate hinge will form near the center of the column (assuming no other structural details that would alter the symmetry of the column response). If the column is in the near field, and subjected to blast from a package bomb, the blast pressures will be concentrated at the base of the column, and move the intermediate hinge downward.

The P-Delta moment generated by the inelastic response must be addressed. Of critical concern is whether the column can develop inelastic response and maintain sufficient moment capacity to resist the associated P-Delta effect. FEMA-350, *Recommended Seismic Design Criteria for New Steel Moment-Frame Buildings* (Federal Emergency Management Agency 2000), addresses this concern for structural steel seismic special moment frames by requiring the moment capacity at 0.03 radians of inelastic response to be greater than the nominal plastic moment, M_p. A similar rationale can be adopted for reinforced concrete members, and a synthesis study of available reinforced concrete SMF data indicates that this detailing provides inelastic response of 0.03 radians at the connections. This level of response would equate to the Moderate to Heavy criteria which are typically associated with LOP III and II, respectively.

Detailing of beam-column joints is discussed in Section 14.3.5 of this chapter, and per the above discussion it is an essential component of achieving P-Delta resistance by providing lateral restraint at frame points.

P-Delta resistance within the clear height of the column is achieved with proper reinforcement detailing so that inelastic response in this region does not unacceptably reduce the flexural resistance of the column. Detailing for columns should be based on SMF provisions in ACI 318-08, Section 21.6. The following subsections present the primary issues to address.

Longitudinal Reinforcement Section 21.6.3.1 of ACI 318-08 limits the area of longitudinal reinforcement to between 1% and 6% of the gross area. The lower limit provides that the column will achieve the yield moment before the cracking moment. The upper limit addresses congestion and avoiding development of high shear stresses. To minimize the number of iterations, it is typically best to begin the design process with reinforcing steel at the 6% limit.

Transverse Reinforcement (Diagonal Shear and Confinement) Transverse reinforcement serves the equally important purposes of resisting diagonal tension and providing confinement. Confinement from closely spaced hoops protects the core concrete from being shattered by high, directly applied blast pressures, or crushed by large deflections and stresses at locations of large inelastic flexural response. Maintaining the integrity of the core concrete is critical for lateral restraint of the longitudinal reinforcement and allows the concrete to contribute to the shear capacity of the section. Hoops and ties for confinement are particularly critical in regions of inelastic response. Per the previous discussion of intermediate hinge location, unless that position can be reliably predicted, it is recommended that transverse reinforcement be provided per Section 21.6.4 of ACI 318-08 over the clear height of the column, rather than only over a distance l_0, as defined in Section 21.6.4.1. Where steel jacketing is provided, transverse reinforcement can be reduced, since the steel jacket will protect against spalling and provide the necessary confinement.

Transverse reinforcement spaced to satisfy confinement requirements (Section 21.6.4) will likely satisfy shear requirements for diagonal tension; however, the requirements of Section 11.4 should always be checked, particularly when the confinement requirements of Section 21.6.4 are not provided over the clear height of the column.

Direct Shear Columns must be designed to resist the effects of direct shear as well as flexure and diagonal tension, although in most cases the direct shear will not control.

Resistance to direct shear from blast loading can be treated in a manner similar to that of Section 11.6 of ACI 318-08, which provides minimum requirements for shear friction reinforcement. The shear friction concept presumes a crack at the critical shear plane, such that sufficient reinforcing steel is required to maintain interlock of the rough, mating surfaces. Since the function of the shear friction reinforcement is to resist the tendency of the shear plane to open, it is the tensile capacity (not shear) that is of interest; as long as there is sufficient steel, and it does not yield, the mating surfaces will remain engaged and resist the shear. Presuming sufficient shear friction reinforcement (as defined ACI 319, Section 11.6), the direct shear capacity of the concrete and any inclined bars that cross the shear plane can be calculated as

$$V_n = 0.18 f'_{cd} bd + A_s f_{yd} \cos\alpha$$

Where, f'_{cd} is the product of the concrete compressive strength, DIF and SIF; A_s is the area of inclined steel bars; f_{yd} is the product of the steel yield stress, DIF and SIF; and α is the angle formed by the plane of the inclined reinforcement and the surface of the shear plane. Note also that since the direct shear phenomenon occurs very early during the blast application, and before the development of flexural response, Type I cross section is used. Type I cross section is defined as the uncracked cross section; that is, the complete concrete cross section can be counted for resisting the direct shear.

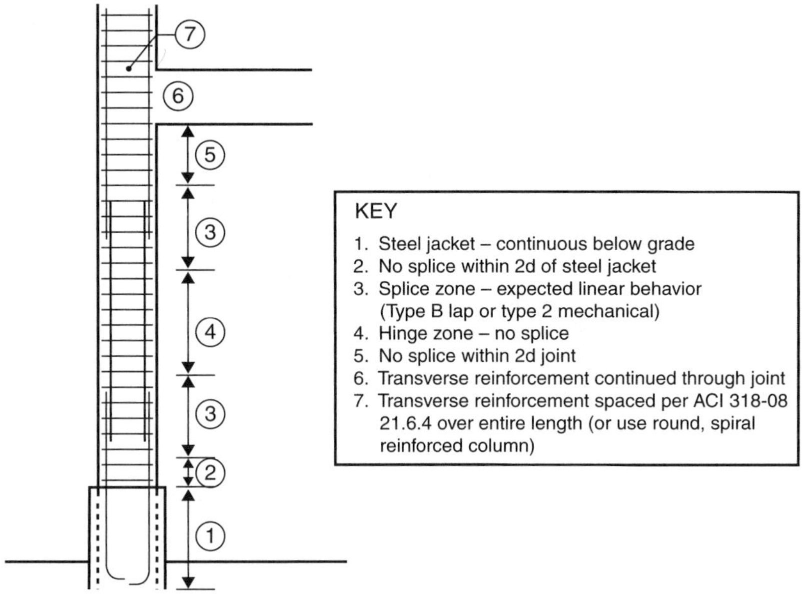

Figure 14.2 Schematic of Column Detailing Recommendations

Splice Locations Even though hinge location can be difficult to estimate, as discussed previously, the attempt should be made so that splices can be avoided at those locations. Splices should never be located in a region that does not have transverse reinforcement placed per Section 21.6.4 of ACI 318-08.

Additionally, splices should not be located within a joint or within a distance of twice the member depth from the face of a joint.

Unbraced Length Designing the column to support some or all of the design load after loss of intermediate lateral support is an indirect means of providing substantial buckling resistance. This equates to a scenario of the loss of bracing from floor framing, and essentially requires designing the column for an unbraced length equal to two story heights.

This is not a blast-resistant design method per se, although it clearly relates to the way the building will resist progressive collapse. However, this is such an important design consideration that it should be carefully considered, even if formal progressive collapse design is not utilized. And, this is an essential design concept whenever a strong-column, weak-beam approach is used.

14.3.4 Beams

This section addresses beams with clear spans greater than four times the beam effective depth. Beams shorter than this should be treated as deep beams with shear capacity determined by methods in other sources. Use of shear beams for blast-resistant design should be carefully considered due to the potentially brittle failure modes associated with them.

Exterior beams subjected directly to lateral air-blast pressures experience large shear forces, out-of-plane deformations that can lead to the development of plastic hinges, and reflected pressures that may have the potential to shatter concrete. Interior beams may be subjected to either direct blast pressures in the form of uplift (for situations in which the interior of the building becomes pressurized) and forces transmitted from other structural members.

Plastic hinge development for exterior beams is similar to that for columns, as previously discussed. Plastic hinge zones can develop at fixed supports and at intermediate points along the length of a beam. The exact location of an intermediate plastic hinge can be determined by analysis and depends on the location of the blast source as well as the span and properties of the column. In lieu of such complex analysis, two general cases can be considered to bound the problem. Beams subjected to far-field blast load will most likely develop hinges at supports and at midspan. This assumes that the load applied to the beam from a far-field blast will be nearly uniform over the column length. Sources of exceptions to this assumption could include very long span beams and beams with intermediate framing. Beams subjected to near-field blast load will most likely develop the middle plastic hinge at a location biased to the position of the blast source.

Simply supported beams can be used in blast-resistant structures if they can be shown to satisfy the required level of performance. However, the designer must be aware that the simple supports limit the available potential energy dissipation and additional margin against unanticipated blast effects.

Detailing for beams should be based on SMF provisions in ACI 318-08, Section 21.5. The following subsections present primary issues to address during the design process.

Longitudinal Reinforcement Longitudinal reinforcement should comply with ACI 318-08, Sections 21.5.2.1 and 21.5.2.2. Beams subjected to uplift should have reinforcement designed to resist both the uplift and rebound.

Transverse Reinforcement (Diagonal Shear and Confinement) Transverse reinforcement serves the equally important purposes of resisting diagonal tension and providing confinement. Confinement from closely spaced hoops protects the core concrete from being shattered by high, directly applied blast pressures, or crushed by large deflections and stresses at locations of large inelastic flexural response. Maintaining the integrity of the core concrete is critical for lateral restraint of the longitudinal reinforcement, and allows the concrete to contribute to the shear capacity of the section. Hoops and ties for confinement are particularly critical in regions of inelastic response. Transverse reinforcement should be spaced to satisfy 21.5.3.2 and provided over the clear span of the beam, unless specific analysis of yield zone location supports alternate placement. Transverse reinforcement spaced to satisfy 21.5.3.2 will likely satisfy shear requirements for diagonal tension; however, the requirements of Section 11.4 should always be checked, particularly when the confinement requirements of Section 21.5.3.2 are not provided over the clear span of the column.

Single-piece hoops as shown in Detail A of Figure R21.5.3 of ACI 318-08 are preferred to hoops built up from multiple members.

Direct Shear The discussion and recommendations for direct shear in columns can be directly applied to beams.

Splice Location Even though hinge location can be difficult to estimate, as discussed previously, the attempt should be made so that splices can be avoided at those locations. Splices should never be located in a region that does not have transverse reinforcement placed per Section 21.5.4 of ACI 318-08.

14.3.5 Beam-Column Joints

Joint design is critical to providing an engineered response to blast loading. The development of inelastic response, which typically accounts for the majority of energy dissipation for LOP I–III, is directly related to connection performance. The performance intent of LOP IV is essentially elastic response to the *design* blast load; achieving this intent, and also providing margins against unanticipated load magnitudes, requires equally careful joint detailing.

The performance expectations of the various levels of protection can potentially be achieved with a simple joint between beams and columns, but the trade-off of this design approach is potentially massive beams.

The monolithic nature of reinforced concrete can be used to great advantage in the development of robust beam-column joints. Reinforced concrete joints inherently provide substantial deformation continuity between beams and columns. To provide the level of performance demanded for blast resistance, measures must be taken to ensure ductility and to protect the core concrete. These goals are primarily achieved through detailing considerations for the beam longitudinal reinforcement (ductility) and column transverse, confinement reinforcement (protection of core concrete).

Longitudinal Reinforcement Wherever possible, beam longitudinal reinforcement should be continuous through joints. Splices should not occur within joints, or within a distance of twice the beam depth from the face of the joint.

Reinforcement that terminates at a joint should comply with Section 21.7.2.2, which requires the reinforcement to be anchored at the far face of the confined column core.

Transverse Reinforcement Column transverse reinforcement should be continued through the joint, as described in Section 21.7.3. This section provides consideration of the similar function that is provided by continuous beam longitudinal reinforcement for joints that have beams framed to all four faces. The

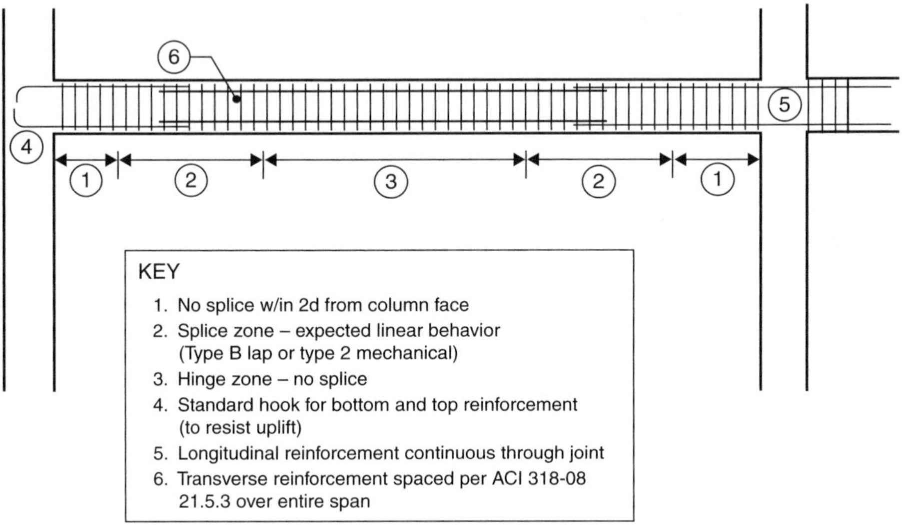

Figure 14.3 Schematic of Beam Detailing Recommendations

important consideration is to have confining reinforcement *around* the column longitudinal reinforcement within the joint.

Shear Strength Shear strength should be computed per Section 21.7.4. Also of importance in this section is the definition that a beam is considered to provide confinement to the joint if a least three-quarters of the face of the joint is covered by the beam. This definition applies to the calculation of shear resistance. However, relative to the consideration of transverse reinforcement discussed previously, the most important consideration is enclosure of the column longitudinal reinforcement that continues through the joint. Thus, a beam that satisfies 21.7.4 for shear capacity of the joint should not necessarily negate the use of transverse reinforcement, spaced as previously discussed, to provide additional protection of the joint core.

14.3.6 Slabs

The designer must first determine the structural behavior anticipated for the slab. Slabs can serve numerous roles, including:

Flexural element: This is a conventional role for a building slab. With respect to blast resistance, a slab will respond as a flexural element to directly applied blast forces. This can occur with explosions initiated within the building, or when the interior becomes pressurized by failure of exterior elements. Blast pressures acting on the soffit of the slab will generate uplift and stresses opposed to gravity loads.

Diaphragm: Slabs that are intended to act as diaphragms for lateral response to conventional loads can also be used to efficiently distribute blast forces through the structure, and thus dissipate the blast effects by engaging additional structural elements and building mass.

Local Restraint: Slabs that are not explicitly intended to perform as a diaphragm can still be used to provide local lateral restraint to beams that are directly exposed to transverse blast forces. This can be a very effective way to support beams against the weak-axis bending.

Membrane: Slabs that develop extreme deformations from large, directly applied blast loading may be designed to transition from a flexural response to a tension membrane response. In addition to careful detailing of the slab, this mode of response requires careful consideration of the large in-plane forces the slab will transmit to the surrounding structural elements.

The following subsections are concerned with slab-detailing considerations.

Flexural Reinforcement The flexural reinforcement ratio should not be less than 0.0018. This is the limit for temperature and shrinkage, when using 60 ksi reinforcement, given in Section 7.12. In practice, the reinforcement ratio will almost always exceed this.

Flexural reinforcement should be provided at top and bottom throughout the slab. For blast scenarios in which uplift is expected, the top reinforcement should be explicitly designed to resist the uplift forces, and the bottom reinforcement to resist rebound. In the event that design blast loads will not subject the slab to uplift, but the slab is part of the structural system that will resist blast effects, it is good practice to provide approximately equal amounts of top and bottom reinforcement. This approach will provide reserve capacity for events such as an unanticipated reversal of curvature or loss of a support.

Bottom reinforcement should be continuous across interior supports. This is particularly important in the column strip of two-way flat-slab floor systems. At exterior supports, the reinforcement should be fully developed. This bottom reinforcement detailing is a precaution against the loss of a support, providing the structural potential for the slab to bridge across the lost support.

Connected Elements Structural elements connected to a slab should generally be designed to establish the full capacity of the slab (i.e., balanced design). This must be carefully considered where the slab is intended to develop tension membrane behavior, because the in-plane forces that will need to be equilibrated by the surrounding structure will be many times larger than the flexural forces transferred from the slab.

Slabs that serve as lateral restraint for beams directly subjected to blast pressures must have sufficient connectivity for the beam to engage the mass and in-plane stiffness of the slab.

Fully distributed force transfer between the slab and connected elements is the preferred mechanism for all situations. For structures with concrete framing members, this is readily achieved. With steel-framed structures, the best method is to cast the member into the concrete slab. For example, for a steel girder supporting a concrete deck, the top flange of the girder should be cast into the deck, if possible, rather than using shear studs.

14.3.7 Walls

As with slabs, the designer must first determine the structural behavior anticipated for the wall. Walls can serve numerous roles, including:

Gravity System: As part of the gravity load system, walls may experience uplift from internal pressurization, or, in rare instances, increased compressive loads from internal overpressures at higher floors.

Flexural Element: A wall will have to be able to respond in flexure in the event of blast forces applied directly to one face, or unbalanced blast forces applied to both faces. This can occur with explosions that are set off within a building, or if the interior becomes pressurized from failure of exterior elements.

Diaphragm: Walls that are intended to act as diaphragms for lateral response to conventional loads can also be used to efficiently distribute blast forces through the structure, and thus dissipate the blast effects by engaging additional structural elements and building mass.

Local Restraint: Walls that are not explicitly intended to perform as a diaphragm can still be used to provide local lateral restraint to columns that are directly exposed to transverse blast forces.

The following subsections are concerned with wall-detailing considerations.

Reinforcement The minimum web reinforcement ratio should be 0.0025, which will satisfy requirements for both in- and out-of-plane loads. Reinforcement should be placed in at least two curtains and fully developed at the wall boundaries. Section 21.9 of ACI 318-08, "Special Structural Walls," contains additional useful design recommendations and commentary.

Connected Elements Walls that may be subjected directly to transverse blast pressure and that also provide gravity support for elements in higher stories should be constructed with columns at each end and a beam spanning between the columns. The purpose of this is to provide a gravity load path in the event that the wall is compromised by the transverse blast load. The beams, columns, and joints should be designed per previous sections of this chapter.

REFERENCES

This chapter is adapted in large part from information in the PCA's *Blast Resistant Design Guide for Reinforced Concrete Structures* (Portland Cement Association 2009), with the permission of the Portland Cement Association.

American Concrete Institute. 2008. *Building Code Requirements for Structural Concrete* (ACI 318-08) *and Commentary* (ACI 318R-08). Farmington Hills, MI: American Concrete Institute.

American Welding Society. 2005. *Structural Welding Code—Reinforcing Steel* (ANSI/AWS D1.4). Miami, FL: American Welding Society.

ASTM. 2006. *Specification for Low-Alloy Steel Deformed Bars for Concrete Reinforcement* (A706). West Conshohocken, PA: ASTM International.

_______. 2008. *Specification for Deformed and Plain Billet-Steel Bars for Concrete Reinforcement* (A615). West Conshohocken, PA: ASTM International.

Bilow, D. and M. Kamara. 2007. Capacity of joints to resist impact loads in concrete moment-resisting frame buildings. Presented at the Structural Engineering Institute Structural Congress, Long Beach, CA, May 16–19, 2007.

Departments of the Army, the Navy and the Air Force (DOANAF). 1990. *Structures to Resist the Effects of Accidental Explosions* (Department of the Army Technical Manual TM 5-1300). Washington, DC: Departments of the Army, the Navy and the Air Force.

Federal Emergency Management Agency. 2000. *Recommended Seismic Design Criteria for New Steel Moment-Frame Buildings* (FEMA 350). Washington, DC: Federal Emergency Management Agency, Department of Homeland Security.

Hayes, J. et al. 2005. Can strengthening for earthquake improve blast and progressive collapse resistance? *Journal of Structural Engineering* 131 (8): 1157–1177.

Portland Cement Association. 2009. *Blast Resistant Design Guide for Reinforced Concrete Structures*. Skokie, IL: Portland Cement Association.

Unified Facilities Criteria Program. 2008. *Structures to Resist the Effects of Accidental Explosions* (UFC 3-340-02). Washington, DC: U.S. Department of Defense, Unified Facilities Criteria Program.

15 Blast-Resistant Design Concepts and Member Detailing: Steel

Charles Carter

15.1 GENERAL

This chapter provides guidance on the design of steel structures for blast loads. It is adapted from information in two references: "Defensive Design" (Shipe and Carter 2003) and *Defensive Design of Structural Steel Buildings* (Gilsanz et al. 2009), with the permission of the American Institute of Steel Construction (AISC).

Considerations for defensive design usually fall into one of three general categories: typical building designs, prescriptive building designs, and performance-based building designs. Some designs combine prescriptive and performance-based approaches.

15.1.1 Typical Building Designs

The majority of buildings receive no special treatment other than judicious attention to redundant configurations and robust connection designs so that the structural (and nonstructural) components are tied together effectively. The typical details used in steel buildings inherently provide for redundancy and robustness, with the capability for load redistribution through alternative load paths. This fact has been demonstrated repeatedly when steel buildings have been subjected to abnormal loadings.

In such cases, the following ideas can be beneficial:

- Configure the building's lateral systems to provide multiple load paths from the roof to the foundations. Multiple lines of lateral framing distributed throughout the building are generally better than fewer isolated systems.
- Provide horizontal floor and roof diaphragms to tie the gravity and lateral framing systems together.
- Minimize framing irregularities in both horizontal and vertical framing when possible. Horizontal and vertical offsets with copes and/or

eccentricities can reduce the available strength at member ends—or require extensive reinforcement to maintain that strength.

- Use multiples of the same shape, rather than changing girder and column sizes. The additional strength in girders and columns that are heavier for convenience could cost less or be free. The use of a smaller number of different shapes in the building means a labor savings in fabrication and erection, and often more than offsets the cost of the additional steel weight, which is only around 25% of the cost of the structure once it is fabricated and erected.
- Remember that serviceability limit states indirectly add significant structural redundancy to steel framing. Usually, beams and girders are sized for deflection or floor-vibration criteria, and girders and columns are commonly sized for drift control. As a result, these elements have significant reserve strength.
- Use typical shear, moment, and/or bracing connections judiciously. Reserve strength is gained at low cost if connection details are clean. It costs little to fill the web of a girder with bolts using a single-plate or double-angle connection.
- Recognize other sources of redundancy and robustness inherent in steel buildings, including: the common overstrength in the steel materials and connecting elements, membrane action in the floor and roof diaphragms, and the strength and stiffness contributions of nonstructural components.

With little—and sometimes no—modification, steel framing provides redundancy and robustness.

15.1.2 Prescriptive Building Designs

Some buildings receive special treatment through the application of prescriptive criteria for design that go beyond those in the basic building code. While there are a variety of prescribed criteria, such as the removal of a single column, there is usually no attempt to characterize the exact effects of the blast. Instead, the goal could be to reduce the probability of progressive collapse in areas not directly affected by the blast.

In such cases, several strategies can be employed:

- Use a perimeter moment frame. This can result in a system that is significantly robust. As an extreme example, the perimeter moment frame in the tube structure of each of the World Trade Center towers spanned a hole about 140 ft wide before succumbing to the combined effects of structural damage and fire, each likely at their lifetime maximum values. More likely, the prescribed criterion will be the removal of a single column. In some cases, the framing will have enough redundancy to accept column removal without modification. If not, the column spacing can be reduced or the framing hardened by increasing size or switching to composite construction.

- Use a strong story or floor. This solution can be a truss system with diagonals or a Vierendeel truss system, incorporated into a single story or multiple stories in the building. Additionally, a single, strong floor with heavier framing and moment connections throughout could carry or hang at least 10 floors. The World Trade Center towers also demonstrate this solution. The damaged core columns in each tower to some extent hung from the hat trusses, which created strong-story framing at the top of each tower. The link between the core gravity columns and the perimeter tube framing carried more than 10 stories in one tower and more than 25 stories in the other. Nonetheless, exercise caution when considering hat-truss framing to create a strong story. Unless specifically designed for progressive-collapse resistance, hat trusses normally reduce the level of reserve strength and redundancy because of the efficiency they allow in the structural system.
- Consider other innovative solutions. One particularly innovative solution was used in the U.S. Federal Courthouse in Seattle, WA. The building has a steel-framed composite core and gravity steel framing around it. The perimeter has steel cables banding it to prevent progressive collapse should a column be lost. The result is one of the most open and inviting blast-hardened buildings built to date.

There are other potential prescriptive criteria, but structural solutions usually flow from criteria like those described above. Most often, there is no attempt to rigorously assess the actual blast effects and the emphasis is on arresting the effects of the blast.

15.1.3 Performance-Based Building Designs

Performance-based criteria are used for a small number of buildings, normally ones that are government-related, high-risk, or high-profile. Performance criteria vary, but generally require that the building withstand the effects of the blast, protect the occupants of the building, and/or maintain a defined level of operability. The nature and characteristics of the threats are identified realistically and modeled in the design. Note that the performance criteria affect more than the structural frame, and often require nonstructural elements, such as blast-resistant windows, special site layouts, and site perimeter protection.

Performance-based criteria are used for a comparatively small number of buildings, where the nature and characteristics of the threat are realistically identified and modeled in the design. When the threat, building characteristics, and performance criteria are known, the solution is a matter of design for the dynamic loading of the blast, including mitigation of damage and the associated progressive collapse potential. The following sections illustrate this approach.

15.2 BLAST EFFECTS ON STRUCTURAL STEEL AND COMPOSITE STRUCTURES

Steel's response to the characteristics of a blast affects the loading for several reasons, discussed in the following subsections.

15.2.1 Member Ductility

Structural steel has a linear stress-strain relationship up to the yield stress, but then can undergo extreme elongation without an increase in stress—about 10 to 15 times the amount needed to reach yield. Stress then increases in the "strain hardening" range until a total elongation of about 20 percent to 30 percent is reached. This response has benefits beyond routine design-level forces for resisting the effects of a blast. The ductility ratio m, defined as the maximum deflection over the elastic deflection, commonly is used to account for this effect.

15.2.2 Connection Ductility

Steel connections can be extremely ductile. As in seismic design practice, connections can be configured so that yielding limit states will control over fracture limit states. However, there is still significant ductility in many limit states that involve fracture. Block shear rupture, for example, is a fracture limit-state accompanied by significant deformation prior to the actual failure.

Additionally, some limit states have "failures" that, in blast applications, can be considered benign or even beneficial. For example, there is little structural consequence to connection slip or bearing deformations at bolt holes due to blast effects. Moreover, these phenomena consume energy when they occur, progressively reducing the blast effect through the structure.

15.2.3 Overstrength

Structural steel usually is stronger than the specified minimum strength. For example, ASTM A992, *Standard Specification for Structural Steel Shapes* (ASTM 2006) has a specified minimum yield strength $F_y = 50$ ksi. From the American Institute of Steel Construction (AISC) *Seismic Provisions for Structural Steel Buildings* (American Institute of Steel Construction 2005a), it has an expected yield strength F_{ye} of 55 ksi. This supports the practice endorsed by the Department of Defense Explosives Safety Board, which allows the use of an average yield strength increase factor (generally a 10 percent increase in the yield strength) for blast design.

15.2.4 Beneficial Strain-Rate Effects

Structural steel benefits from an increase in apparent strength when the rate of loading is rapid. The yield point increases substantially by a factor that is called the Dynamic Increase Factor for yield stress. The ultimate tensile stress also increases, but not as greatly as the yield stress. The total elongation at failure typically remains unchanged or decreases slightly because of the increased strain rate. The modulus of elasticity is unaffected by the strain rate.

15.2.5 Beneficial Effects of Composite Construction

The use of composite construction can have benefits for blast-design applications due to the mass effect of composite systems with steel elements (490 lb/ft^3) and concrete elements (150 lb/ft^3). The inelastic action in a composite system generally will limit deflections and local deformations, and partially mitigate rebound effects through the damping effect of concrete cracking.

15.2.6 Perimeter Column Design

At the building perimeter, the columns will be loaded by the blast through the facade. A critical design decision is the selection of an appropriate tributary area for the column, which must be based upon the expected performance of the facade in the blast. Will the facade and its connection to the structure be such that the full blast load will be delivered to the framing, or will the facade components be shattered in the blast before the load can be delivered through them?

The tributary area is then used with the maximum total dynamic pressure and the dynamic load factor to find an equivalent static load on the column. Either a rigorous beam-column design approach or the simplified approach suggested in UFC 3-340-02 (Unified Facilities Criteria Program 2008) can be used:

- Support conditions for the column are taken as if the column were a simple-span beam.
- The moment perpendicular to the plane of the facade is taken as $wl^2/8$ based upon the blast loading.
- The moment in the plane of the facade is a function of the magnitude of the strong-axis moment and the angle of incidence of the blast to the column.
- The axial load applied concurrently with the blast load is based upon the full dead load and one-quarter of the live load.

Based upon these loadings, UFC 3-340-02 recommends checking limit states like flange-local buckling, web-local buckling, shear yield, and interaction between flexural and axial forces.

15.2.7 Perimeter Girder Design

The consequences of a girder failure normally are not as high as the consequences of a column failure. When designing the girder to the same performance level as the columns, use the foregoing approach to column design for the girder design, without the axial load. Otherwise, a less restrictive approach can be used, permitting inelastic deformations. It is important to consider the differing support conditions between the top and bottom flanges of the girder. Often this can be mitigated by orienting the infill beams to provide restraint to the girder bottom flange along the span. Also, the girder end connections often will be best configured as moment connections.

15.2.8 Slab Design

The blast pressure could subject the slab to significant uplift, depending upon the expected performance of the facade in the blast. As with girder design, the consequences of a local slab failure normally are not as high as the consequences of a column failure, so it could be sufficient to utilize the typical reinforcing steel in the upper portion of the slab. If a higher performance objective is established, the blast pressure can be applied to the slab and the reinforcement selected to be appropriate for the loading.

15.3 ANALYSIS AND DESIGN OF STRUCTURAL MEMBERS

Failure modes determine the mechanism of collapse for the structural element. Deformation criteria classify the response of the element. The primary failure modes used to design structural elements are breaching, tension-compression, bending, shear, and axial-bending interaction.

Member failure is defined through support rotation and ductility ratios for use with single-degree-of-freedom (SDOF) and simplified multi-degree-of-freedom (MDOF) systems. Strain-based failure criteria may also be justified if strains are calculated using finite element methods that characterize the material properties, the details of construction, and many of the possible failure mechanisms. Ductility is the ratio between the maximum deflection and the maximum elastic deflection. This parameter is smaller than one (1) if the behavior is elastic, and larger than one (1) if the behavior is plastic.

Energy dissipated along the load path is ignored, and it is assumed that the elements are absorbing all of the energy, which is a conservative assumption. The designer should give attention to the reactions and the connections along the load path from one member to another one, as these will be required to transfer the full energy of the system.

15.4 STEEL MATERIAL PROPERTIES FOR BLAST DESIGN

The properties used in conventional design of steel buildings are modified when considering blast loadings to account for actual strength level and dynamic effects.

15.4.1 Strength Increase Factor (SIF)

For steel grades with $F_y = 50$ ksi or less, the average yield stress in current production is approximately 10% larger than this minimum specified value. Therefore, for blast design the yield stresses should be multiplied by a Strength Increase Factor (SIF) of 1.10. For higher-strength grades, this average is smaller than 5% larger; therefore, no factor is used on those grades. The tensile strength F_u is not factored in any case.

15.4.2 Dynamic Increase Factor (DIF)

Steel mechanical properties vary with the time rate of strain. As compared with the static values normally used in design, the following properties vary for dynamic loading:

- The yield point increases substantially.
- The ultimate tensile strength increases slightly.
- Modulus of elasticity does not vary, and the elongation at rupture either remains constant or is slightly reduced.

The factor used to modify the static stress due to dynamic load is the Dynamic Increase Factor (DIF). These factors are defined in *Design of Blast Resistant Buildings in Petrochemicals Facilities* (American Society of Civil Engineers 1997) and UFC 3-340-02.

The values summarized in Table 15.1 are based upon an average strain ratio of 0.10 in./in./sec, which is characteristic of low-pressure explosions. Higher values of strain ratio will result in larger values of DIF. UFC 3-340-02 provides values of DIF for different average strain rates.

Common grades in use today, such as ASTM A992, A500, and A53 (ASTM 2006, ASTM 2007b, ASTM 2007a) are not addressed in the aforementioned

Table 15.1 Dynamic Increase Factors (DIF) for Structural Steel

	DIF (Low Pressure)		
	Yield Stress		
Material	Bending/Shear	Tension/Compression	Ultimate Stress
A36	1.29	1.19	1.10
A588	1.19	1.12	1.05
A514	1.09	1.05	1.00
A446	1.10	1.10	1.00

references. All are normal construction grades, similar to ASTM A36 (ASTM 2008), and suitable for use with the same DIF values.

15.4.3 Dynamic Design Stress

Based on the expected ductility ratio and/or the damage allowed in the structural element, different dynamic yield points are defined by UFC 3-340-02. If the ductility ratio is smaller than or equal to 10, the dynamic design stress is:

$$f_{ds} = f_{dy} = SIF \cdot DIF \cdot f_y \tag{15.1}$$

If the ductility ratio is greater than 10, the dynamic design stress is:

$$f_{ds} = f_{dy} + \frac{f_{du} - f_{dy}}{4} \tag{15.2}$$

where

$$f_{du} = DIF \cdot f_u \tag{15.3}$$

The dynamic design stress for shear is

$$f_{dv} = 0.55 f_{ds} \tag{15.4}$$

For typical design, unless governed by more detailed requirements such as *Design of Blast Resistant Buildings in Petrochemical Facilities* (ASCE 1997), UFC 3-340-02, or other established criteria, a simplified value may be used as follows:

$$f_{ds} = 1.3 f_y \tag{15.5}$$

$$f_{du} = 1.05 f_u \tag{15.6}$$

15.5 DESIGN CRITERIA FOR BLAST DESIGN

15.5.1 General

Due to the nature of blast loading, plastic design is required, and is measured by support rotations and ductility. Different codes can be used to define the strength of the elements, but no safety factor should be used in those calculations. Local and global stability should be addressed in those elements where plasticity must be achieved and ductility criteria must be used.

Design and failure criteria are based on the results of explosive tests conducted by the U.S. government and reported in publications such as the *Tri-Service*

Manual by the Department of Defense, and the *Protective Design Manual* by the U.S. Air Force.

The designer should determine, at the beginning of the project, the acceptance criteria to be used based on the level of protection desired for the building. Most projects have predefined criteria that must be used.

15.5.2 Load Combinations

In the absence of other governing criteria, the following load combination should be used:

$$1.0D + 0.25L + 1.0B \tag{15.7}$$

where D is the dead load
L is the live load
B is the blast load

15.5.3 Resistance Factor and Factor of Safety

For Load and Resistance Factor Design (LRFD), the resistance factor ϕ is not used in blast design. For Allowable Stress Design (ASD), the safety factor Ω is not used in blast design. Thus, the strength is simply taken as the nominal strength, without reduction.

15.5.4 Local Buckling

Blast design is based on the ultimate strength of the elements and ductility of the system. Structural members subject to blast loads should be capable of undergoing plastic deformation. To allow hinges to form in the elements, sections must be flexurally compact in accordance with criteria in Chapter B of the AISC *Specification for Structural Steel Buildings* (American Institute of Steel Construction 2005b), unless the effects of local buckling can be tolerated in the response. The dynamic design stress should be used in the calculation of the limiting slenderness to establish the local buckling criteria.

15.5.5 Lateral-Torsional Buckling

Flexural members should be braced at an interval that is small enough to permit plastic hinge formation. Essentially, this removes lateral-torsional buckling as the controlling limit state in beams.

15.5.6 Deformation Criteria

Support rotation is defined as the tangent angle at the support formed by the maximum beam deflection. In the plastic range, this value, neglecting the elastic deformation, can be related to the plastic hinge rotation. Note that if the hinge is

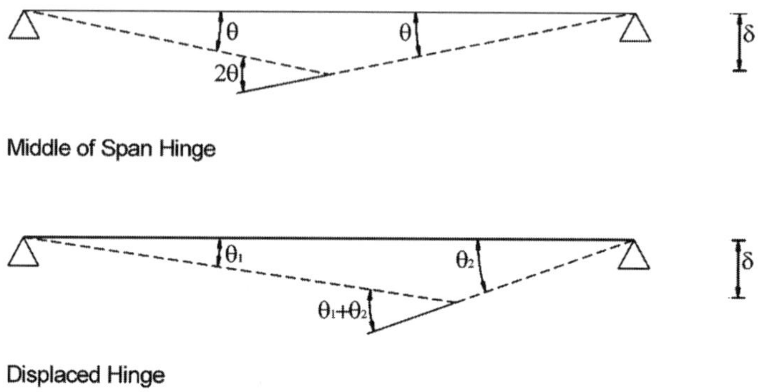

Figure 15.1 Hinge Rotation versus Connection Rotation

not formed at the center of the beam, the support rotations are different at both sides, and the maximum rotation should be considered as shown in Figure 15.1.

Ductility, μ, as used in blast design, is defined as the ratio between the maximum deflection (Δ_m) and the elastic deflection Δ_E).

$$\mu = \frac{\Delta_m}{\Delta_E} \tag{15.8}$$

ASCE's *Design of Blast Resistant Buildings in Petrochemical Facilities* (American Society of Civil Engineers 1997) classifies the deformation range in three different damaged stages: low, medium, and high response as a function of the damage in the building. UFC 3-340-02 also classifies the response as a function of the protection provided by the structural elements. This section is based upon a low level of protection that implies a high response as the intent is to provide a design that avoids imminent collapse.

For different structural elements, this guide follows the response criteria as shown in Table 15.2. The rotation criteria in Table 15.2 refer to support rotation. The criteria are defined for the behavior of a single element. These criteria apply to the design of the members, and the connection between the members may have different criteria.

Table 15.2 Response Criteria for Structural Steel for a Low Level of Protection

Element	Response Ratio (Ductility)	Rotation
Open-web steel joists[1]	$\mu < 2$	$\theta < 3°$
Steel beams	$\mu < 20$	$\theta < 10°$
Steel columns	$\mu < 5$	$\theta < 2°$

[1]Response ratio (ductility) controlled by downward loading, and rotation controlled by upward loading

As an example, assume a beam with a 30-ft span that starts to yield at midspan when the deflection is 4 in. A support rotation of 10° implies a deflection of 15 × tan(10°) = 31 in. A ductility of 1 implies a deflection of 4 in. Thus, a ductility of 20, again taken from Table 15.2, would require a deflection of 80 in. Obviously, this level of ductility is not achievable, and the support rotation controls.

Other more restrictive criteria exist, such as Corps of Engineers Program Design and Construction (COE PDC) criteria, and are usually used in commercial and governmental buildings.

Ductility limits are linked to a given mode of response. A flexural ductility is different from shear or tension. For steel elements not shown in Table 15.2, other sources can be consulted, including UFC 3-340-02 and *Design of Blast Resistant Buildings in Petrochemicals Facilities*. Although it is intended for seismic design, FEMA 356 (Federal Emergency Management Agency 2000) provides useful information.

UFC 3-340-02 defines the blast criteria and the structural properties used in Figure 15.2.

15.5.7 Detailing for Specific Failure Modes:

Failure modes include breaching, tension, compression, shear, flexure, and combined loading.

- *Breaching*: Blast loads in contact with or in very close proximity to an element. May cause failure before a more typical structural response like flexure can occur, due to the high pressure produced by the explosion. Generally, if the scaled distance is below two, it is possible for the element to breach before the overall response of the structural element starts.

  For very close charges, in addition to the air blast, temperature and the shock wave are important. An explosion in direct contact with a floor or wall will interact directly with it and induce a shock wave inside it. The speed and magnitude of the shock waves can cause the floor or wall to crack internally. For example, when the shock wave reaches the opposite face of a concrete wall, a section of concrete may spall or separate from the wall because the energy of the shock wave exceeds the tensile strength of the material.

  This behavior can be estimated using information in Figure 15.3, from the *Protective Construction Design Manual*, ESL TR 87-57 (Air Force Engineering and Services Center 1989), based upon the thickness (T, ft), the standoff distance (R, ft), and the charge (W, equivalent lbs TNT). Depending on the performance requirements, the slab may be designed to allow a local breach in the bay closest to the blast, yet withstand the blast in adjacent bays. Typically, the standoff and charge weight are specified, given the threat analysis. Other breaching curves can be found in ESL-TR-87-57 or UFC 3-340-01 (Unified Facilities Criteria Program 2002).
- *Tension*: According to AISC's *Specification for Structural Steel Buildings*, the maximum strength for elements under tension is based on yielding of

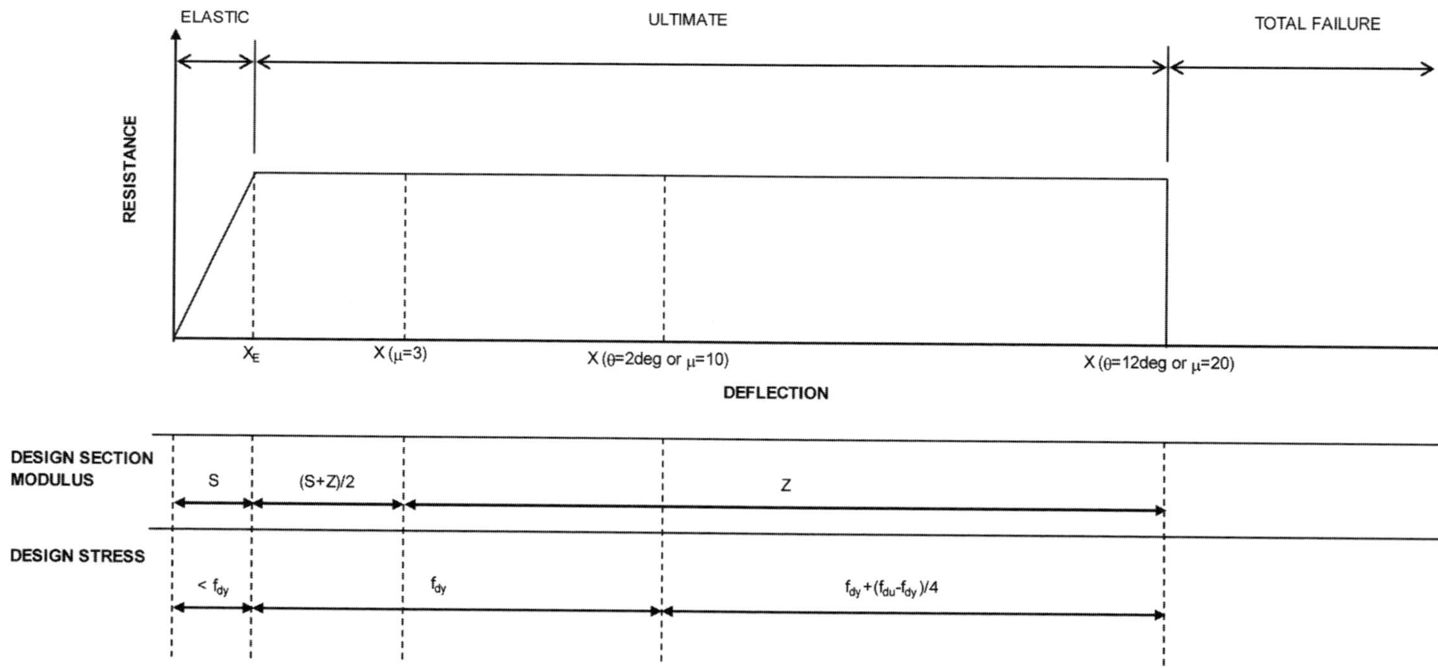

Figure 15.2 UFC 3-340-02 Criteria

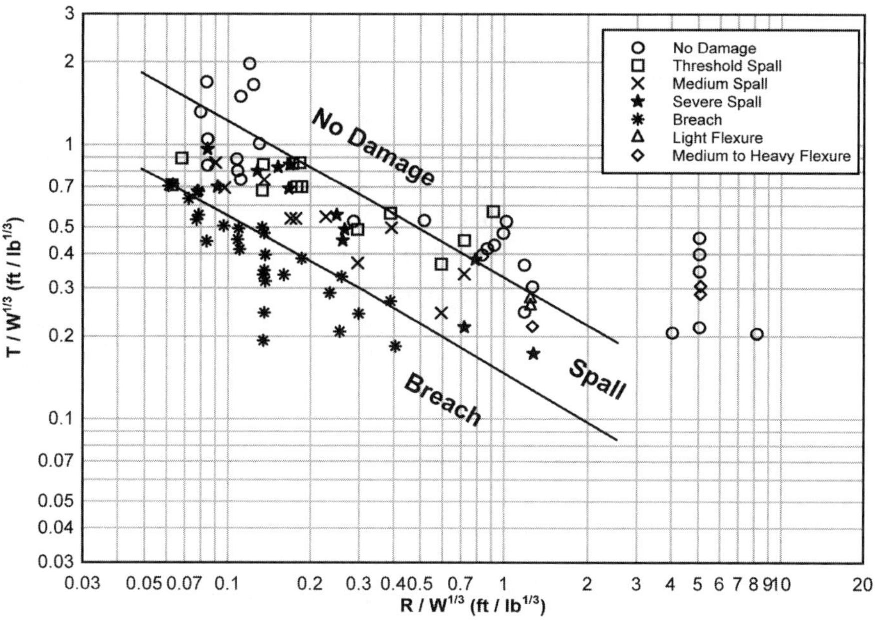

Figure 15.3 Protective Construction Manual Breach Chart

the member and is defined by

$$P_u = \phi f_y A_g \tag{15.9}$$

where A_g = the gross area of the element
f_y = the material yield stress
ϕ = the resistance factor, 0.90

For blast loading, ϕ is not used, and $f_y = f_{ds}$ is the dynamic design stress defined in Equations 15.1 and 15.2.

Axial hinge properties for these kinds of elements are elastic-plastic, based on the ultimate load defined above. If necessary, for computational stability, a small strain hardening slope, such as 0.1%, can be used in the plastic area.

- *Compression*: The strength of compression elements, based on the AISC *Specification for Structural Steel Buildings,* is governed by the following equation:

$$P_u = \phi f_{cr} A_g \tag{15.10}$$

where A_g = the gross area of the element
f_{cr} = the critical stress
ϕ = 0.90

For blast loading, ϕ is not used, and f_{cr} = the critical stress by the appropriate AISC equation substituting f_{ds} for f_y. For a tension-compression hinge, the recommendations of FEMA 356 can be used, if no other proper criterion is defined.

- *Shear*: For I-shapes it is assumed that moment is carried primarily by the flanges, while shear is carried primarily by the web. Thus, moment-shear interaction is neglected. The yield strength of elements in shear, based on the AISC *Specification*, is governed by the following equation:

$$V_p = \phi f_v A_w \tag{15.11}$$

where A_w = the area of the web
$f_v = 0.6 f_y$
$\phi = 1.00$ for most W-shapes

For blast loading, ϕ is not used, A_w is the area of the web, and $f_v = f_{dv}$ is the dynamic yield stress; $f_v = 0.55 f_y$ in UFC 3-340-02 (Unified Facilities Criteria Program 2008). Note that connection strength is, in most of the cases, more critical than the shear strength of the beam.

- *Flexure*: Based on plastic response allowed for blast design, a plastic hinge will be allowed to form in the structural element. The assumption that the plastic hinge is concentrated at a section will be taken as adequate for practical purposes, even though deflection values may not be accurate.

The definition of plastic moment is a function of the ductility ratio expected in the structural element. Based on Figure 15.7 of UFC 3-340-02, if the ductility ratio is smaller than 3, the full plastic moment cannot be developed by the section. Thus, an average between the elastic and plastic section modulus is used:

$$M_p = f_{ds}\left(\frac{S+Z}{2}\right) \tag{15.12}$$

If the ductility ratio expected is larger than 3, the full plastic moment can be developed. Thus,

$$M_p = f_{ds} Z \tag{15.13}$$

For hand calculations, the behavior of the section is assumed elastic-perfectly plastic. For hand calculations in this guide, the plastic moment strength of the section is given by Equation 15.13. To avoid computational instabilities, the hinge properties used for computer calculations should include plastic hardening, where the yield moment is defined by Equation 15.12 and the ultimate moment is defined by Equation 15.13 as recommended in UFC 3-340-02. A criterion for lateral bracing is also given in UFC 3-340-02 although not discussed here.

- *Combined Forces*: Provisions for combined axial and bending effects are presented in the AISC *Specification*. The strengths and resistance factors defined above are used in the appropriate interaction equations.

15.6 EXAMPLES

15.6.1 Example 1—Determining Capacities

The strength of the elements determined in these examples is based on UFC 3-340-02. For element design, it is assumed that there is no energy dissipation along the load path. A more accurate and less conservative procedure is to use the dynamic reactions of one member on another along the load path.

When using member reactions to load following members, natural frequencies should be compared. If the natural frequencies are close, a simultaneous solution is required to account for the interaction between the two members. If the period of the primary element (i.e., beam) is at least twice the period of the secondary element (i.e., girder), they can be treated as individual SDOF structures on unyielding supports. If not, an MDOF solution of the same system should be used.

Material Properties All calculations for this example are done for steel with the following properties:

- $F_y = 50$ ksi
- $F_u = 70$ ksi

For a ductility ratio $\mu \leq 10$, the dynamic design stress, defined in Equation 15.5, is:

$$f_{ds} = f_{dy} = 1.3F_y = 1.3 \times 50 = 65 \text{ ksi}$$

For a ductility ratio $\mu > 10$, the dynamic design stress, defined in Equation 15.2, is:

$$f_{ds} = f_{dy} + \left(\frac{f_{du} - f_{dy}}{4}\right) = 65 + \left(\frac{1.05 \times 70 - 65}{4}\right) = 67 \text{ ksi}$$

where: $f_{du} = 1.05f_u$ from Equation 15.6.

The dynamic design shear stress, defined in Equation 15.4, for a ductility ratio $\mu \leq 10$ is:

$$f_{dv} = 0.55 \times f_{ds} = 0.55 \times 65 = 36 \text{ ksi}$$

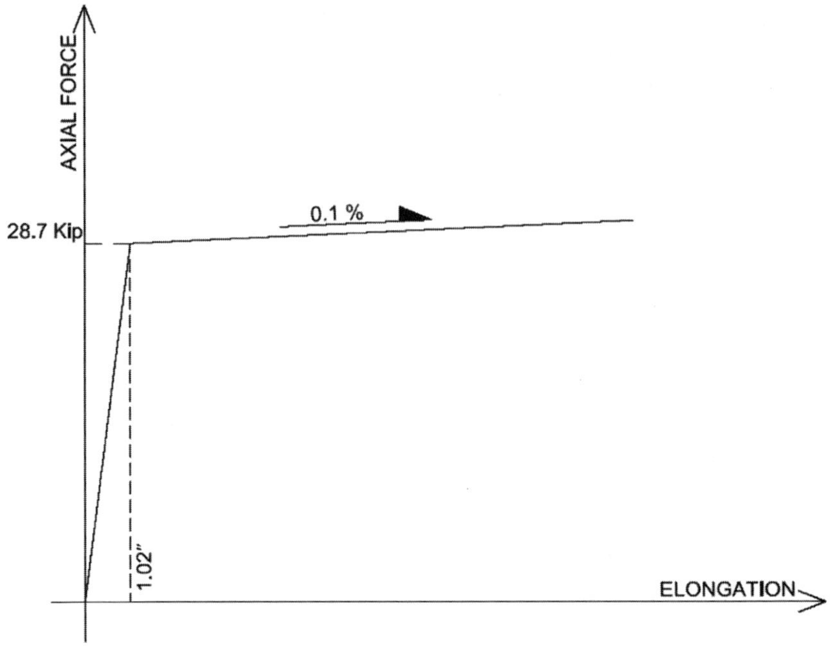

Figure 15.4 Tension Hinge Property

Tension-Only Rod Brace A tension rod with $^3/_4$-in. diameter is used. Because the rod has upset ends, we can use the full cross section area. The tension strength is obtained as:

$$T_{\max} = A\ f_{ds} = \pi \times \left(\frac{3/4}{2}\right)^2 \times 65 = 28.7 \text{ kips}$$

The hinge property for this rod follows the elastic-plastic curve shown in Figure 15.4. A slight slope is included in the plastic region for computational convergence purposes.

Given a maximum horizontal displacement of 4.2 in., the maximum rotation at the base of the column for this displacement is:

$$\alpha = \text{atan}\left(\frac{\Delta}{H}\right) = \text{atan}\left(\frac{4.2}{15 \times 12}\right) = 1.34\,\text{deg} < 2\,\text{deg}$$

For a small-displacement formulation, the elongation in the rod can be determined. First, the angle of the diagonal is determined:

$$\theta = \text{atan}\left(\frac{H}{L}\right) = \text{atan}\left(\frac{15}{35}\right) = 23.2\,\text{deg}$$

Then, the elongation, assuming the top of the column moves horizontally:

$$\Delta L = \Delta \cos\theta = 4.2 \times \cos 23.2 = 3.86 \text{ in}$$

Thus, the elastic deformation of this rod is:

$$\Delta L_{el} = \frac{f_{ds} L}{E} = \frac{65 \times 38 \times 12}{29000} = 1.02 \text{ in}$$

Therefore, the ductility demand in the rod is:

$$\mu = \frac{3.86}{1.02} = 3.78$$

Diagonal Brace (in Tension and Compression) An ASTM A500 Grade C HSS6 × 6 × 3/4 is used. This section has an area of 5.24 in.^{2} and a radius of gyration of 2.34 in. The brace length is 17 ft.

The slenderness of this element is:

$$\lambda = \frac{KL}{r} = \frac{1 \times 17\text{ft} \times 12^{\text{in.}}/_{\text{ft}}}{2.34 \text{ in}} = 88 < C_c = \sqrt{\frac{2\pi^2 E}{f_{ds}}} = 94$$

The buckling stress is obtained from UFC 3-340-02 as:

$$F_a = \frac{\left(1 - \dfrac{\left(\dfrac{Kl}{r}\right)^2}{2C_c^2}\right) f_{ds}}{\dfrac{5}{3} + \dfrac{3}{8}\dfrac{\left(\dfrac{Kl}{r}\right)}{C_c} - \dfrac{1}{8}\dfrac{\left(\dfrac{Kl}{r}\right)^3}{C_c^3}} \tag{15.14}$$

Substituting values, the nominal buckling stress is:

$$F_a = \frac{\left(1 - \dfrac{88^2}{2 \times 94^2}\right) \times 65}{\dfrac{5}{3} + \dfrac{3}{8} \times \dfrac{88}{94} - \dfrac{1}{8} \times \dfrac{88^3}{94^3}} = 19.2 \text{ ksi}$$

The denominator in the above equation (which equals 1.92), is the factor of safety. Hence, the compression capacity without the factor of safety is:

$$P_n = 1.92 A F_a = 1.67 \times 5.24 \times 19.2 = 193 \text{ kips}$$

The tension capacity for this element is:

$$P_u = A f_{ds} = 5.24 \times 65 = 341 \text{ kips}$$

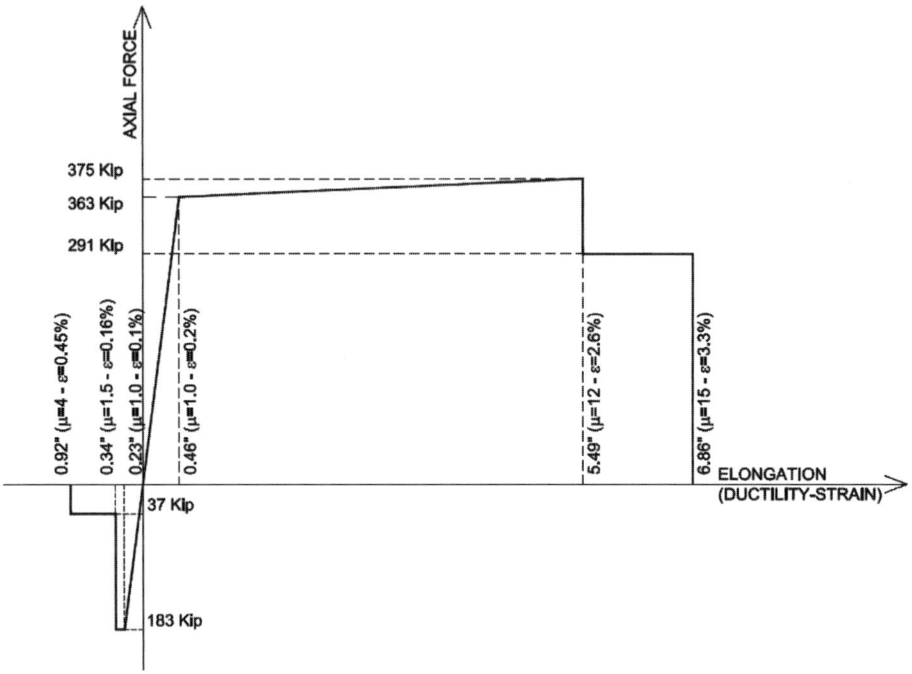

Figure 15.5 Tension-Compression Hinge Properties

Using information from FEMA 356 (Federal Emergency Management Agency 2000), a tension-compression hinge can be developed, as shown in Figure 15.5. This asymmetric curve includes the buckling ductility. In the tension zone, the hardening slope is modified to account for the dynamic yield stress for the yield axial force and the dynamic ultimate stress for the ultimate axial force. In the compression zone, a hardening slope of 0.1% is introduced to avoid computational instabilities.

Column Assume the column is a W12×53 with an effective length $KL = 15$ ft. The section properties are:

$$A = 15.6 \text{ in.}^2 \quad r_{\min} = r_y = 2.48 \text{ in.}$$

For this column:

$$C_c = \sqrt{\frac{2\pi^2 E}{f_{ds}}} = \sqrt{\frac{2 \times \pi^2 \times 29000}{65}} = 94$$

The maximum slenderness of this element is:

$$\frac{KL}{r_y} = \frac{15 \times 12}{2.48} = 72.6 < 94$$

The buckling stress is:

$$F_a = \frac{\left(1 - \frac{72.6^2}{2 \times 94^2}\right) \times 65}{\frac{5}{3} + \frac{3}{8} \times \frac{72.6}{94} - \frac{1}{8} \times \frac{72.6^3}{94^3}} = 24 \text{ ksi}$$

And the compression capacity without the factor of safety is:

$$P_u = 1.90 A F_a = 1.90 \times 15.6 \times 24 = 711 \text{ kips}$$

The tension capacity of this element is:

$$T_u = A \ f_{ds} = 15.6 \times 65 = 1014 \text{ kips}$$

Beam Assume the beam is a W12×35, 24-ft long, and it has chevron braces connected to it at midspan. This shape has:

$$I = 285 \text{ in}^4 \quad S = 45.6 \text{ in}^3 \quad Z = 51.2 \text{ in}^3$$

It is modeled as a simply supported beam with a concentrated load at midspan, where the braces meet.

For ductility ratios smaller than 3, the elastic-plastic bending moment is given by

$$M_{pl} = \frac{S + Z}{2} f_{ds} = \frac{45.6 + 51.2}{2} \times \frac{65}{12} = 262 \text{ kip} - \text{ft}$$

The elastic deflection for this moment due to the concentrated load at the midspan is

$$\Delta = \frac{M_{pl} L^2}{12EI} = \frac{262 \times 12 \times (24 \times 12)^2}{12 \times 29{,}000 \times 285} = 2.63 \text{ in.}$$

This deflection gives an elastic rotation:

$$\theta = \arctan\left(\frac{\Delta}{L/2}\right) = \arctan\left(\frac{2.63}{24 \times 12/2}\right) = 1.05 \text{ deg}$$

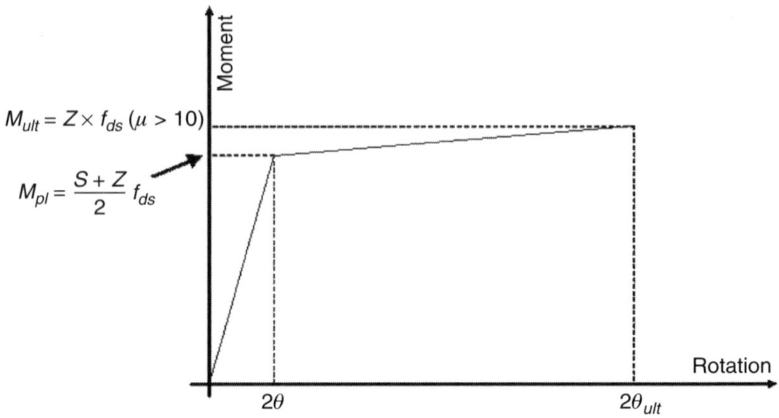

Figure 15.6 Beam Hinge Moment-Rotation

Note that the hinge rotation is twice this support rotation. These parameters define an elastic-perfectly plastic moment-rotation curve. Since many programs have convergence problems with a perfectly plastic zone, a sloped line should be introduced in this plastic region. This slope is obtained from Figure 15.6 for ductility greater than 10. Therefore, the ultimate bending moment is computed as:

$$M_{ult} = Z \times f_{ds} \, (\mu > 10) = \frac{51.2 \times 67}{12} = 286 \text{ kips} - \text{ft}$$

The maximum rotation from UFC 3-340-02 is:

$$\theta_{ult} = \min(10 \deg, \ 20 \times \theta) = \min(10 \deg, \ 26 \deg) = 10 \deg$$

Again, note that hinge rotation is twice this support rotation.

15.6.2 Example 2—Design and Analysis for Blast Loads on Members

Figures 15.7 and 15.8 illustrate the method used in this example for a simply supported beam with load and mass uniformly distributed.

The same transformation as shown in Figure 15.7 can be done by multiplying only the mass by the load-mass factor.

$$M_{SDOF} = M K_{LM} = M \frac{K_M}{K_L}$$

The K_{LM} approach is simpler because it only uses one transformation factor. This example uses the load factor K_L and the mass factor K_M because they have a more physical interpretation.

EQUIVALENT SDoF PROCEDURE

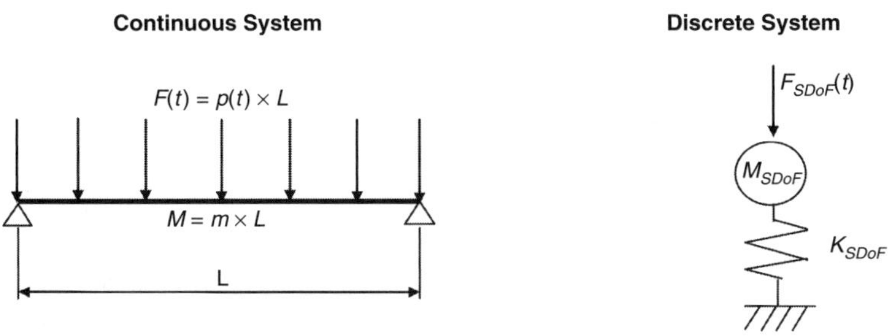

Mass = $M = m \times L$ ⟶ Equivalent Mass = $M_{SDoF} = M \times K_M$

Load = $F(t) = p(t) \times L$ ⟶ Equivalent Load = $F_{SDoF}(t) = F(t) \times K_L$

Elastic Stiffness = $K = \dfrac{384 \times E \times I}{5 \times L^3}$ ⟶ Equivalent Stiffness = $K_{SDoF} = K \times K_L$

Plastic Stiffness = $K_{pl} = 0$

Yield Force = $R_{yield} = \dfrac{8 \times M_p}{L}$ ⟶ Equivalent Yield Force = $R_{yield,\ SDoF} = R_{yield} \times K_L$

Figure 15.7 MDOF to SDOF Simplification

Wind Girt The wind girt shown in Figure 15.9 has been designed as an MC8 $\times$ 20 (ASTM A572 Grade 50; $F_y = 50$ ksi; $F_u = 65$ ksi) for wind effects with a deflection limitation of $L/260$. The blast deflection criteria given in Table 15.2 show that the ductility should be less than 20, and support rotation should be less than 10 degrees.

The duration of the blast is 6.2 milliseconds, and the peak load for the 79.5 psi peak blast pressure is:

$$F_{peak} = \frac{79.5 \times 12 \times 12}{1000} \times 5 \times 25 = 1431 \text{ kips}$$

Also given, the load factor $K_L = 0.50$ and the mass factor $K_M = 0.33$.

As a preliminary design, the support rotation criterion is used because it does not imply the knowledge of the actual section used. Rigid-perfectly plastic behavior is assumed, and the element is sufficiently braced against lateral-torsional buckling.

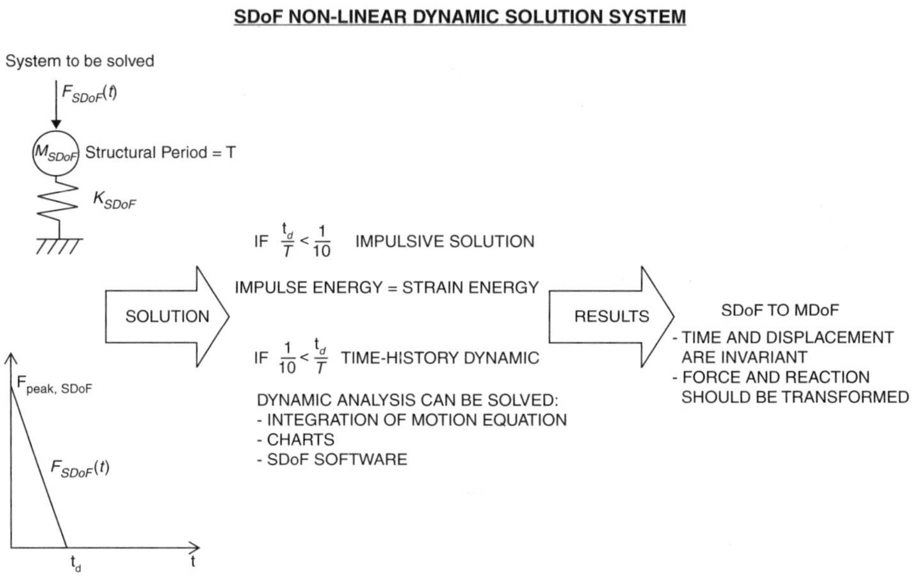

Figure 15.8 SDOF Solution

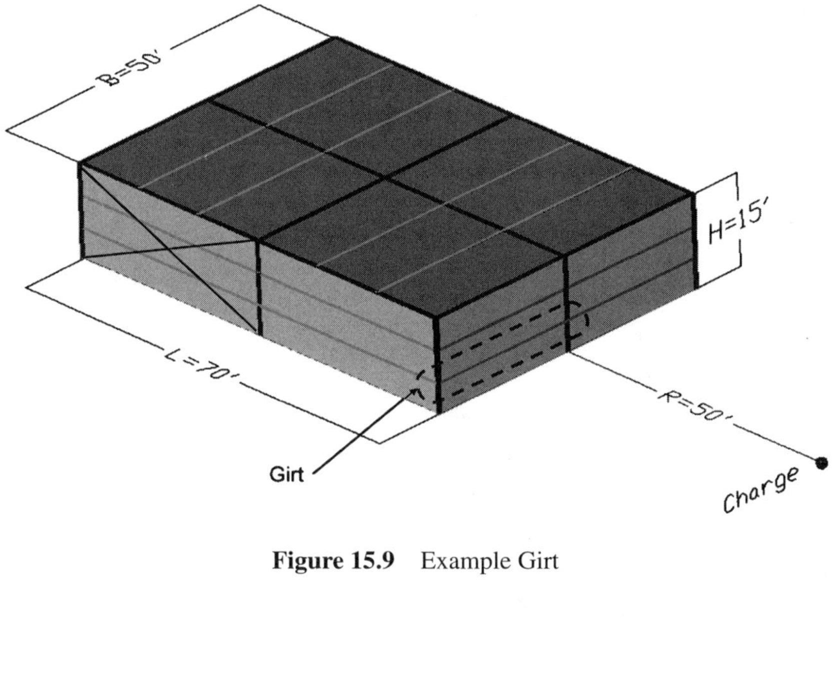

Figure 15.9 Example Girt

For this rotation criterion, the maximum displacement allowed is:

$$\Delta_{\max} = \frac{25 \times 12}{2} \times \sin(10\,\text{deg}) = 26 \text{ in.}$$

The self weight of the facade is 40 psf. The total weight of the system is:

$$w = \frac{40 \times 25 \times 5}{1000} = 5 \text{ kips}$$

Based on the single-degree-of-freedom simplification used for mass and loads uniformly distributed in the plastic range, the load and stiffness are multiplied by the load factor $K_L = 0.50$, and the mass is multiplied by the mass factor $K_M = 0.33$. Therefore, the parameters used in the discrete system are:

$$F_{peak,SDOF} = F_{peak}K_L = 1431 \times 0.5 = 715 \text{ kips } w_{SDOF}$$
$$= wK_M = 5 \times 0.33 = 1.65 \text{ kips}$$

The equivalent impulse in this single-degree-of-freedom system is:

$$I_{SDOF} = \frac{F_{peak,SDOF}\, t_{blast}}{2} = \frac{715 \times 6.2 \times 10^{-3}}{2} = 2.2 \text{ kips} - \text{sec}$$

The total energy produced by the blast load in the single-degree-of-freedom is:

$$W_{\text{Impulse},SDOF} = \frac{I_{SDOF}^2}{2m_{SDOF}} = \frac{2.2^2}{2 \times \dfrac{1.65}{386}} = 566 \text{ kips} - \text{in}$$

To limit the maximum displacement to comply with the support rotation criteria, the single-degree-of-freedom yield force should be:

$$R_{yield,SDOF} = \frac{W_{\text{Impulse},SDOF}}{\Delta_{\max}} = \frac{566}{26} = 21.8 \text{ kips}$$

The yield force for the continuous system is:

$$R_{yield} = \frac{21.8}{0.5} = 43.6 \text{ kips}$$

For this yield force, the maximum moment is:

$$M_{pl} = \frac{R_{yield}L}{8} = \frac{43.6 \times 25}{8} = 136 \text{ kips} - \text{ft}$$

For a maximum tensile strength of 65 ksi, the plastic modulus should be greater than:

$$Z_{\min} = \frac{136 \times 12}{65} = 25 \text{ in}^3$$

Based on this preliminary design, there is no MC8 strong enough to support the blast. There are several possible modifications to improve the behavior of the system: Increase the excited mass, increase the strength-stiffness of the system, or decrease the blast load by integrating a variable blast pressure that is a function of the distance to the charge along the girt.

In this example we will increase the steel section to a W8 × 28 (ASTM A992; $F_y = 50$ ksi; $F_u = 65$ ksi):

$$Z_x = 27.2 \text{ in}^3 \quad M_{pl} = Z\ f_{ds} = \frac{27.2 \times 65}{12} = 147 \text{ kip} - \text{ft}$$

$$I_x = 98 \text{ in}^4$$

$$\Delta_{el} = \frac{5}{48}\frac{M_{pl}L^2}{EI} = \frac{5}{48}\frac{147 \times 12 \times (25 \times 12)^2}{29{,}000 \times 98} = 5.82 \text{ in}$$

The strength parameters to use in the dynamic calculation are:

$$R_{yield} = \frac{8M_{pl}}{L} = \frac{8 \times 147}{25} = 47.0 \text{ kips for the SDOF}$$

$$R_{yield,SDOF} = R_{yield}K_L = 47.0 \times 0.5 = 23.6 \text{ kips}$$

$$K = \frac{R_{yield}}{\Delta_{el}} = \frac{47}{5.82} = 8.08 \text{ kips/in for the SDOF}$$

$$K_{SDOF} = KK_L = 8.08 \times 0.5 = 4.04 \text{ kips/in}$$

The mass and load do not change from the latest calculations.

The period of the system is:

$$T = 2\pi\sqrt{\frac{m_{SDOF}}{K_{SDOF}}} = 2 \times \pi \times \sqrt{\frac{1.65}{386 \times 4.04}} = 0.20 \text{ sec}$$

The structural period (0.20 sec) is more than 10 times longer than the load duration (0.0062 sec), hence the assumption of impulsive load is correct.

Figure 15.10 shows the displacement computed using SDOF software. The maximum deflection is 28.8 in. > 26.0 in.; hence, the rotation criterion is not met by a slight margin.

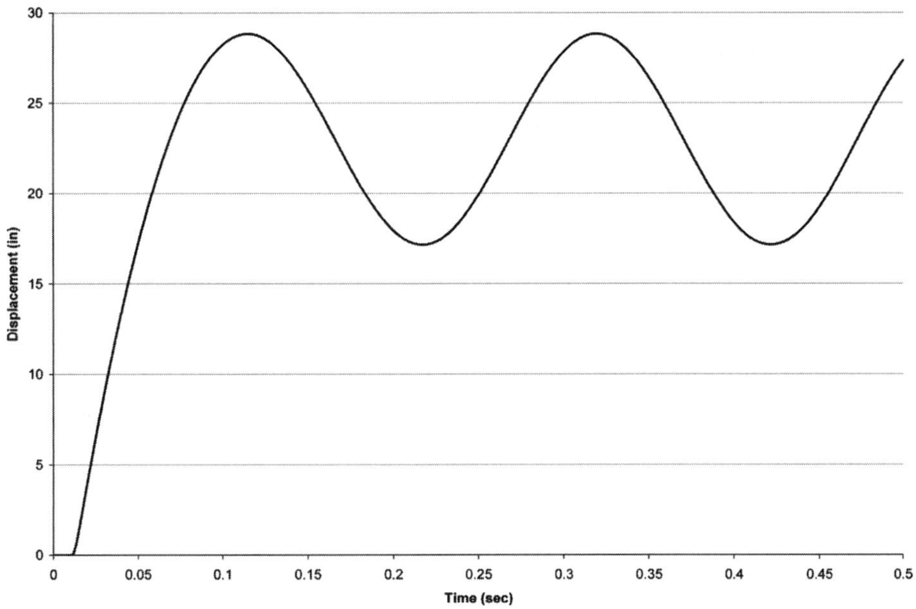

Figure 15.10 SDOF Displacement

The elastic displacement for this beam is 5.82 in.; therefore, the ductility ratio for this system is:

$$\mu = \frac{28.83}{5.82} = 5 < 20$$

The maximum reaction happens at the time of the maximum response. In an impulse load, at the time the response is maximum, the load is zero. For a longer blast duration, the reaction is a combination of the direct load and the capacity of the resisting element. According to UFC 3-340-02, the maximum shear force in plastic design can be obtained as the shear reaction that corresponds to the elastic limit of the section:

$$V = \frac{R_{yield}}{2} = \frac{47.2}{2} = 23.6 \text{ kips}$$

The dynamic shear capacity is:

$$V_p = f_{dv} A_w = 36 \times 2.30 = 82.8 \text{ kips} > 23.6 \text{ kips}$$

where f_{dv} is the blast shear capacity.
A_w is the area of the web

Hence, the section can support the shear. The connection should be designed for this capacity.

Facade Column The column shown in Figure 15.11 has been designed as a W12 × 53 (ASTM A992; $F_y = 50$ ksi; $F_u = 65$ ksi) for wind effects with a deflection limitation of *L*/260. The properties of this section are:

$$A = 15.6\,\text{in.}$$
$$I_x = 425\,\text{in}.^4 \quad Z_x = 77.9\,\text{in.}^3$$
$$r_y = 2.48\,\text{in.} \quad r_x = 5.23\,\text{in.}$$

As for the wind girt, the duration of the blast is 6.2 milliseconds, and the peak load for the 79.5 psi peak blast pressure is:

$$F_{peak} = \frac{79.5 \times 12 \times 12}{1000} \times 5 \times 25 = 1431\ \text{kips}$$

Also given, the load factor $K_L = 0.50$ and the mass factor $K_M = 0.33$.

Columns are designed to remain elastic; therefore, the maximum bending capacity should not be reached. As a preliminary design, the column will be designed without axial compression. Combined bending compression formulation will be used to check the adequacy of this element.

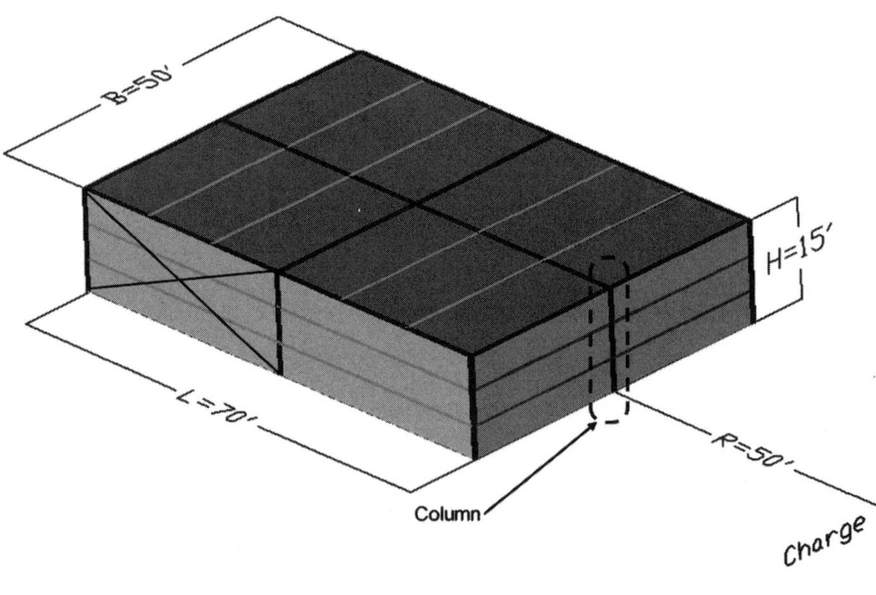

Figure 15.11 Example Column

As a preliminary and conservative assumption, the girt is assumed infinitely rigid, and all the blast pressure is absorbed by the column.

The self weight of the facade is 40 psf. The total weight of the system is:

$$w = \frac{40 \times 25 \times 15}{1000} = 15 \text{ kips}$$

The peak load for the 79.5 psi peak blast pressure is:

$$F_{peak} = \frac{79.5 \times 12 \times 12}{1000} \times 15 \times 25 = 4293 \text{ kips}$$

Based on the single-degree-of-freedom simplification used for mass and loads uniformly distributed in the plastic range, the load and the stiffness are multiplied by the load factor $K_L = 0.50$, and the mass is multiplied by the mass factor $K_M = 0.33$.

Therefore, the parameters used in the discrete system are:

$$\begin{aligned} F_{peak,SDOF} &= 0.5 \times 4293 = 2146 \text{ kips} \\ w_{SDOF} &= 0.33 \times 15 = 4.95 \text{ kips} \end{aligned}$$

The equivalent impulse in this single-degree-of-freedom system is:

$$I_{SDOF} = \frac{F_{peak,SDOF}\, t_{blast}}{2} = \frac{2146 \times 6.2 \times 10^{-3}}{2} = 6.65 \text{ kips} - \text{sec}$$

The total energy produced by the blast load in the single-degree-of-freedom is:

$$W_{\text{Impulse},SDOF} = \frac{I_{SDOF}^2}{2\, m_{SDOF}} = \frac{6.65^2}{2 \times \dfrac{4.95}{386}} = 1724 \text{ kip} - \text{in.}$$

For this W12 × 53, assuming a uniformly distributed load, the following properties define the structural behavior:

$$M_{pl} = Zf_{ds} = \frac{77.9 \times 65}{12} = 422\ kip - ft$$

$$\Delta_{el} = \frac{5}{48}\frac{M_{pl}L^2}{E\ I} = \frac{5}{48}\frac{422 \times 12 \times (15 \times 12)^2}{29{,}000 \times 425} = 1.38 \text{ in.}$$

The parameters to use in the dynamic calculation are:

$$R_{yield} = \frac{8M_{pl}}{L} = \frac{8 \times 422}{15} = 225\,\text{kips, for the SDOF}$$

$$R_{yield,SDOF} = R_{yield}K_L = 225 \times 0.5 = 112.5\text{ kips}$$

$$K = \frac{R_{yield}}{\Delta_{el}} = \frac{225}{1.38} = 163\ kips/in.,\ \text{for the SDOF}$$

$$K_{SDOF} = 163 \times 0.5 = 81.5\ kips/in.$$

The period of the system is:

$$T = 2\pi\sqrt{\frac{m_{SDOF}}{K_{SDOF}}} = 2 \times \pi \times \sqrt{\frac{4.95}{386 \times 81.5}} = 0.08\ \text{sec}$$

The column period (0.08 sec) is more than 10 times the load duration (0.0062 sec); hence, the assumption of impulsive load can be assumed. The period of the column is smaller than half of the beam period (0.08 sec $< \frac{0.20}{2} = 0.1$ sec); hence, the system is uncoupled and can be modeled separately.

Assuming elastic behavior, the maximum elastic energy that can be adsorbed by the SDOF system is given by the equation:

$$W_{S,el,\max} = \frac{R^2_{yield,SDOF}}{2K_{SDOF}} = \frac{112.5^2}{2 \times 81.5} = 77.6\text{ kips} - \text{in.} \ll 1724\text{ kips} - \text{in.}$$

This energy is smaller than the energy induced by the impulse; hence, this element will achieve plastic behavior. There is not an economical solution to solve this problem because of the element's inability to remain elastic for a directly applied tributary blast load. But the preliminary assumption that based this design in the rigid behavior of the girt is highly conservative; the maximum load that this element is carrying comes from the reaction in the girt.

Assuming this reaction is static, the system to solve is defined in Figure 15.12. The maximum bending moment for this configuration is:

$$M_x = 2 \times 23.6 \times \frac{15}{3} = 236\text{ kips} - \text{ft}$$

The axial load at the column, based on 30 psf roof dead load, is:

$$P = 30 \times \frac{50}{2} \times \frac{70}{4} = 13.1\text{ kips}$$

In the following, the capacity properties of this element are calculated to be used in the interaction axial-bending curves.

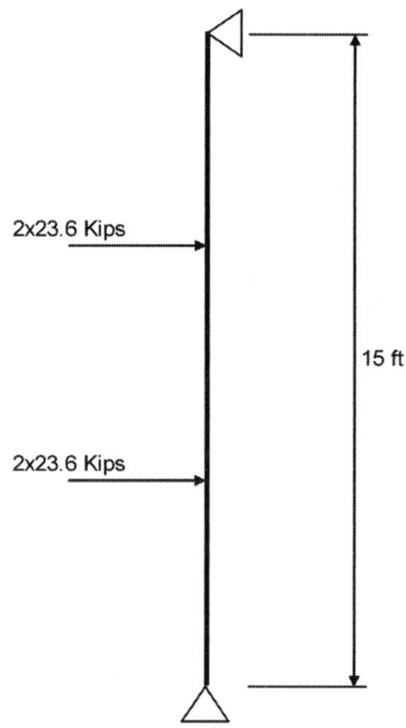

Figure 15.12 Equivalent Static Reaction System

For buckling about the weak axis, the column is assumed unbraced by the girts. The buckling length is $K_y L = 1 \times 15 = 15$ ft; hence, the slenderness is:

$$\frac{K_y L}{r_y} = \frac{15 \times 12}{2.48} = 72.6$$

For the buckling about the strong axis, the column is also unbraced. The buckling length is $K_x L = 1 \times 15 = 15$ ft; hence, the slenderness is:

$$\frac{K_x L}{r_x} = \frac{15 \times 12}{5.23} = 34.4$$

Also,

$$C_c = \sqrt{\frac{2\pi^2 E}{f_{ds}}} = \sqrt{\frac{2 \times \pi^2 \times 29000}{65}} = 94$$

And the allowable compression stress is calculated by Equation 15-14:

$$F_a = \frac{\left(1 - \dfrac{72.6^2}{2 \times 94^2}\right) \times 65}{\dfrac{5}{3} + \dfrac{3}{8} \times \dfrac{72.6}{94} - \dfrac{1}{8} \times \dfrac{72.6^3}{94^3}} = 24 \text{ ksi}$$

The nominal compression capacity results:

$$P_u = 1.90AF_a = 1.90 \times 15.6 \times 24 = 711 \text{ kips}$$

The Euler stress in the x-axis is:

$$F_{ex} = \frac{12\pi^2 E}{23\lambda_x^2} = \frac{12 \times \pi^2 \times 29000}{23 \times 34^2} = 129 \text{ ksi}$$

In this equation, 12/23 is the factor of safety for buckling. And the unfactored Euler load in the x-axis results:

$$P_{ex} = \frac{23}{12} AF_{ex} = \frac{23}{12} \times 15.6 \times 129 = 3857 \text{ kips}$$

The tension capacity is:

$$P_p = A\ f_{ds} = 15.6 \times 65 = 1014 \text{ kips}$$

The plastic moment was calculated before as:

$$M_{pl,x} = Z_x f_{ds} = \frac{77.9 \times 65}{12} = 422\ kip - ft$$

This value was calculated assuming the member was completely braced, but this beam is laterally unsupported because the girts cannot carry any axial load after the blast. Hence, the unbraced flexural strength is:

$$M_{m,x} = \left(1.07 - \frac{L/r_y\sqrt{f_{ds}}}{3160}\right) M_{pl,x} = \left(1.07 - \frac{15 \times 12/2.48 \times \sqrt{65}}{3160}\right)$$
$$\times 422 = 0.88 \times 422 = 371 \text{ kip} - \text{ft}$$

The first interaction curve that the system should satisfy is:

$$\frac{P}{P_u} + \frac{C_{mx} M_x}{\left(1 - \frac{P}{P_{ex}}\right) M_{mx}} = \frac{13.1}{636} + \frac{1 \times 236}{\left(1 - \frac{13.1}{3857}\right) \times 371}$$
$$= 0.015 + 0.64 = 0.65 \leq 1$$

where

$$C_{mx} = 1$$

All the other parameters were defined previously.

The second interaction curve to accomplish is:

$$\frac{M_x}{M_{pl,x}} = \frac{236}{422} = 0.56 \leq 1 \text{ for } \frac{P}{P_p} = \frac{13.1}{1014} = 0.013 < 0.15$$

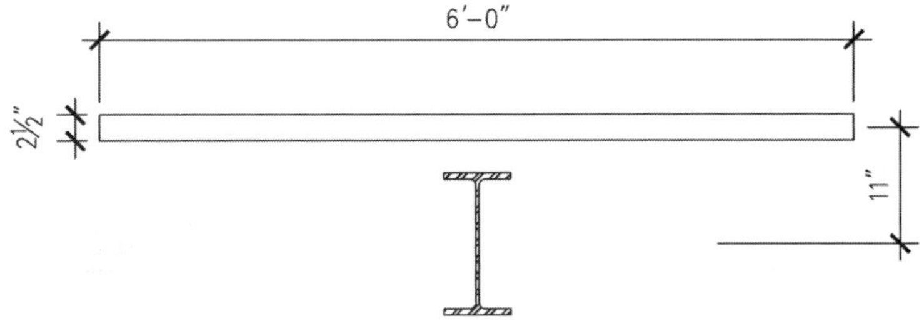

Figure 15.13 Composite Beam Section

This calculation does not consider the reduced stiffness due to the interaction of bending and compression. For higher compression loads and if the moment magnifier is bigger than 10 percent, the analysis should consider second-order effects.

Composite Beam In this example, the composite beam shown in Figure 15.13 will be designed for the roof blast load. The structure is a 25-ft span composite beam, W14 × 22 with a $5^1/_2$ slab with a 3″ metal deck. The concrete strength is 3.0 ksi. For blast design, this strength is multiplied by 1.12. The steel used is Grade 50, as in the previous examples.

The section properties based on the blast strength for the concrete and the steel are:

$$I = 830 \text{ in.}^4$$
$$M_{pl} = 395 \text{ kip} - \text{ft}$$

For rebound purposes, the moment capacity of the steel, based on $Z = 33.2 \text{ in}^3$, is:

$$M_{pl,steel} = Zf_{dy} = \frac{33.2(65)}{12} = 180 \text{ kip} - \text{ft}$$

The load to be used for this example is shown in Figure 15.14.

Finally, assume that the existing dead load is 60 psf.

The total weight for the beam is:

$$w = \frac{60 \times 25 \times 6}{1000} = 9 \text{ kips}$$

The peak load for the 12.2 psi peak blast pressure is:

$$F_{peak} = \frac{12.2 \times 12 \times 12}{1000} \times 6 \times 25 = 263 \text{ kips}$$

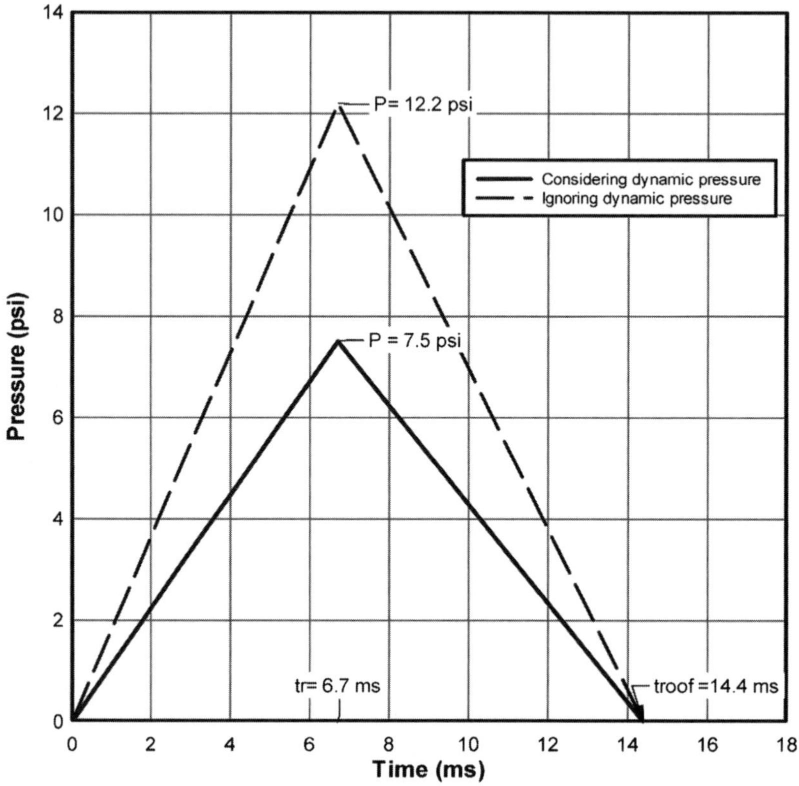

Figure 15.14 Roof Blast Load

Based on the single-degree-of-freedom simplification used for mass and loads uniformly distributed in the plastic range, the load and the stiffness are multiplied by the load factor $K_L = 0.50$, and the mass is multiplied by the mass factor $K_M = 0.33$. Therefore, the parameters used in the discrete system are:

$$F_{peak,SDOF} = 0.5 \times 263 = 131.5 \text{ kips}$$
$$w_{SDOF} = 0.33 \times 9 = 2.97 \text{ kips}$$

The maximum elastic deflection is:

$$\Delta_{el} = \frac{5}{48}\frac{M_{pl}L^2}{EI} = \frac{5}{48}\frac{395 \times 12 \times (25 \times 12)^2}{29000 \times 830} = 1.85 \text{ in.}$$

The parameters to use in the dynamic calculation are:

$$R_{yield} = \frac{8M_{pl}}{L} = \frac{8 \times 395}{25} = 126 \text{ kips}$$

For the design of this beam, the existing dead load is applied simultaneously with the blast load and will reduce the beam capacity.

$$R_{yield,reduced} = 126 - 9 = 117 \text{ kips. For the SDOF}$$

$$R_{yield,SDOF} = 117 \times 0.5 = 58.5 \text{ kips}$$

$$K = \frac{R_{yield}}{\Delta_{el}} = \frac{126}{1.85} = 68 \text{ kips/in. For the SDOF}$$

$$K_{SDOF} = 68 \times 0.5 = 34 \text{ kips/in.}$$

The period of the system is:

$$T = 2\pi\sqrt{\frac{m_{SDOF}}{K_{SDOF}}} = 2\pi\sqrt{\frac{w_{SDOF}}{gK_{SDOF}}} = 2\pi\sqrt{\frac{2.97}{386 \times 34}} = 0.09 \text{ sec}$$

For this example, the beam period is less than 10 times the load duration,0.09 sec $< 10(0.0144) = 0.144$ sec; hence, the impulse formulation cannot be used.

For the rebound

$$R_{yield,steel} = \frac{8M_{pl,steel}}{L} = \frac{8 \times 180}{25} = 57.6 \text{ kips}$$

The elastic deflection for the noncomposite steel beam is:

$$\Delta_{el,steel} = \frac{40}{384}\frac{M_{pl,steel}L^2}{E\,I_{steel}} = \frac{40}{384}\frac{180 \times 12 \times (25 \times 12)^2}{29000 \times 199} = 3.50 \text{ in.}$$

The maximum negative force is:

$$R_{\text{Rebound},SDOF} = 0.5 \times 57.6 \text{ kips} = 28.8 \text{ kips}$$

Therefore, the elastic stiffness at the rebound is:

$$K_{\text{Rebound},SDOF} = \frac{R_{\text{Rebound},SDOF}}{\Delta_{\text{el,steel}}} = \frac{28.8}{3.50} = 8.22 \text{ kips/in.}$$

The period at the rebound is:

$$T = 2\pi\sqrt{\frac{m_{SDOF}}{K_{\text{Rebound},SDOF}}} = 2 \times \pi \times \sqrt{\frac{2.97}{386 \times 8.22}} = 0.19 \text{ sec}$$

For the rebound, the beam period is greater than 10 times the load duration,0.19 sec > 0.0144 sec; hence, the impulse formulation can be used for the rebound response.

The parameters for the load used in the analysis were previously calculated:

$$w_{SDOF} = 2.97 \text{ kips} \quad F_{peak,SDOF} = 131.5 \text{ kips}$$

The load is applied follow the timing shown in Figure 15.14.

The yield and stiffness were obtained previously as:

$$R_{yield,SDOF} = 58.5 \text{ kips} \quad K_{SDOF} = 34 \text{ kips/in.}$$

And the rebound properties are:

$$R_{\text{Rebound},SDOF} = 28.8 \text{ kips} \quad K_{\text{Rebound},SDOF} = 8.22 \text{ kips}/\text{in}$$

Using nonlinear SDOF software, the maximum displacement is (see Figure 15.15):

$$\Delta_{SDOF} = 1.84 \text{ in.} \quad \Delta_{SDOF,\text{Rebound}} = 3.61 \text{ in.}$$

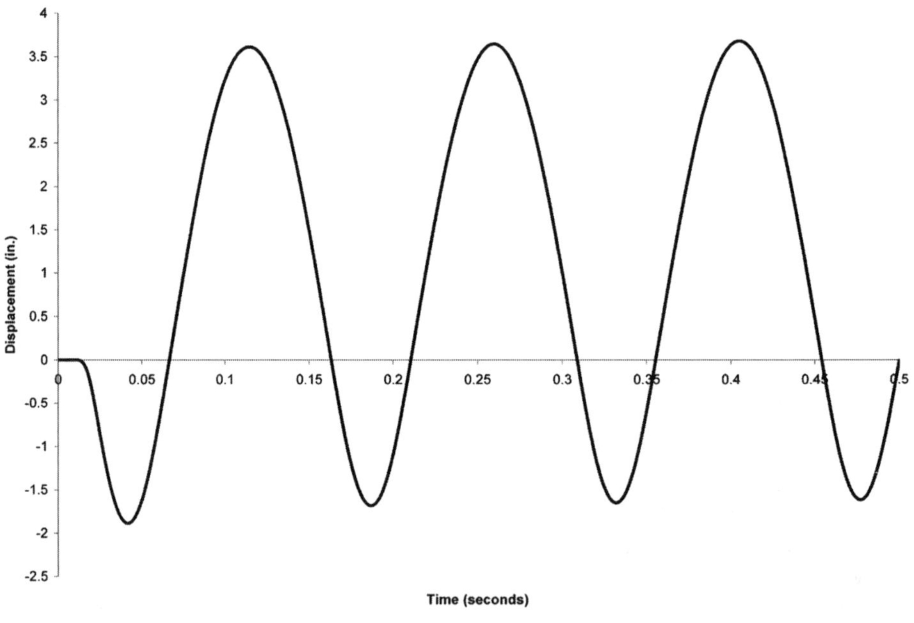

Figure 15.15 SDOF Displacement

The dead load deflection should be added to the solution obtained by SDOF.

$$\Delta_{dead} = \frac{5}{384}\frac{WL^3}{EI} = \frac{5}{384}\frac{9 \times (25 \times 12)^3}{29000 \times 830} = 0.13 \text{ in.}$$

The maximum deflection results:

$$\Delta_{\max} = \Delta_{SDOF} + \Delta_{dead} = 1.84 + 0.13 = 1.97 \text{ in.}$$

Hence, the ductility results:

$$\mu = \frac{\Delta_{\max}}{\Delta_{el}} = \frac{1.97}{1.85} = 1.06$$

For the rebound, the ductility is:

$$\mu = \frac{3.61}{3.50} = 1.03$$

For the SDOF solution the element behaves as a composite member and yields during the rebound. Figure 15.16 shows the force resultant from the SDOF system.

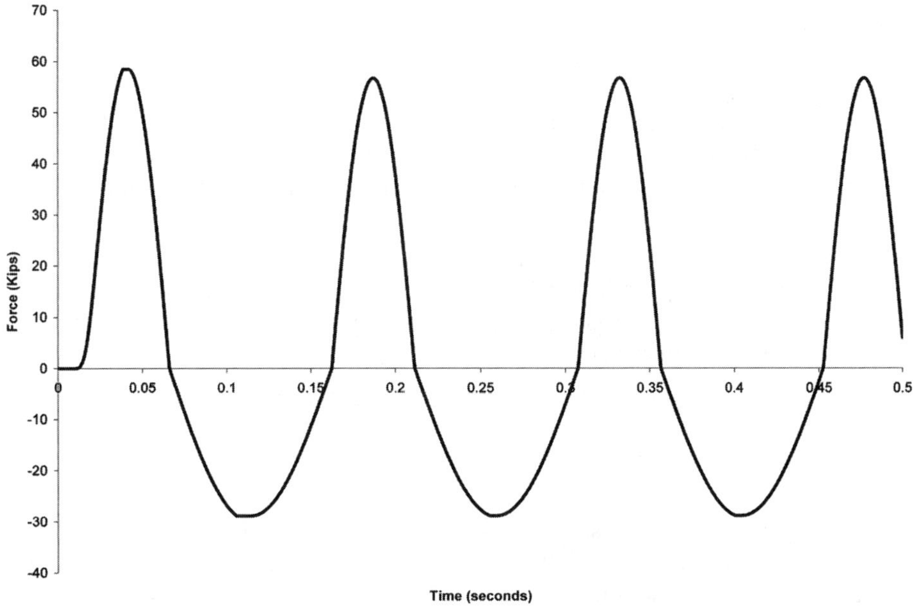

Figure 15.16 SDOF Force

The maximum shear force in plastic design can be obtained as the shear reaction that plastifies the section:

$$V = \frac{R_{yield}}{2} = \frac{126}{2} = 63 \text{ kips}$$

The dynamic shear capacity is:

$$V_p = f_{dv} A_w = 36 \times 2.76 = 99.4 \text{ kips} > 63 \text{ kips}$$

where f_{dv} is the blast shear capacity
A_w is the area of the web

Hence, the section can support the shear. The connection should be designed for this capacity

15.7 DESIGN OF CONNECTIONS

Connections require as much attention as members, especially in blast design. Failures of steel structures under overload can initiate at the connections of members, as opposed to the members themselves, for many reasons:

- The design of members is often controlled by considerations other than strength, including deflection, vibration control, and architectural requirements. Such members are often substantially stronger than required to resist the design forces. Connections, however, typically can be designed based only on considerations of strength. When this is the case, there may be less overstrength or reserve capacity to match the overstrength of the members.
- Connections may be controlled by limit states with varying levels of ductility. When less ductile limit states, such as tension rupture or block shear rupture control, there is less capacity for redistribution of forces and the mobilization of ductile behavior.
- Connections tend to be of limited size and, therefore, even when exhibiting ductile behavior, they can only accommodate limited plastic deformation before reaching their ultimate capacities

It is important in the design of blast-resistant structures, particularly those expected to be loaded into the inelastic range of behavior, that sudden failure modes be avoided so that the plastic response of the structure can be mobilized. It is also important to remember that blast loading may result in load reversal,

Table 15.3 Dynamic Increase Factors for Connection Design

Connection Material	Action	Yield Strength	Ultimate Strength
Plate, A36, A572	Bending/shear	1.3	1.1
	Tension/compression/bearing	1.2	1.1
Plate A588	Bending/shear	1.2	1.05
	Tension/compression/bearing	1.12	1.05
Bolts	Shear/tension	1.0	1.0
Weld metal	Same as base metals		

sometimes in the form of rebound. Connections should be designed with equal strength under load reversal, unless the following apply:

1. Member strength is limited in one direction of load application by considerations of buckling, as will occur for braces in compression and flexural members with one flange braced and the other unbraced.
2. Nonlinear dynamic analysis is performed to determine the maximum connection loads in both positive and negative loading applications.

Connections should be designed to develop the full plastic capacity of the supported members, so that the plastic response of the structure can be mobilized in resisting blast-induced stresses. It should be noted that the dynamic plastic capacity of an element, loaded briefly by impulsive loading, is often greater than the static plastic capacity. Alternatively, when nonlinear dynamic analysis of structures and structural elements under blast loading is performed, it is permissible to design the connections for the peak forces obtained from the analysis.

The specified yield and tensile strength of the connection material, including bolts, welds, angles, and plates, may be increased in accordance with Table 15.3 to account for the Dynamic Increase Factor (DIF). Note that Strength Increase Factors (SIF), discussed previously for members, are not used in the design of connections as an additional means of ensuring that connections will be capable of developing the strength of the member.

REFERENCES

Air Force Engineering and Services Center. 1989. *Protective Construction Design Manual*. (ESL-TR-87-57). Tyndall Air Force Base, FL: Air Force Engineering and Services Laboratory.

American Institute of Steel Construction. 2005a. *Seismic Provisions for Structural Steel Buildings*. Chicago, IL: American Institute of Steel Construction.

________. 2005b. *Specification for Structural Steel Buildings*. Chicago, IL: American Institute of Steel Construction.

American Society of Civil Engineers, Petrochemical Committee, Task Committee on Blast Resistant Design. 1997. *Design of Blast Resistant Buildings in Petrochemical Facilities*. New York: American Society of Civil Engineers.

ASTM. 2006. *Standard Specification for Structural Steel Shapes* (A992). West Conshohocken, PA: ASTM International.

________. 2007a. *Standard Specification for Pipe, Steel, Black and Hot-Dipped, Zinc-Coated, Welded and Seamless* (A53). West Conshohocken, PA: ASTM International.

________. 2007b. *Standard Specification for Cold-Formed Welded and Seamless Carbon Steel Structural Tubing in Rounds and Shapes* (A500). West Conshohocken, PA: ASTM International.

________. 2008. *Standard Specification for Carbon Structural Steel* (A36). West Conshohocken, PA: ASTM International.

Departments of the Army, the Navy and the Air Force (DOANAF). 1990. *Structures to Resist the Effects of Accidental Explosions* (Department of the Army Technical Manual TM 5–1300). Washington, DC: Departments of the Army, the Navy and the Air Force.

Federal Emergency Management Agency. 2000. *Prestandard and Commentary for the Seismic Rehabilitation of Buildings* (FEMA 356). Washington, DC: Federal Emergency Management Agency.

Gilsanz, Ramon *et al.* 2009. *Defensive Design of Structural Steel Buildings* (AISC Design Guide No. 27), Chapters 6 and 7. Chicago, IL: American Institute of Steel Construction.

Shipe, James A. and Charles J. Carter. 2003. Defensive design. *Modern Steel Construction* 43 (11): 25–31.

Unified Facilities Criteria Program. 2002. *Design and Analysis of Hardened Structures to Conventional Weapons Effects (FOUO)* (UFC 3-340-01). Washington, DC: U.S. Department of Defense, Unified Facilities Criteria Program.

________. 2008. *Structures to Resist the Effects of Accidental Explosions* (UFC 3-340-02). Washington, DC: U.S. Department of Defense, Unified Facilities Criteria Program.

16 Blast-Resistant Design Concepts and Member Detailing: Masonry

Shalva Marjanishvili

Prior to the twentieth century, unreinforced masonry was the primary building material for both residential and commercial construction. As a result, unreinforced masonry structures constitute a large portion of existing buildings, many of which have historical and architectural importance.

Beginning in the early 1980s with terrorist attacks in Lebanon and Kuwait, high-profile and high-risk civilian federal buildings have been designed to resist the effects of explosive attacks. With each major attack on U.S. interests, more attention has been paid to this low probability/high consequence threat scenario. As a result, recent work has been dedicated to the design and analysis of masonry structures to resist air-blast loads.

When a wall is first subjected to an air-blast load, it experiences out-of-plane flexure, producing tensile strains on the interior face of the wall and compressive strains on the exterior. Once the maximum positive deflection associated with the out-of-plane flexure has been attained, the wall will begin to vibrate due to rebound forces created by negative blast pressures. The wall undergoes negative deflection and curvature, causing tensile strains to develop on the exterior face of the wall, and increasing shear stresses at wall supports.

Concrete masonry unit construction is primarily used for walls. These walls, when properly designed, will provide economical resistance to relatively low air-blast pressures. Limitations of masonry walls include development of large deformations and rebound response.

Most masonry design codes advocate elastic design to specified stress levels, even for dynamic loads. This approach is uneconomical and potentially unsafe because of its inability to predict structural response beyond the allowable stress limit. Masonry walls, when subjected to air-blast loads, are expected to undergo large post-elastic response, and therefore, they should be designed using the strength design methods.

Below are two primary assumptions made when assessing flexural strength of reinforced masonry:

Assumption 1.

Singly reinforced masonry units rely on composite action between compression in concrete and tension in reinforcement.

a. Plain sections remain plain during bending. At large deflections, significant deviation from this assumption may occur. This can be corrected by including both bending and shear deformations in the analysis.
b. There is a perfect bond between grout and reinforcement.
c. Concrete tension strength is neglected.
d. Concrete stress block is idealized as a rectangle.
e. Reinforcement stress/strain is idealized as elastic-perfectly plastic with appropriate strain-rate hardening effects.
f. Flexural strength is attained when either tension reinforcement or extreme fibers of concrete reach their ultimate strains.

Assumption 2.

Under extreme loading such as air blast, it is expected that the compressive strength of masonry will be exceeded. When this happens, the resistance is provided purely by compression and tension in the reinforcement. This approach requires that masonry be reinforced with two curtains of reinforcement. Therefore, doubly reinforced masonry units, under this assumption, rely on compressive and tension forces in the reinforcement only.

a. Plain sections remain plain during bending.
b. Reinforcement is assumed to be held in its original position, as if the masonry were undamaged.
c. Concrete tension and compression strengths are neglected.
d. Reinforcement stress/strain relationship is idealized as elastic-perfectly plastic with appropriate strain hardening effects.
e. Maximum flexural strength is attained when either curtain of reinforcement yields either in tension or in compression.

Assumption 1 is appropriate only for level of protection (LOP) I, the superficial damage level. For other damage levels (LOP II–LOP IV), it is necessary to use Assumption 2. It is worth noting that reinforced masonry will often behave under Assumption 2.

When a masonry wall is subjected to relatively small air-blast pressures with low performance requirements (LOP I), ACI 530-08, Section 3.3.5 (American Concrete Institute 2008) can be used to estimate out-of-plane moment capacity of the wall, with the following modifications:

1. ϕ strength reduction factors shall be taken as one (i.e., $\phi = 1.0$).
2. Yield strength of reinforcement shall be taken as dynamic (i.e., $f_y = f_{dy}$).
3. Compressive strength of masonry shall be taken as dynamic (i.e., $f'_m = f'_{dm}$).

UFC 3-340-02 (Unified Facilities Criteria 2008) suggests the following equation for estimating ultimate moment capacity under Assumption 2:

$$M_u = A_s f_{dy} d_c \qquad \text{(UFC 3-340-02, Equation 6.2)}$$

where: A_s = area of joint reinforcement at one face
f_{dy} = dynamic yield strength of reinforcement
d_c = distance between centroids of compression and tension reinforcement

16.1 GENERAL CONSIDERATIONS

Reinforced masonry subjected to service design loads (excluding air-blast) can be designed using both allowable and strength design methodologies. For air-blast design, only strength design methodology should be used, since reinforced masonry elements subjected to air-blast loads are likely to experience large nonlinear deformations with relatively large strain and ductility demands. Therefore, it is necessary to employ a design methodology that can account for response well beyond linear limit states. Working stress design, although much simpler then strength design, only accounts for response below nonlinear range, and implies that all reinforced masonry should be designed linearly. Furthermore, air-blast design requires that structural elements respond in a ductile manner, with large nonlinear deformations. For these reasons, allowable stress design should not be used for air-blast design of reinforced masonry.

Although unreinforced masonry should not be used for new construction, there are many historic buildings with existing unreinforced masonry walls. When existing unreinforced masonry walls cannot be replaced, the stability of existing unreinforced masonry walls should be assessed. The stability assessment should assume that the wall is simply supported if flexible supports are provided. It should be permissible to assume arching action in the masonry wall if rigid supports are provided.

There will be cases when air-blast design requirements contradict or largely exceed the conventional masonry design required by building codes. The design engineer should verify that the final masonry design meets all design load requirements, including blast design requirements. If the blast design requirements countermand or contradict other masonry design constraints, the owner and architect should be appraised of the issue by the design professional, and a decision should be reached by the group in order to resolve the conflicting requirements.

The design engineer should direct the detailing on the design documents such that all requirements of the blast-resistant design reworking and renovation of an existing structure are met. The design engineer should verify that the design documents clearly detail any required repairs of the existing structure, modifications to the existing structure, or further work to the existing structure that is necessary to achieve the LOP required of the modified existing structure.

16.1.1 Masonry

Maximum masonry strength, $f'm$, should not exceed 6,000 psi for any LOP in new construction. This strength limitation is intended to preclude nonductile failure modes, such as shear and/or crushing.

Minimum masonry strength, $f'm$, should not be less than 1,500 psi for any LOP in new construction.

The design engineer should determine the material properties of existing masonry structural elements to ensure proper design to achieve the LOP required for the existing masonry structure.

UFC 3-240-02 suggests the following equations to determine effective moment of inertia:

$$I_{eff} = I_a = \frac{I_n + I_{cr}}{2} \qquad \text{(UFC 3-340-02, Equation 6-6)}$$

$$I_{cr} = 0.005bd_c^3 \qquad \text{(UFC 3-340-02, Equation 6-7)}$$

While ACI 530-08 suggests that I_{eff} is approximately equal to one-half of gross moment of inertia, it also suggests the following equation for a more accurate estimate of effective moment of inertia:

$$I_{eff} = I_n \left(\frac{M_{cr}}{M_a}\right)^3 + I_{cr}\left[1 - \left(\frac{M_{cr}}{M_a}\right)^3\right] \leq I_n \leq 0.5I_g$$

(ACI 530-08, Section 3.5.3)

This handbook recommends the ACI approach when using Assumption 1, and the UFC approach when using Assumption 2. This recommendation is based on the understanding that the UFC document covers reinforced masonry response due to extreme loading conditions, while ACI-350-08 is concerned mainly with elastic response.

16.1.2 Reinforcement

All reinforcement should comply with ASTM A615 (Grade 60) or A706 for all LOPs in new construction (ASTM 2006, ASTM 2008a).

It is desirable to use reinforcement with high ductility demand capability. Currently only ASTM A706 reinforcement is considered to have high ductility capacity. However, according to UFC 3-340-02, it is sufficient to use ASTM A615 (Grade 60).

The modulus of elasticity of reinforcement should be taken as required by ACI 350-08.

At strain rates characteristic of air-blast response, reinforcing steel exhibits a significant increase in yield strength above static test values. For 60-ksi reinforcement, UFC 3-340-02 recommends a dynamic increase factor of 1.17, that is:

$$f_{dy} = 1.17 f_y = 1.17 \times 60 \text{ ksi} = 70.2 \text{ ksi}$$

Reinforcement splices should be limited to zones that remain elastic during blast response. Furthermore, not more than 50% of splices should be located along a single line. That is, splices should be staggered along the length or height of the wall and away from potential yield lines.

Splices should develop full tensile capacity of reinforcement and should conform to ACI 350-08. It is possible to use mechanical splices; however, the use of mechanical splices is limited by the geometric constraints of concrete masonry units. Use of splices should be limited.

16.1.3 Mortar

Mortar is a mixture of cementitious materials, aggregates and water. All mortars should comply with ASTM Standard C270 (ASTM 2000). Mortars with lime should be avoided, as lime decreases the strength of mortar.

16.1.4 Grout

Grout provides bonding of longitudinal reinforcement and adds strength, rigidity, and mass to the wall system. All grouts are required to conform to ASTM Standard C476 (ASTM 2008b). It is recommended that blast masonry walls be fully grouted.

16.1.5 Construction Methods

Two commonly used types of masonry construction are running bond and stack bond.

With running bond, masonry units are normally staggered by at least one-quarter of the length of the unit. With typical 8-in. masonry units, the joints are offset by at least 4 in., and since the length of the units is 16 in., it is common to offset units by half the length of the unit.

With stack bond, joints in each successive course line up vertically. This type of construction is not desirable in air-blast-resistant construction, since there is little or no continuity across the head joints (the head joint is the joint that runs up and down the wall). The only continuity across the head joints is provided by longitudinal reinforcement and grout.

Conventional construction methods of reinforced masonry walls include running bond and stacked bond with the maximum spacing of vertical reinforcement at 48 in. on center (o.c.). This implies that not every cell is grouted.

For air-blast-resistant design, it is recommended that every cell be grouted and at least one reinforcing bar be placed in each concrete masonry unit. For example, with 16-in. masonry units the maximum reinforcement spacing should not exceed 16 in. o.c., and every cell should be grouted. Since concrete masonry units (CMUs) are of modular construction with a modular dimension of 8 in., reinforcement must be placed in increments of 8 in.

When designing with reinforced masonry, dimensional constraints need to be accounted for as follows:

1. Actual dimensions of masonry units are 3/8 in. less than the nominal dimensions.
2. Reinforcement should be placed at increments of 8 in. o.c. (i.e., 8 in., 16 in.).
3. Since the largest CMU block has a length of 16 in., it is recommended that at least one reinforcing bar be placed in each masonry unit, resulting in a maximum spacing of 16 in. o.c.
4. Masonry unit wall thickness will dictate the minimum cover dimension. It may be very difficult to achieve double curtain reinforcement in 4-in. masonry units.
5. Double curtain reinforcement can be achieved with 8-in. masonry units by staggering longitudinal reinforcement as shown in Figure 16.9

16.2 FAILURE MODES

Failure modes of masonry walls are categorized as either ductile or brittle. Ductile failure is preferred, and is generally characterized by bending failure caused by yielding of tensile reinforcement. Brittle failure modes include shear, buckling, and crushing. Figures 16.1 though 16.3 are photos depicting failures of masonry construction. Figure 16.1 shows infill masonry wall performance when structural collapse is initiated. Figure 16.2 depicts infill masonry wall failure without collapse of bearing elements. In this case, the masonry wall could potentially become heavy debris thrown into the occupied rooms. Both Figures 16.1 and 16.2 show masonry wall performance under extremely high air-blast loads, while Figure 16.3 shows reinforced masonry wall performance under moderate to low air-blast pressures. Although the masonry wall shown in Figure 16.3 experienced some cracking and stucco spalling, it performed satisfactorily. These figures illustrate how reinforced masonry walls could provide economical protection under relatively moderate air-blast pressures.

Table 16.1 correlates qualitative damage description with the damage levels superficial, moderate, heavy, and hazardous and their corresponding levels of protection from IV though I.

The first step in air-blast design of masonry walls is to establish the desirable performance level in terms of acceptable damage levels, as shown in Table 16.1. Then, dynamic analysis is conducted to calculate damage indicators, as described under each column for corresponding damage levels. If it is determined that the damage indicators are exceeded for a desired performance level, the

Figure 16.1 Failure of Building with Masonry Infill Walls

Figure 16.2 Failure of Masonry Infill Wall

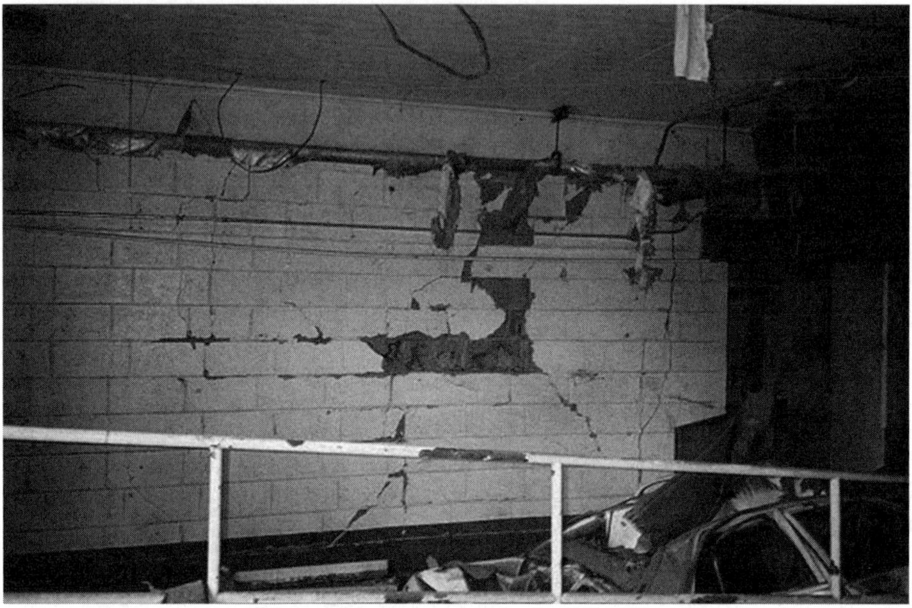

Figure 16.3 Crack Pattern on Masonry Infill Wall

wall must be redesigned and re-analyzed. Thus, air-blast design is an iterative process.

To simplify air-blast analysis and design procedures, often a single-degree-of-freedom (SDOF) analysis is conducted with a predetermined deflected shape of the wall. The results of this SDOF analysis are usually obtained as ductilities and support rotations. These calculated ductilities and rotations are then compared to allowable values, to determine damage limits (see Chapter 3, section 3.5 of this book for allowable ductilities and rotations for reinforced masonry).

Table 16.1 also lists performance indicators for unreinforced masonry. Although unreinforced masonry is not recommended in new construction, there are many existing buildings with unreinforced masonry facades which may require evaluation under air-blast loads.

16.2.1 Flexure

Flexure provides a ductile failure mode. This is the preferred failure mechanism and occurs in properly designed and supported masonry. This can be achieved by designing masonry for the balanced condition whereby reinforcement will yield just before the masonry concrete begins crushing. Figure 16.4 depicts typical deflected shape due to flexural response.

When air-blast pressures are relatively low, it may be possible to design a reinforced masonry wall with just one curtain of reinforcement located in the middle.

Table 16.1 Qualitative Damage Expectations for Masonry

Element	Limit State	Superficial LOP IV	Moderate LOP III	Heavy LOP II	Hazardous LOP I
Reinforced	Multiple	Masonry is effective in resisting moment and shear. The cover over the reinforcement on both surfaces of the wall remains intact. System should essentially remain elastic. Axial load-bearing elements maintain bearing capacity. No connection or shear failure. Compressive strain in masonry does not exceed 0.003.	Masonry is crushed and not effective in resisting moment; however, shear resistance remains though interlocking. Compression reinforcement equal to the tension reinforcement is required to resist moment. The cover over the reinforcement on both surfaces of the wall remains intact. No spalling or scabbing of masonry. Axial load-bearing elements maintain bearing capacity. No connection or shear failure. Compressive strain in masonry should not exceed 0.007.	The concrete cover over the reinforcement on both surfaces of the element is completely disengaged. Equal tension and compression reinforcement that is properly tied together are required to resist moment Limited spalling and scabbing of masonry. Axial load-bearing elements maintain bearing capacity. No connection or shear failure. Compressive strain in masonry should not exceed 0.010.	Masonry is completely disengaged. Significant spalling and scabbing of masonry. Axial load-bearing elements maintain bearing capacity. No connection or shear failure. Compressive strain in masonry exceeds 0.010.
Unreinforced	Multiple	Not permitted	Not permitted	Significant spalling and scabbing of masonry. No connection or shear failure. Compressive strain in masonry does not exceed 0.003.	Significant spalling and scabbing of masonry. No connection or shear failure. Compressive strain in masonry does not exceed 0.007.

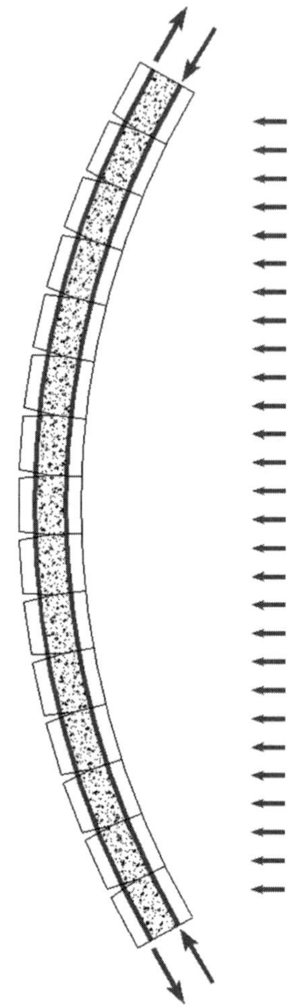

Figure 16.4 Flexural Response Mode of Reinforced Masonry

In this case, Assumption 1 is used to determine the bending capacity of the masonry. The wall should still be designed so that tensile yielding of reinforcement occurs before the compressive block of masonry starts crushing. This will result in a relatively weak but ductile wall.

When air-blast pressures are relatively high (in excess of 10 psi), it may be necessary to design the reinforced masonry wall with a double curtain of reinforcement, using Assumption 2. In this case, it is desirable to place equal amounts of reinforcement on each side of the wall. Furthermore, it may be necessary to increase the thickness of the masonry wall to accommodate two curtains of reinforcement. This can, however, result in a stronger and more rigid wall, which may become susceptible to diagonal and direct shear failure. Thus, there should be an analysis check, and if necessary additional design iterations, to confirm that the wall is designed in such a way that it reaches the flexural capacity limit before the shear capacity limit, when subjected to load distribution similar to air-blast load.

16.2.2 Diagonal Tension Shear

Diagonal tension shear failure occurs when a masonry wall response reaches the diagonal tension resistance limit before the bending capacity is exhausted. Figure 16.5 depicts typical deflected shape corresponding to diagonal tension shear response. Diagonal tension failure exhibits very little ductility capacity and therefore is brittle in nature. Because of this, diagonal tension failure should

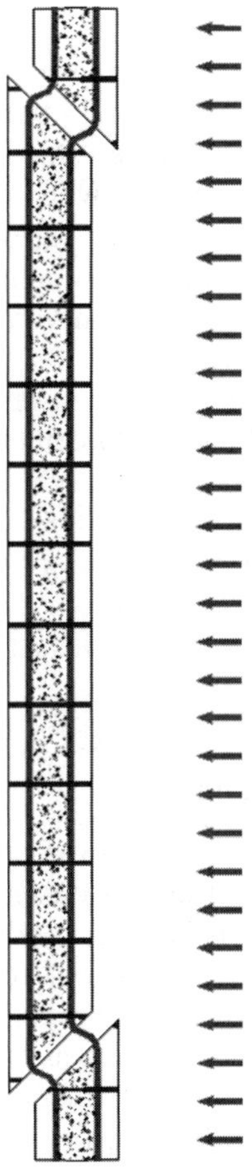

Figure 16.5 Diagonal Shear Response Mode of Reinforced Masonry

be avoided in blast-resistant masonry walls. One method to avoid diagonal tension failure mode is to decrease the flexural capacity of the wall in such a way that the wall undergoes flexural yielding before shear capacity is exhausted. However, the flexural resistance of the wall should not be reduced when doing so will compromise the desired level of protection. Alternatively, the diagonal tension capacity of the masonry wall can be increased by increasing the thickness of the masonry, by grouting more cells, or by placing shear reinforcement. Placement of shear reinforcement is feasible only with two curtains of reinforcement. If shear reinforcement is required, it should be placed in such a way that it crosses expected diagonal shear crack between the two curtains of reinforcement.

Diagonal tension capacity for out-of-plane bending of reinforced masonry walls can be assessed using ACI-350-08 strength design procedures.

16.2.3 Direct Shear

Direct shear occurs when explosion occurs very close to the wall and when mortar joints are too weak to provide shear stress flow continuity across the masonry units. Figure 16.6 depicts typical deflected shape consistent with diagonal shear response. Direct shear may be of concern when scaled distance is less then 4. At this standoff it is expected that air-blast pressures will be of very high magnitude and extremely short duration. If the air-blast pressures are of extremely short duration, there will be no time for the masonry wall to respond in flexure mode, and direct shear response will dominate.

Direct shear capacity depends on friction resistance across the mortar joint and dowel action of the longitudinal reinforcement. Direction shear strength generally exceeds diagonal shear strength. The direct shear failure mode is almost always very brittle and should be avoided. One way to avoid direct shear failure is to increase the standoff between the masonry wall and the expected explosion source. Furthermore, since reinforced masonry walls do not perform well under extremely high air-blast pressures, direct shear failure modes should be avoided whenever possible.

16.2.4 Breach and Spall Phenomena

Breach occurs when a weapon is placed directly against or very close to the masonry wall surface. Detonation at this close proximity causes shattering of the masonry units. Breach analysis is often conducted using computational fluid dynamic codes with appropriate equations of state, or by experimental studies. Breach effects can be mitigated by masonry wall thickness, proper confinement, and application of anti-spall laminates such as fiber-reinforced polymer (FRP).

Spall typically occurs on the back side of the wall when a weapon is placed at the breaching distance but does not cause full breach. Spall occurs due to a compression wave traveling through the thickness of the masonry.

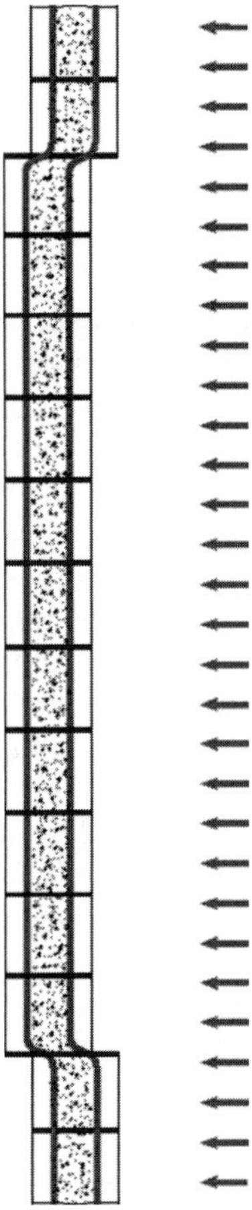

Figure 16.6 Direct Shear Response Mode of Reinforced Masonry

When reflected from the back surface of the wall, this causes through-thickness tensile stresses at the back face that can exceed the tensile strength of the masonry.

Both breach and spall will cause fragmentation on the back side of the wall. Spall can be mitigated by application of shielding (laminate material) such as FRP or so-called "catchment systems" such as geotextile fabric to stop fragments. Laminate shielding is permanently adhered to the wall surface (on the protected side of the wall) and may act compositely with the masonry, while catching systems are basically a curtain or fabric located on the protected side of the wall to catch debris generated by failed masonry. Both systems provide protection by the development of membrane resistance, and therefore they should be connected to supports (normally floor slabs) in a way that allows full development of the membrane capacity of the base material. This is done to avoid premature failure of the system. Fragmentation effects are discussed in the detail in Chapter 8 of this book.

16.3 REINFORCED MASONRY DETAILING

The detailing guidelines presented in this section are intended to promote the desired structural behaviors discussed in Chapter 9.

Since masonry walls resist air-blast loads primarily though nonlinear deformations, it is important that wall supports develop the full flexural capacity of the wall. Reinforced and unreinforced walls can be analyzed using single-degree-of-freedom (SDOF) methods or pressure-impulse (P-I) diagrams. Axially loaded walls should be designed considering slenderness effects.

Non-load-bearing partition walls should be reinforced and grouted, as required, to resist the applied blast loads. Alternatively, debris catch systems should be used to minimize the debris impact hazard. The effectiveness of the debris catch systems should be determined through applicable explosive test data or advanced finite element analysis.

The design shear forces should not be less than the shear forces associated with the ultimate flexural strength of the element. The shear capacity of reinforced masonry wall structures can be increased by placement of shear reinforcement. Shear reinforcement should be detailed in accordance with UFC 3-340-02.

Masonry wall structures should be sufficiently reinforced to resist the in-plane and out-of-plane response to dynamic blast loads. It should be permissible to consider arching action for the evaluation of wall response and to use empirical P-I charts based on masonry wall tests; however, the impulse asymptote "layover" (the "bump" at the extreme asymptote) associated with the beneficial effects of negative phase loading may be nonconservative, and should not be considered for design. The use of P-I charts should be limited to flexural modes in response to external blast loads and to wall types that are within the P-I database. Diagonal tension and direct shear capacity must be verified independently. Walls should be designed to develop the shear forces associated with plastic hinging. Connections should be designed, detailed, and constructed such that the governing failure mode of the connection is a ductile mode related to yielding of the steel elements of the connection, such as angles, plates, and rebar. Fracture of bolts and welds should not be the governing failure mode of connections.

This section applies to non-load-bearing masonry walls only. Reinforced concrete masonry units should allow full development of the reinforcing steel. Reinforced concrete masonry unit walls should be designed as either one- or two-way systems using cell and joint reinforcement. Shear resistance can be provided by the concrete masonry unit in combination with steel reinforcement, although the inclusion of this behavior should be considered in light of the assumptions identified at the beginning of this chapter.

16.3.1 General

All concrete masonry units should be fully grouted for LOP III and LOP IV.

All lap splices should be tension lap splices, as specified in ACI 530 and ACI 530.1 (American Concrete Institute 2008).

Use of mechanical splices should be limited to zones that remain elastic during blast response, unless the mechanical splices can be shown to develop the expected tensile strength of the bar under conditions generated by blast loads. All mechanical splices should meet or exceed the specifications in ACI 530 and ACI 530.1.

Use of welded splices should be limited to zones that remain elastic during blast response, unless the welded splices can be shown to develop the expected tensile strength of the bar under conditions generated by blast loads. All welded splices should meet or exceed the specifications in ACI 530 and ACI 530.1.

16.3.2 Longitudinal Reinforcement

For air-blast-resistant design, it is desirable to place longitudinal reinforcement in each cell in fully grouted concrete masonry units. This will result in reinforcement placed at 8 in. o.c. This detailing approach is beneficial since every grouted cell adds weight, rigidity, and shear strength, and closely spaced bars may reduce the size of fragments should the wall fail under air-blast loads. Figures 16.7 though 16.10 provide sample detailing options for reinforced masonry.

16.3.3 Horizontal Reinforcement

A bond beam course with two horizontal reinforcing bars should be placed all around the cap course of the wall. Bars should be spliced and the course grouted, per requirements of ACI 530.

All new construction should have a bond beam course with a minimum of two horizontal reinforcing bars. Reinforcing bars should be located at the level of the lowest attachment of the masonry wall to the floor diaphragm. Bond beam should be placed at any horizontal diaphragm or floor system supported by the wall. If the attachment band in the diaphragm is thicker than the bond beam course, then the bond beam course should be placed on the bottom of the diaphragm, and all cells in courses to the top of the attachment band should be fully grouted. Bars should be spliced and the course grouted per requirements of ACI 530. See Figure 16.11 for examples of bond beam reinforcement.

A bond beam course with two horizontal reinforcing bars should be placed at all lintel locations; the bar should be either hooked around vertical reinforcement on either side of the opening or continuous along the wall, with the course grouted per requirements of ACI 530.

Control joints for all LOPs should be designed, at a minimum, with one vertical bar in a fully grouted cell column on each side of the control joint.

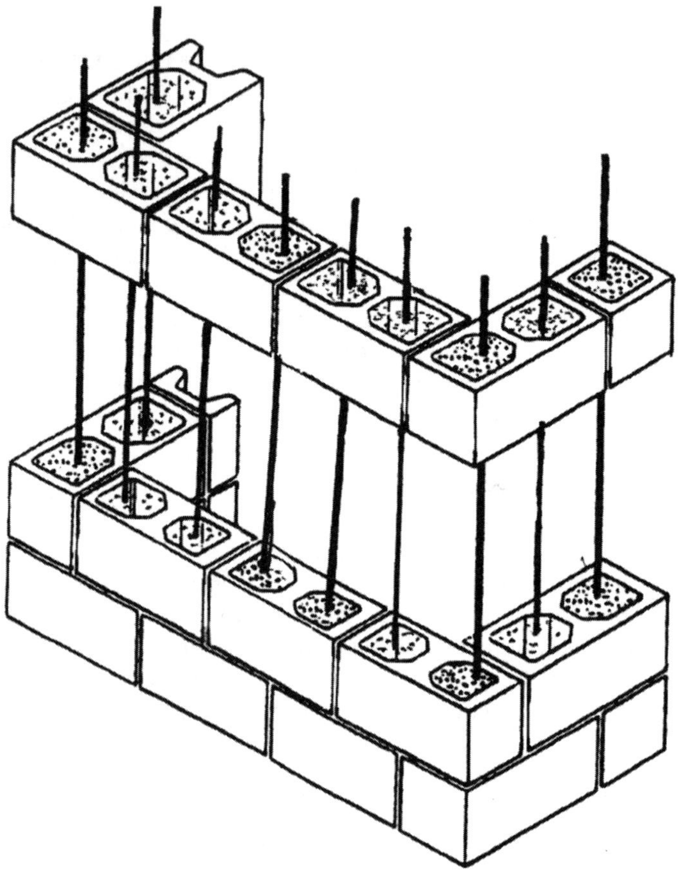

Figure 16.7 Fully Grouted Reinforced Masonry Wall

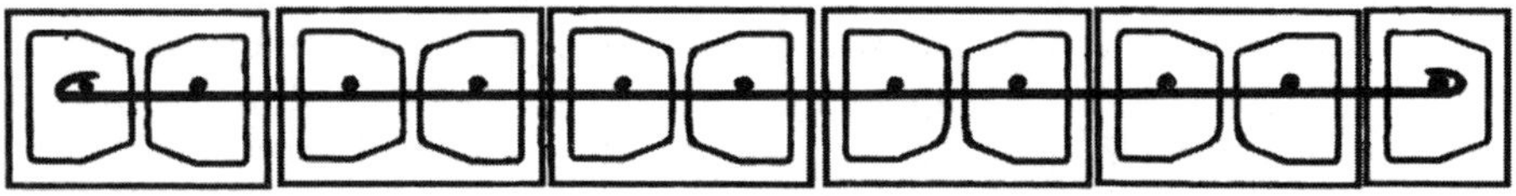

Figure 16.8 Singly Reinforced Masonry Wall

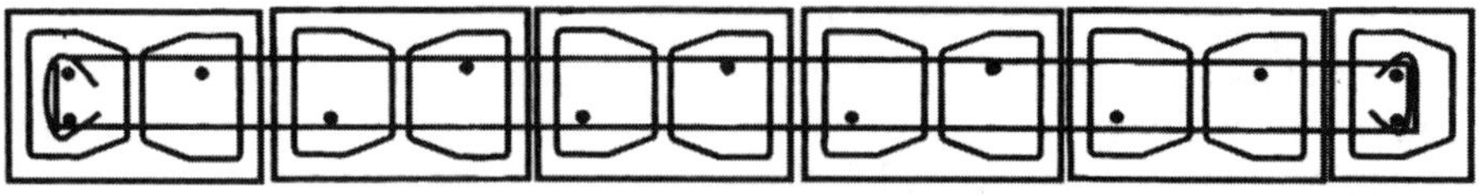

Figure 16.9 Masonry Wall Doubly Reinforced by Staggering Longitudinal Reinforcement

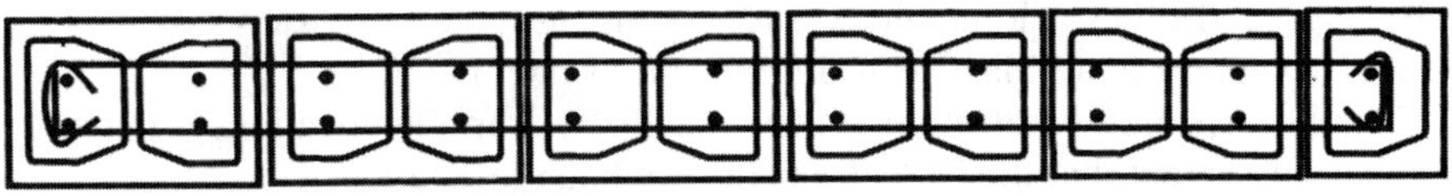

Figure 16.10 Doubly Reinforced Masonry Wall – Two Bars per Each Cell

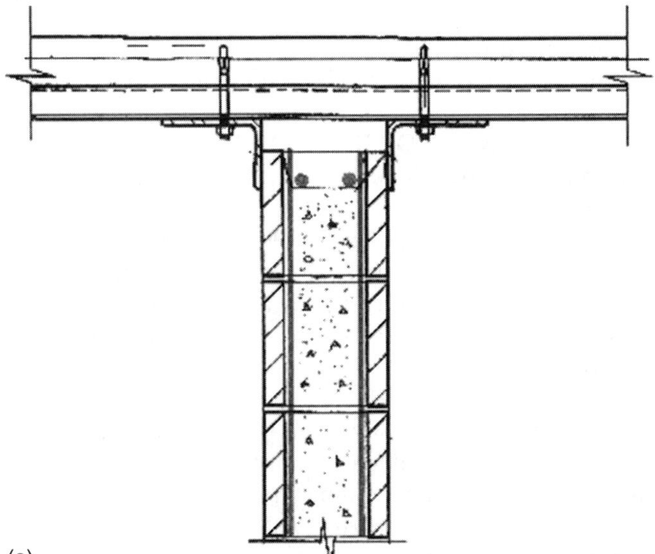

(a)

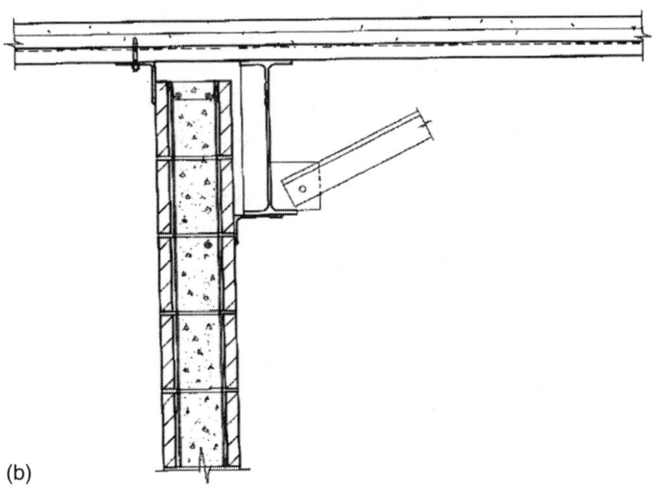

(b)

Figure 16.11 Examples of Bond Beam

Code-approved control joint block and expansion material systems should be used as necessary in all blast-resistant masonry systems.

The design professional should determine the appropriate locations for control joints, to the extent that their locations impact blast resistance.

The design professional should specify continuity and/or end conditions of horizontal bars and bond beams at control joints.

16.3.4 Walls

All elements constructed to LOPs I and II should have at least one continuous vertical reinforcing bar in a fully grouted cell column, with spacing not to exceed 48 in. o.c.

All elements constructed to LOP III and IV should have at least one continuous vertical reinforcing bar per block in a fully grouted cell column, with spacing not to exceed 16 in. oc.

In addition, at least one vertical bar should be placed in a fully grouted cell column at all inside or outside corners, alongside any openings, and in the cell column on each side of control joints, if they are used in the structure

All elements constructed to LOP III and IV should have a minimum vertical reinforcement ratio of 0.0025.

Bearing Walls Bearing walls are subjected to constant axial load in addition to blast loading. The presence of axial load in reinforced masonry increases bending capacity up to a so-called balanced point on a P/M diagram, after which axial load causes rapid deterioration of bearing wall air-blast-resistance capacity, due to buckling. Additionally, the presence of axial load induces larger stresses on block masonry under compression, and may cause the masonry wall to fail in concrete crushing mode rather than through the yielding of the reinforcement (i.e., compression-controlled section).

Careful attention must be paid during the structural analysis of bearing walls to avoid these undesirable failure modes (crushing and buckling).

Slender Walls Slender walls normally carry some axial load, which affects the overall resistance of the wall and should be accounted for in the analysis. All slender wall design should follow ACI 350-08, Section 3.3.5

Infill Walls Infill masonry walls typically act as partitions and are supported by floor slabs or beams. If properly designed they can provide physical separation and protection against air-blast waves and fragments.

16.3.5 Support Connections

Since masonry walls are primary blast-load-resisting systems, and since masonry walls are normally expected to respond to air-blast load in nonlinear regime, it is necessary to provide sufficient support to allow masonry walls to develop full plastic mechanism. This can be achieved by fully developing longitudinal reinforcement into the supports.

16.4 UNREINFORCED MASONRY

Unreinforced masonry, due to its brittle nature, is not recommended in new construction. However, there is a vast amount of unreinforced masonry in existing and historic structures. Sometimes, existing reinforced masonry walls contain reinforcement that provides negligible improvement over the unreinforced masonry. Lightly reinforced masonry walls of this type are usually analyzed using the principles of unreinforced masonry discussed in this section.

Methods have been established for assessing the stability of unreinforced masonry walls under blast loading. The phenomenon of arching action is one of the primary mechanisms by which unreinforced masonry resists complete failure. However, even slight in-plane vertical movements of "rigid" supports negate the large benefits predicted by the assumed arching action.

16.4.1 Performance Evaluation

When an external explosion imposes an air-blast load on a heavy external wall system, the air-blast pressure must overcome the inertial forces of the wall itself before putting the system into motion and causing damage. In the case of thick, unreinforced masonry walls, these forces are very large, and therefore the resulting damage is less than one might intuitively imagine.

Informative data for masonry can be obtained from the Concrete Masonry Unit Database Software (CMUDS) program. The full-scale equivalent dimensions for the data have spans of 8 to 11 feet and thicknesses of 6 to 8 inches. CMUDS also provides data on reinforced masonry, with reinforcement ratios ranging from 0.15% to 00%, although most of the data have reinforcing ratios from 0.15% to 0.30%. The data include full-scale shock tube testing and high explosives testing on full- and quarter-scale walls. Elastic-perfectly plastic analysis should not be used for unreinforced masonry elements, as this will lead to a nonconservative calculation of the blast load that such elements can resist.

Wall Analysis A common configuration for windows in historic unreinforced masonry buildings, especially those constructed in the late 1800s and the early 1900s, is to have two or three windows per structural bay. This creates masonry piers between individual window openings. These piers tend to be critical wall-support elements.

It is best to begin calculations assuming that the windows can be anchored to the walls at all four sides, distributing the forces from the windows to the supporting walls at the head, sill, and jambs. Walls should have sufficient dynamic capacity to allow window glazing to break first. This approach, commonly known as balanced design, has been shown to reduce the weight (the size and the amount) of debris that may enter the occupied space. There is a direct correlation between injuries and size and velocity of debris.

Two computational models for analyzing wall response are described below.

Simply-Supported Wall A conservative approach to modeling the resistance of an unreinforced masonry wall for an air-blast load is to represent the wall as a slender member subjected to a uniformly applied dynamic load. It is assumed that there is no contribution of axial compressive forces at the top and bottom of the wall, and the wall is analyzed as a simply-supported beam. The flexural resistance of the wall is dependent on the compressive stresses from its self weight, and the tensile strength as defined by the mortar's modulus of rupture. Because unreinforced masonry has little tensile resistance, the flexural capacity of the wall is minimal, resulting in extensive cracking and brittle failure at the interior face of the wall.

Rigid Supports Another model, based on UFC 3-340-02 (the recognized air-blast-resistance design manual throughout the United States), assumes that the top and bottom supports are completely rigid and provide restraint against elongation. Strength comes from compressive blocks that form at the top, bottom, and mid-height hinge locations of the wall, behavior known as a compressive membrane phenomenon. This model also assumes that the mortar joints will be cracked, disregarding what little flexural tensile strength remains in the masonry. The deflected wall "wedges" itself between floor slabs and also forms a plastic hinge at mid-height. This model takes into account that masonry infill walls will often have a gap at the top, either from a mortar joint or from a design that accounts for flexure of the floor systems above.

16.4.2 Retrofit Recommendations

Retrofit masonry walls can be costly, but may be unavoidable due to various constraints such as the historic value of a building. Masonry wall retrofits are usually required when an existing wall cannot be replaced and does not meet the desired performance level of protection. Retrofit recommendations provided in this section apply equally to unreinforced and reinforced masonry walls.

Catchment Systems Or Curtains Depending on the performance intent, an alternative to structural retrofit of unreinforced masonry walls is to design geotextile fabric blast curtain located on the inside face of a non-blast-mitigating masonry wall. This curtain will provide protection from masonry fragments. The geotextile fabric and connections should be designed for membrane action of the fabric. The fabric connections should transfer loads into the floor system without causing fabric tears from stress concentrations. The performance of a specified geotextile fabric and its connections should be verified by blast testing. Generally, blast testing, in addition to demonstrating acceptable performance for design loads, should also demonstrate the post-failure behavior of the retrofit system, to ensure that the failure of the system does not create brittle and hazardous failure.

Secondary Walls A secondary blast-mitigating wall system installed on the inside face of an existing wall system, in addition to being designed to withstand fragmentation and impact effects (due to the frangible wall in front), should also be designed as a new wall, in accordance with this chapter.

Steel Supporting Frames The most straightforward retrofit is to install steel frames, interior to the existing walls, that support the wall and the window system. This solution is appropriate where the wall piers are not able to support the additional air-blast load imposed by the windows. This retrofit is often unacceptable to architects, as it either creates an uneven wall surface or requires reducing the overall room size by furring out the walls.

Shotcrete Walls In some cases, other loading condition requirements may be combined with air-blast requirements to develop multipurpose solutions. Where progressive collapse or seismic design requirements would benefit from shotcrete walls, this may be used as an air-blast retrofit of unreinforced masonry walls as well. Window systems could be mounted in the new concrete wall system. This would be an especially useful technique for relatively larger blast loads.

Fiber-Reinforced Polymer (FRP) Another retrofit method is to apply vertical FRP composite strips along the inside face of the masonry wall. FRP composites have significant tensile strength in the direction of the fibers and can be highly effective in increasing the wall's out-of-plane flexural capacity. Additional advantages of this method are that it is relatively unobtrusive to the architectural detailing of the wall and existing structure, and its application process is not labor-intensive, requiring little disruption to building occupants.

The flexural capacity of an FRP-retrofitted unreinforced masonry wall can be assessed using basic moment-equilibrium relations, as with a steel-reinforced masonry wall. The tensile strength of the masonry is disregarded, and all tensile resistance is assumed to be provided by the FRP. The flexural resistance can then be computed as a function of the compressive strength of the masonry, the tensile strength of the FRP, the thickness of the masonry panel, and the value of the applied axial force.

Blast tests have demonstrated that it is unlikely that the wall's predominant failure mode will be failure of the FRP. Also, tests of unreinforced masonry units to out-of-plane air-blast loading have demonstrated that shear failure at the supports is a predominant failure mechanism. The FRP reinforcing provides no additional shear resistance to the wall section, and if connections at the wall supports do not allow transfer of excessive shear forces to other structural elements, shear failure at the wall supports can occur. Also, connections at wall supports are instrumental in preventing out-of-plane flexural collapse towards the front side of the wall due to rebound forces.

REFERENCES

American Concrete Institute. 2008. *Building Code Requirements for Masonry Structures and Commentary* (ACI 530-08/ACI 530.1-08/ASCE 5-08/TMS 602-08). Farmington Hills, MI: American Concrete Institute.

American Society of Civil Engineers, Petrochemical Committee, Task Committee on Blast Resistant Design. 1997. *Design of Blast Resistant Buildings in Petrochemical Facilities*. New York: American Society of Civil Engineers.

ASTM. 2000. *Standard Specification for Mortar for Unit Masonry* (C270). West Conshohocken, PA: ASTM International.

______. 2006. *Specification for Low-Alloy Steel Deformed Bars for Concrete Reinforcement* (A706). West Conshohocken, PA: ASTM International.

______. 2008a. *Specification for Deformed and Plain Billet-Steel Bars for Concrete Reinforcement* (A615). West Conshohocken, PA: ASTM International.

______. 2008b. *Standard Specification for Grout for Masonry* (C476). West Conshohocken, PA: ASTM International.

Brandow G. E., G. C. Hart, and A. Virdee. 1997. *Design of Reinforced Masonry Structures*. Citrus Heights, CA: Concrete Masonry Association of California and Nevada.

Departments of the Army, the Navy and the Air Force (DOANAF). 1990. *Structures to Resist the Effects of Accidental Explosions* (Department of the Army Technical Manual TM 5−1300). Washington, DC: Departments of the Army, the Navy and the Air Force.

Federal Emergency Management Agency. 2003. *Primer for Design of Commercial Buildings to Mitigate Terrorist Attacks* (FEMA 427). Washington, DC: Federal Emergency Management Agency, Department of Homeland Security.

Paulay, T. and M. J. N. Priestley. 1992. *Seismic Design of Reinforced Concrete and Masonry Buildings*. Hoboken, NJ: John Wiley and Sons.

Protective Design Center. 2008. *Single Degree of Freedom Structural Response Limits for Antiterrorism Design* (PDC TR-06-08). Omaha, NE: U.S. Army Corps of Engineers, Protective Design Center.

Unified Facilities Criteria Program. 2002. *Design and Analysis of Hardened Structures to Conventional Weapons Effects* (UFC 3-340-0). Washington, DC: U.S. Department of Defense, Unified Facilities Criteria Program. For Official Use Only.

Unified Facilities Criteria Program. 2006. DoD Minimum Antiterrorism Standards for Buildings (UFC 4-010-01). Washington, DC: U.S. Department of Defense, Unified Facilities Criteria Program.

Unified Facilities Criteria Program. 2008. *Structures to Resist the Effects of Accidental Explosions* (UFC 3-340-02). Washington, DC: U.S. Department of Defense, Unified Facilities Criteria Program. Supersedes TM 5-1300.

U.S. Army Corps of Engineers. 1999. *Airblast Protection Retrofit for Unreinforced Concrete Masonry Walls* (ETL 1110-3-494). Washington, DC: U.S. Army Corps of Engineers.

U.S. Department of State, Overseas Buildings Operations. 2002. *A&E Design Guidelines for U.S. Diplomatic Mission Buildings*. Washington, DC: U.S. Department of State, Overseas Building Operations.

17 Retrofit of Structural Components and Systems

John E. Crawford and L. Javier Malvar

17.1 INTRODUCTION

Structural components such as columns, beams, floors, and walls are likely to be vulnerable to blast loads. This is important because the resulting damage to these components could produce partial or complete collapse of the structural system, or produce large volumes of injurious debris. For example:

- The blast at the Alfred P. Murrah Federal Building in Oklahoma City led to the shearing of main load-bearing columns and resulted in the collapse of nearly half of the building's structure, which resulted in a large number of casualties. In this incident, it was estimated that 87% of the people in the collapsed portion died (153 out of 175), whereas only 5% of the people in the uncollapsed portion died (10 out of 186) (Federal Emergency Management Agency 1996), indicating that, had the structure better resisted the blast loads and not experienced a collapse, the casualties would have been a lot fewer.
- Even when the structural system survives, blast loads can cause the disintegration of walls and windows, creating secondary fragments that will generate fatalities and injuries in the adjacent rooms.

In this chapter, methods are described for retrofitting structural components to enhance their blast resistance and for assessing the protection the retrofits provide.

Much of this chapter is devoted to describing retrofit techniques to protect reinforced concrete (RC) columns. This was done because so much can be done to prevent the catastrophic damage observed for some types of RC columns when subjected to blast. This information, along with that for the other components, is presented in a summarized form due to space limitation.

17.2 RETROFIT OF COLUMNS

Typically, columns are the component whose integrity is key to sustaining the capability of a building's structural system to survive a blast load. As such, attention to ensuring their survival should have the highest priority when the blast resistance of a building is to be upgraded.

17.2.1 Reinforced Concrete Columns

Soon after the bombing of the Alfred P. Murrah Building, it was shown that high-fidelity physics-based (HFPB) finite element (FE) models would have predicted the failure of the bearing columns nearest the point of detonation, and the resulting progressive building collapse (Crawford, Wesevich, et al. 1995; Crawford, Malvar, et al. 1996; Crawford, Bogosian, and Wesevich 1997). These models also suggested that the use of composite wraps, or steel jackets, could have increased the columns' blast resistance sufficiently to prevent their failure, and thus, the building's collapse.

The concept of jacketing columns had also been suggested by the Federal Emergency Management Agency (Federal Emergency Management Agency 1996), which indicated that retrofit techniques used to resist seismic loads could be extended to resist blast loads. But it was not until 1999 (Crawford 2001) that, in a series of three blast tests conducted by the Defense Threat Reduction Agency (DTRA), it was demonstrated that steel and fiber-reinforced polymer (FRP) jacketing could be used to significantly enhance the blast resistance of RC columns.

Test Data These first three blast tests were conducted on columns located on the ground floor of a full-scale four-story RC building (Figure 17.1); they are designated as Tests 1 to 3 in the figure. The same blast load on identical columns (i.e., except for the retrofit) was used in all three tests, the results being shown in Figure 17.2.

The Test 1 column's performance under these conditions is shown in Figure 17.2a, where a diagonal shear failure can be observed—which is not surprising, given the relatively wide tie spacing of this conventionally designed column. However, when this type of column design is retrofitted with a steel jacket (Test 2), or carbon FRP (CFRP) hoop wraps and vertical CFRP strips (Test 3), it can survive the same load in nearly pristine condition (Malvar, Morrill, and Crawford 1999).

Based on these initial results, an extensive series of full-scale column components was tested using various types of FRP retrofit designs. The results from the blast tests were augmented with a companion program of ten laboratory tests of column specimens identical to those tested by blast, but loaded statically with hydraulic rams (Figure 17.3). These rams were applied in such a way as to approximate the deformation patterns observed in the blast tests. Aramid FRP (AFRP), glass FRP (GFRP), and CFRP composites were used successfully in these tests, although CFRP has shown better durability characteristics (American Concrete

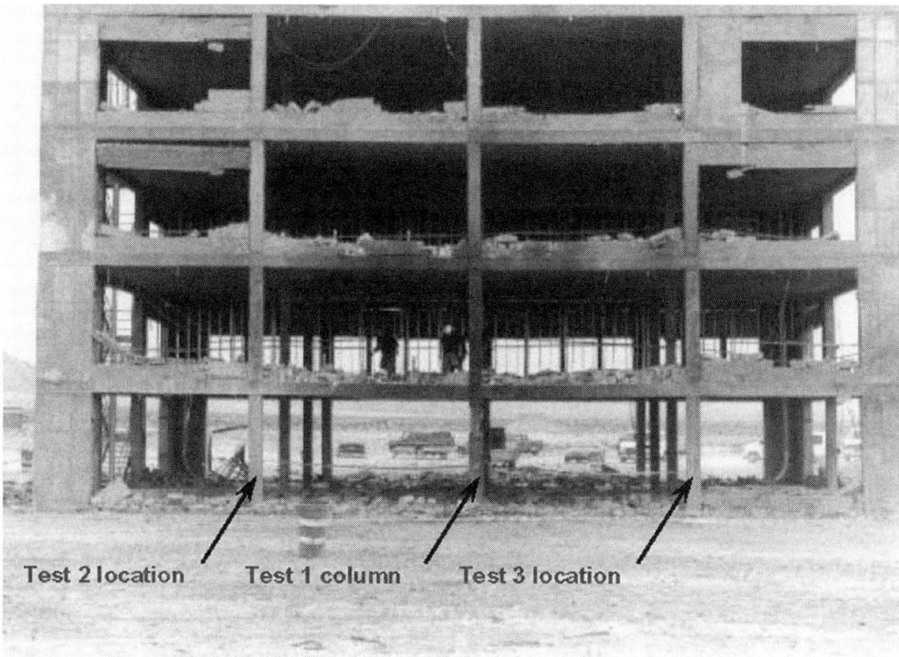

Figure 17.1 Location of initial blast tests of unretrofitted and retrofitted RC columns (tests were conducted using a four-story building constructed using a cast-in-place conventionally designed RC flat-slab framing system; figure shows structure after Test 1 (Photo courtesy of DTRA)

Institute 2002, Malvar 1998a), and its higher stiffness can provide more efficient confinement with less material (Malvar, Morrill, and Crawford 2004).

Design/Analysis Tools One of the major outcomes of the test program was the development and validation of a procedure for the design of FRP wrap and steel jackets to retrofit RC columns to enhance their blast resistance. This procedure can quickly assess the performance of both conventional and retrofitted RC columns subjected to blast loads, and it provides an easily used means to design an FRP wrap or steel jacket to retrofit an RC column so that it can survive a specific blast threat (Malvar, Morrill, and Crawford 1999; Morrill, Malvar, and Crawford 1999; Morrill, Malvar, and Crawford et al. 2000; Morrill, Malvar, and Crawford et al. 2004; Perea and Morrill 2002; Kersul and Sunshine 2002). The retrofit design procedure was embodied in the code CBARD (Crawford, Malvar, and Ferritto et al. 2003). For composites, the design procedure determines the number of FRP hoop wraps needed (and possible vertical FRP reinforcement) for a given blast load. For steel jackets, it determines the required jacket thickness, given the proposed jacket geometry.

(a) (b) (c)

Figure 17.2 Depiction of results from full-scale blast testing of retrofitted and unretrofitted RC columns (Photo courtesy of DTRA): (a) Unretrofitted column (Test 1), (b) Steel jacket column (Test 2), (c) FRP wrapped column (Test 3).

Figure 17.3 Quasi-static laboratory test demonstrating the highly ductile behavior achievable for an RC column retrofitted with CFRP wrap to prevent shear failure (Photo courtesy of DTRA)

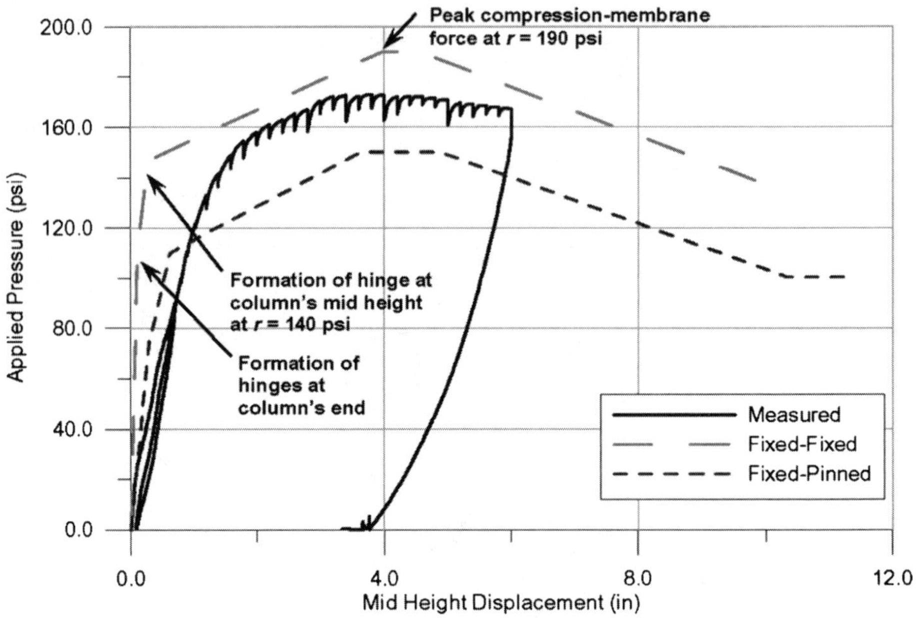

Figure 17.4 Measured versus predicted resistance function for an RC column (as computed by CBARD), which is given in the form of lateral pressure applied versus lateral deflection at mid-height (the added capacity beyond the third hinge formation is provided by compression-membrane behavior)

In addition to the blast tests, as mentioned, static laboratory tests (Figure 17.3) with loading conditions that simulated the deformations observed in the blast tests were conducted to measure the flexural resistance of the columns (Morrill et al. 2001). Static testing cannot replicate the part of the behavior related to the initial highly transient response driven by the short-duration pressure loading of the blast (e.g., the apparent material strengthening at high strain rates), or the wrapping of the blast loads around the column, but it can yield data that can be used to verify the blast design procedure. For example, these data provide demonstrative evidence of the importance of compression-membrane (C-M) forces (Figures 17.4 and 17.5), demonstrate the effectiveness of FRP wrap at enhancing shear resistance (Figure 17.5), and provide load-deflection information for validating design tools (e.g., CBARD).

Although retrofitting columns with steel jackets or composite wraps is reminiscent of seismic retrofits, there are basic differences. For example, apparent material strengthening occurs at the high strain rates induced by the blast loading (Malvar 1998b, Malvar and Ross 1998), a phenomenon that, for concrete, has been linked to moisture content at lower strain rates (below about 1/s) and to inertial effects above that, and that is typically neglected during seismic loading. Also, the deformed shape of building columns under blast loads usually includes

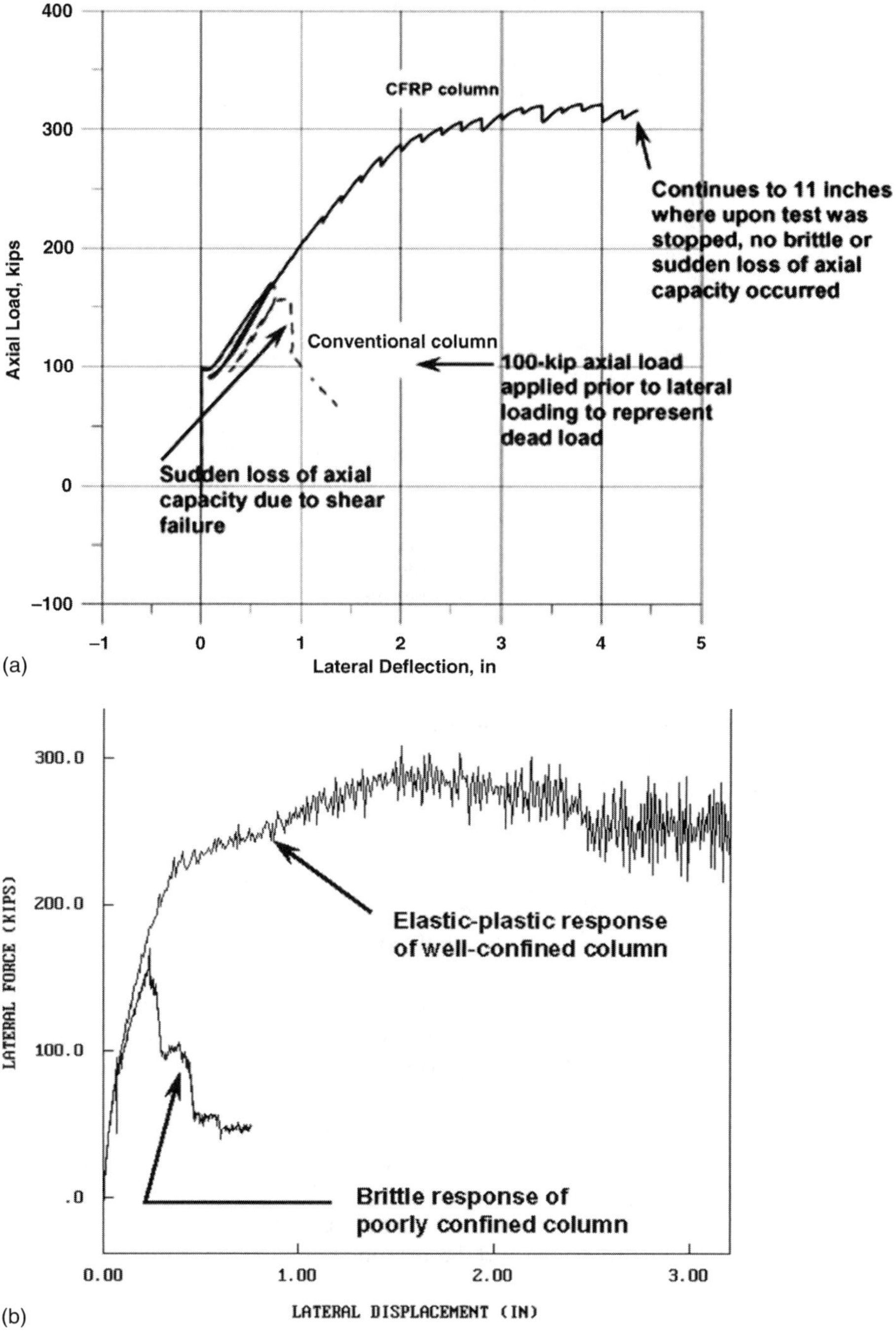

Figure 17.5 Response of conventional and CFRP hoop retrofitted 14-inch square RC columns (the axial load in the CFRP column initially increases with increasing lateral deflection due to compression membrane then decreases to 100 kips when the test was stopped): (a) Axial load, (b) Lateral load

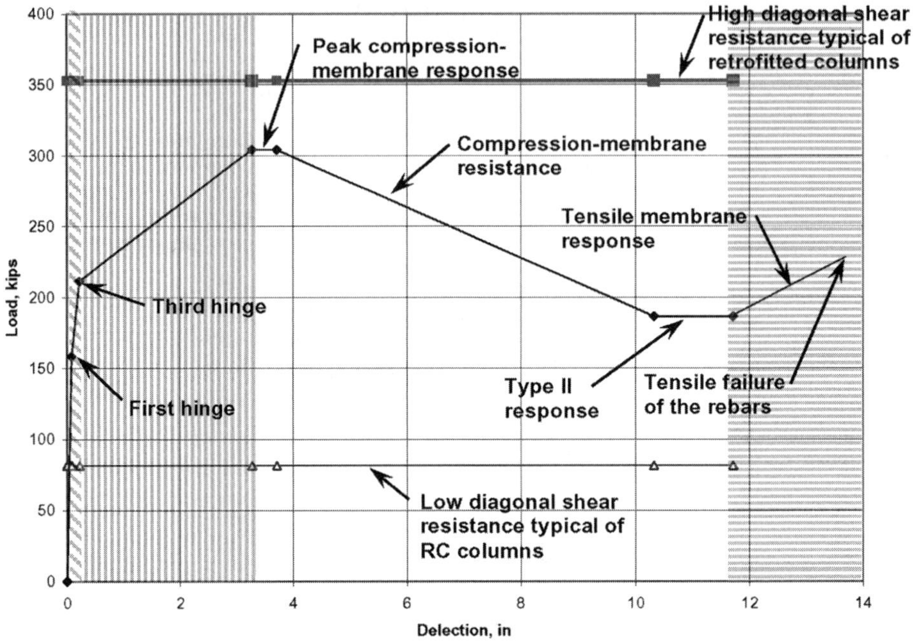

Figure 17.6 Classic form for resistance function (represented by the middle line) pertaining to the fully ductile response of an RC column; this resistance only develops if the diagonal shear capacity always exceeds it

two inflection points and three plastic hinges (Figure 17.3), as compared to a single inflection and two plastic hinges for seismic loading. This is an important difference, as the deformed shape with two inflexion points allows for the development of C-M forces, which can significantly increase the lateral resistance of the column, typically anywhere between 20% and 100%, depending on the column's geometry and the rigidity of the boundary conditions (Crawford, Malvar, and Morrill et al. 2001). This C-M behavior develops due to the vertical (axial) constraint at the top of the column provided by the inertia of the supported structure under high loading rates. Figure 17.6 shows schematically the flexural resistance after the formation of three hinges and the increase provided by the C-M resistance developed subsequently. C-M behavior is typically not present during seismic events; it is a phenomenon associated with both the form and rapidity of the blast response and the presence of significant mass above the column.

Procedure for Retrofitting RC Columns The key concept employed in retrofitting an RC column pertains to preventing shear and other brittle forms of failure under the blast loading, by ensuring that the shear capacity of the column exceeds its flexural strength (i.e., including its C–M capacity), and therefore ensuring that the column dissipates the most energy possible via flexure. For example, in Figure 17.6, if the shear capacity is depicted by the top line (demarked with the solid squares), then the complete flexural behavior will be allowed to develop. This latter situation ensures that the brittle shear failure response is avoided altogether (i.e., over the whole range of the column's response), and instead ensures a ductile flexural response up to the point where P-δ failure occurs. When the shear capacity of the column is too low (e.g., if in Figure 17.6 the shear capacity is represented by the lowest line, demarked by the open triangles), this can be remediated by employing steel jackets or composite wraps to shift this line upward—for example, to the location of the high shear resistance line. In the case of composites, hoop wraps in general are sufficient to provide the required shear capacity and ensure ductile behavior, and vertical sheets or strips are not needed. Moreover, vertical composite reinforcements tend to rupture at low column lateral displacement, often before compression membrane is developed—that is, at a deflection of half the column depth, or less; see DSWA (Defense Special Weapons Agency 1998) or Malvar and Wesevich (1996)—whereas the hoop wraps can remain intact during large lateral deflections (see Figure 17.3).

Description of Design Methodology. This methodology contains a number of unique features that distinguishes it from other design procedures, such as those for retrofitting columns to improve their seismic capacity and traditional American Concrete Institute (ACI) design. These features include:

- Recognition that some RC columns, even those designed for regions of high seismicity, are prone to fail in shear and/or experience catastrophic loss of axial capacity under a blast load.
- Using resistance functions (e.g., as shown in Figure 17.4) specially tailored to the column's geometry and the retrofit concepts employed.
- Focusing the retrofit design on achieving a fully ductile response mechanism (e.g., as demonstrated in Figure 17.5) for the lateral response of the column. A primary goal in designing the retrofit is to have a column that is fully ductile under lateral load.
- Using a best-estimate response-prediction model for selecting design parameters and estimating capacities of retrofitted and unretrofitted columns.
- Using a single-degree-of-freedom (SDOF) model to predict responses to blast loads for both unretrofitted and retrofitted RC columns.

Blast loads may produce several types of response indicating various deficiencies, or lack thereof, in the column's behavior. The deficiencies likely to be encountered include:

- *Lack of Diagonal Shear Capacity*: The column can fail in shear with little plastic behavior, hence dissipating little energy. This is a worst-case condition for a column subjected to a blast load, and will likely be the predominant reason for retrofitting columns. This deficiency is sometimes referred to, although not preferably, as diagonal tension failure rather than diagonal shear failure.

- *Lack of Flexural Ductility*: Good practice would dictate that columns should achieve a full measure of flexural ductility for blast-resistant design—a full measure being reaching their deflection limit, which is related to keeping the column's axial demand/capacity ratio less than 1 (one). Deflections associated with this level of ductility (i.e., full measure) are typically on the order of a large fraction of the column's width. This may not be practical for wide, short columns, in which case, one must be content with prevention of shear and other brittle failure modes.
- *Lack of Direct Shear Capacity*: This condition is uncommon for building columns. It occurs for columns with high diagonal shear capacity and high flexural resistance. This condition is of particular concern for columns having low ratios of height to depth, which is common in parking structures underneath multistory buildings, and for columns with steel jackets. It is often not practical or easy to enhance the direct shear capacity, but significant energy can be dissipated by the direct shear response itself so that an acceptable behavior may be achieved without resorting to any extraordinary retrofit measure.
- *Lack of Moment Capacity*: If the resistance function is low or has little area under it (i.e., its integral is small, indicating little energy absorption in the response), little energy will dissipate, and unacceptable lateral deflections may ensue. This may be mitigated by adding vertical composite strips (FRP retrofits), a steel jacket, or steel strips (steel retrofits). In these cases, close attention should be paid to ensuring that the resulting enhanced diagonal and direct shear demands are met.
- *Lack of Confinement*: This condition is of most concern for large columns having high axial loads that are subjected to a large blast load. In such a situation, the lateral response to the blast is likely to be quite small, but the shock load may be large enough to cause extensive cracking in the concrete. In this case, if the column has insufficient ties to maintain confinement of the concrete and prevent rebars buckling, then the axial resistance of the column may be severely compromised.

The design methodology presented in CBARD encompasses these situations, and provides a means for designing either FRP wrap or steel jackets to address all of these deficiencies.

17.2.2 Steel Columns

The retrofit of steel columns is relatively simple, as compared to the design process employed for RC columns. Generally, the main concerns for steel columns are that they are likely to lack sufficient shear capacity near their supports, and that they may not have enough flexure resistance to preclude excessive lateral deformation. Retrofit concepts for blast can often employ concepts similar to those for enhancing the static load resistance of a steel column.

There are considerable blast effects data for steel columns and subassemblies comprising a column fixed at its base with a set of girders framing into it at a floor level above, and the connections used (Crawford, Magallanes, and Morrill 2005). These data have provided insight into the behaviors that may be cause for concern in a blast environment and that may require mitigation. The data have also been used to validate HFPB finite element models and simplified engineering models for the analysis and design of steel framing systems for blast loads (Morrill, Crawford, and Magallanes 2007, Magallanes and Morrill 2008).

Damage Modes. In retrofitting steel columns to enhance their resistance to blast loads, three basic modes of response should be considered:

- Mode 1: Excessive lateral deformations
- Mode 2: Joint failures, for example, for baseplate connections and column splices that are near the point of detonation
- Mode 3: Fracture or excessive localized deformation of web and flange

The first two modes are relatively straightforward in terms of using retrofit concepts to mitigate the risk of their occurrence, which are primarily related to the lateral loading engendered by the blast. The Mode 3 failure, however, is peculiar to blast where the rapid loading produced by the blast produces a shear failure along the K-line, or causes the web and flanges to deform excessively. Results from a blast test illustrating some aspects of these failure modes, taken from one of a series of tests of wide flange segments, are shown in Figure 17.7.

Retrofit designs for addressing these potential failure mechanisms are likely to require careful consideration, and no presupposed methodology or approach can be considered adequate to ensure that all the salient issues at a particular site are addressed adequately.

Retrofitting. Mode 1 failures can be addressed in relatively conventional ways by strengthening of the member (e.g., by the addition of cover plates). Both local and global deformation of the section may need to be addressed, depending on the blast scenarios considered. Concerns for excessive deformation for light sections may be particularly difficult to address. The retrofit approach would follow much the same strategy as would be used to retrofit a girder to take a higher load. This might involve the use of cover plates, stiffeners, knee braces, and gusset plates. The main concern in developing the retrofit design will be ensuring that ductile behavior is maintained throughout the response, especially with regard to the added shear demand caused by blast type loadings at the column's supports. Alternatively, a blast shield may be installed in front of the column to reduce its response and negate or lessen the need for retrofit.

Figure 17.7 Examples of failure modes for steel columns: (a) Section view, illustrating Mode 1 P-δ failure, (b) Connection view, illustrating concerns with fracture (Mode 3 failure), (c) Baseplate view, depicting a Mode 2 failure (Photo courtesy of DTRA)

Addressing Mode 2 failures is a little less straightforward, but they can be approached in several ways. These failures are related to a general lack of capability in the various sorts of connections used in conjunction with deploying steel columns. Of these, the failure of column base plates and splices is of most concern. The retrofits for addressing Mode 2 failures are likely to take a variety of forms. The general approach employed in reducing the vulnerability to Mode 2 failures is to ensure that the connection does not represent a weak spot. For example, using a cross tie to remove reliance on the base plate's capacity. The other main concern is to ensure that plastic hinges are able to form in the column as a whole, so as to achieve a fully ductile lateral response for the column.

Mode 3 failures are most easily addressed by taking advantage of the short duration of the blast load and the benefits that can be derived from inertia in this situation. A retrofit concept embodying this approach is depicted in Figure 17.8. Here the interstices between the section's flanges are filled with concrete, which mitigates problems associated with localized bending of both the flanges and webs (Crawford and Lan 2005). This may be augmented with additional webs, as shown in the figure, if additional shear capacity is desired.

Another means to protect the column from web breach (i.e., in the case of a wide flange section) is to place a concrete or other similarly substantive panel in front of the web so that the airblast cannot impinge directly on the web. This could be in the form of a conventional cladding (Crawford, Magallanes, and Morrill 2005).

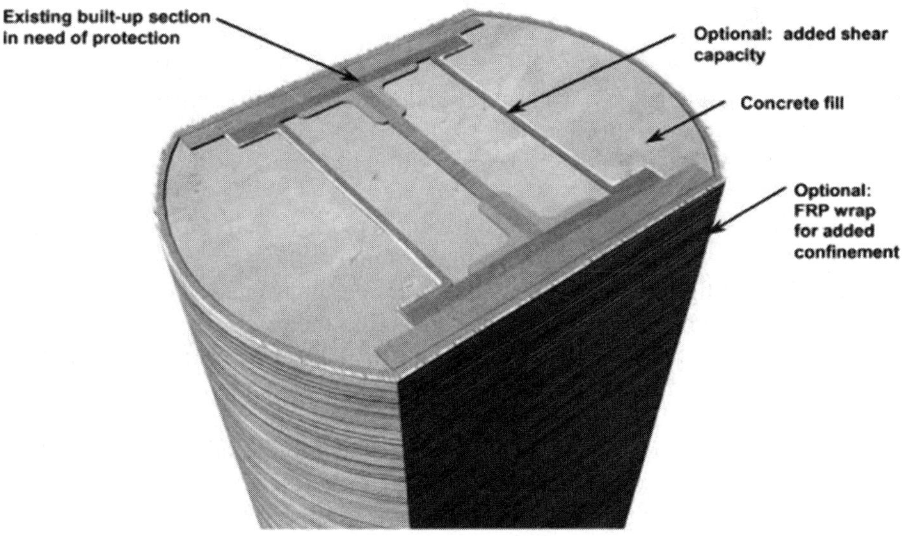

Figure 17.8 Example of the infill concept for preventing web and flange failure (Mode 3) for steel sections

17.3 RETROFIT OF WALLS

Bearing walls, like columns, are a key structural component supporting the gravity loads of a building system, and the loss of bearing walls can also lead to overall collapse. Unlike columns, however, bearing walls can sustain major damage without compromising their function of carrying the gravity loads. These walls, in essence, offer a continuous form of load path redundancy, which allows them to be locally breached by a nearby charge and still perform their primary function of holding the structure up. In contrast to columns, walls can also provide a last line of defense for the building's occupants from the effects of blast, and the failure of both bearing and nonbearing walls can result in deadly secondary fragmentation likely to produce significant casualties as their debris are blown through a building.

Walls are found in a wide variety of strengths, combine a wide range of materials, vary substantially in quality of construction, and often present considerable difficulty in estimating their robustness as a structural system. In the following discussion, the behaviors of walls subjected to out-of-plane static and blast loads are examined. These are different from seismic applications, where walls are generally used more to provide shear resistance, and therefore subjected primarily to in-plane loads.

Composites, polymers, and conventional materials have been used to enhance the blast resistance of masonry walls (Malvar, Crawford, and Morrill 2007). FRPs have been used to enhance the resistance of RC and stud walls. Sheet-metal panels and blast shields are useful in enhancing the blast resistance of any wall. Examples of the use of some of these retrofit concepts are described below.

17.3.1 Masonry Walls

The retrofitting of masonry walls with and without reinforcement to enhance their blast resistance has been widely studied. Those composed of concrete masonry units (CMU) and fired clay bricks will be the focus of the following discussion.

The response of conventional masonry walls to blast loads has been observed in many full and subscale tests. An example of one of these tests is shown in Figure 17.9. In this figure, two side-by-side identical ungrouted CMU walls were tested with a moderate-size blast load. The wall to the right was retrofitted (Figure 17.9b); the one to the left was not. Figure 17.10 shows a snapshot depicting the motions of the unretrofitted wall immediately after a detonation. The performance shown is typical of a masonry wall's response for a moderate blast load, in that the CMU blocks remain relatively intact, while the failure of the wall is largely due to the opening of the masonry's horizontal joints.

Three approaches to retrofitting walls are described. The first two technologies involve bonding reinforced and/or unreinforced layers of polymer to the back (and possibly front) faces of a masonry wall to provide both added strength and ductility for blast load response. The third approach uses some of the same

(a)

(b)

Figure 17.9 Example showing the benefit of using FRP to retrofit a masonry wall to enhance its blast resistance: (a) Control (left) and retrofitted wall (right) prior to blast test, (b) Control (left) and retrofitted wall (right) after the test (Photo courtesy of The Energetic Materials and Research Testing Center)

Figure 17.10 Example typical of the response expected from a CMU wall struck by a blast; at the time shown, the CMU is moving at several tens of feet per second (Photo courtesy of The Energetic Materials and Research Testing Center)

materials as the first two, but in this case they are not bonded to the wall. In this situation, the strengthening elements, such as FRP, are anchored to the floor and act by preventing entry of wall debris rather than by wall strengthening.

Fiber-Reinforced Polymer The use of FRP to enhance the strength and blast resistance of masonry walls has been extensively studied. This would include FRPs such as CFRP, AFRP, and GFRP.

CMU Studies. Muszynski (1998) and Muszynski and Purcell (2003) bonded CFRP and AFRP to masonry walls composed of concrete masonry units. Baylot, Dennis, and Woodson (2000), Baylot et al. (2005), and Dennis, Baylot, and Woodson (2002) conducted experiments on walls using quarter-scale models built with typical 8-inch-wide CMU. In these tests, three types of retrofitting schemes were evaluated for ungrouted and for partially grouted walls. The first retrofit method consisted of 0.04-inch (1 mm) thick GFRP fabric bonded to the back face of the wall. The second retrofit technique consisted of a two-part sprayed-on polyurea product applied to the back face of the wall. The third retrofit method used 20-gauge galvanized steel sheet metal attached to the back of the wall. All the walls failed during the blast testing by separating from the frame, but the GFRP and the polyurea retrofitted walls were successful in preventing loose fragments within the structure, and were, therefore, considered as successful retrofits.

Myers, Belarbi, and El-Domiaty (2003, 2004) tested eight ungrouted, unreinforced masonry (URM) walls retrofitted with FRP (epoxy) laminates and near-surface-mounted (NSM) rods. Two series of walls (100 and 200 mm or 4 and 8 inches thick) with different slenderness ratios and different strengthening schemes were tested at varying blast charge weights and standoff distances. The walls were only supported at the top and bottom, and retrofitted walls failed in shear at those locations. The FRP laminates were able to contain the wall debris.

Carney and Myers (2003, 2005) tested twelve ungrouted URM walls using a pressure bag to apply a uniform loading (Phase I), then field-tested two walls under blast (Phase II). Results of the last four Phase I tests, together with the Phase II tests, are shown in the second paper (Carney and Myers 2005). The walls were reinforced with GFRP sheets or NSM rods, with and without anchorage. The anchorages consisted of wrapping the sheets around NSM rods embedded into the wall's top and bottom support for the first case, and of extending the NSM rod three inches into the wall support for the second. Strengthening the anchorages almost doubled the capacity of the walls. In contrast, unanchored NSM rods did not show any improvement. For FRP reinforced walls, no shear failure was observed during the static tests, while it was apparent in previous blast tests (Myers, Belarbi, and El-Domiaty 2003), indicating a potential limitation in the use of static tests to simulate blast tests.

Stanley, Metzger, and Martinez et al. (2005) used a retrofit system composed of multiple layers of carbon and glass FRP in an epoxy resin (applied with paint rollers), placed on the interior and exterior surfaces of the CMU wall test specimen. This system was successful in containing debris but seemed too labor intensive to apply. Stanley, Metzger, and Morrill et al. (2005a) used two layers of a glass FRP in an epoxy resin applied to the interior surface of the CMU wall test specimen. This wall experienced complete failure, indicating that stiff retrofit systems fail catastrophically once their capacity is exceeded.

Simmons (2005) studied the effectiveness of three types of retrofits (E-glass, 22-gauge steel, and an elastomeric polymer) on CMU walls to increase their resistance to primary fragments from a 120-mm mortar round and a 122-mm rocket; he found that the E-glass significantly reduced the number of fragments that passed through the wall.

Brick Studies. Conley and Mlakar (1999) used CFRP (epoxy) strips to reinforce small brick walls loaded with a gas-loaded piston and subjected to one-way bending. Increases in load capacity of 400% were recorded.

Patoary and Tan (2003) completed static tests using air bags on four built-in solid masonry brick walls retrofitted with three types of FRP systems: glass, carbon, and low-strength fiberglass woven roving. The walls, 120-mm thick, were able to sustain deflections up to 20 mm. Specimens with GFRP or CFRP failed by edge debonding, while specimens with roving failed by fiber rupture. This work was followed by out-of-plane tests on 30 masonry walls reinforced with three different GFRP and CFRP systems (Tan and Patoary 2004).

Retrofitting with Unreinforced or Fiber-Reinforced Polyurea-like Polymers In some situations, the performance of walls reinforced with FRP composites is limited by their peak strain capacity, so materials with higher failure strains might be expected to perform better. Because both fibers and epoxies used for FRP applications have rather low strains to failure, several researchers started looking at polyureas, which are materials with purported failure strains of 100% or more.

These materials have appeared in two forms: one as a spray-on coating that is applied and bonded directly to the wall's surface, the other as a sheet polymer that is placed behind the wall and anchored to the floors. The first system acts in concert with the masonry by adding tensile capacity. The second system acts as a catcher system to contain the wall's debris and is not bonded to the wall. This latter system is described in Section 17.3.1.3.

Unreinforced. Knox et al. (2000) sprayed the back of concrete block walls with various unreinforced polymers, in particular polyurethane and polyurea. These polymers showed exceptional ductility and were able to prevent injurious wall fragmentation. As indicated earlier, Baylot, Dennis, and Woodson (2000) and Baylot et al. (2005) and Dennis, Baylot, and Woodson (2002) conducted experiments on CMU walls using quarter-scale models of typical 8—inch-wide CMU. Among other retrofits, they used a two-part sprayed-on polyurea product applied to the back face of the wall, which was successful in preventing loose fragments within the structure.

Hammons, Knox, and Porter (2002) expanded their application of polyurea to CMU walls and found it very successful in preventing deadly wall fragmentation. Deflections of up to 610 mm (24 inches) were recorded (the nominal CMU wall width was 8 inches, or 200 mm). Other tests, using a high pressure spray–applied two-component polyurea (Stanley Martinez, and Koenig 2005b), and a two-component spray-applied polyurea liner (Stanley, Metzger, and Morrill et al. 2005b) applied to both interior and exterior CMU wall faces, confirmed such findings, although some tearing of the polymer was observed. In these applications with large deflections, the complete resistance function may be mobilized, including the lower resistance beyond compression membrane (due to the crushing of the concrete compression zone), and the subsequent strengthening due to tensile membrane, enabling the dissipation of significant amounts of energy.

Davidson, Fisher et al. (2005) studied 12 polymer-reinforced masonry walls during seven explosive tests. The walls were restrained at the top and bottom only and responded in a one-way fashion. During the tests, arching due to end constraints was suspected to have contributed to the loss of the compression face near the ends. A simple spray overlap of 6 inches (152 mm) was sufficient to transfer the loads to the frame. Compared to stiff FRP fabric reinforcing, the spray materials were cheaper, bonded better to the wall, and were easier to anchor to the frame (see also Davidson, Porter et al. 2004).

Reinforced. These polyurea polymers can also be further strengthened with continuous fibers, which can provide some additional energy dissipation up to fiber rupture. Tests were conducted by Crawford and Morrill (2002) to define the properties of reinforced and nonreinforced polyurea spray-on coatings. These materials were used to retrofit masonry walls subjected to blast loads. This is detailed in several studies (Crawford and Morrill 2003a, 2003b; Johnson, Slawson, Cummins et al. 2003, 2004; Johnson and Slawson 2004; Morrill, Nicolaisen, and Hutchinson 2005; Piepenburg et al. 2002, 2003; Stanley, Metzger, and Morrill et al. 2005c; Stanley, Metzger, and Koenig. 2005a).

Piepenburg et al. (2002) used an aramid grid fabric to reinforce a polyurea spray retrofit of a CMU wall, and were successful in containing the CMU debris. Piepenburg et al. (2003) also used an aramid within two thin sprayed-on polyurea layers on a CMU wall. Crawford and Morrill (2003a) describe a retrofit consisting of an aramid laminate that is secured to the wall with a polyurethane adhesive, and anchored to the diaphragms with energy-dissipating compliant anchorages. Johnson, Slawson, and Cummins et al. (2003, 2004) used an open-weave aramid with different fiber lay-ups with a spray-on polyurea. The capacity of the walls retrofitted with reinforced polyurea was 1.4–2.0 times higher than that of the walls retrofitted with just polyurea, and 5.5–7.5 higher than that of the unretrofitted walls. Morrill, Nicolaisen, and Hutchinson (2005) used a polyurea coating, with and without an aramid fiber mesh, on the back side of brick wall samples tested quasi-statically in a laboratory environment. These tests were useful in determining complete quasi-static resistance functions, including tensile membrane.

Stanley, Metzger, and Morrill et al. (2005c) used a spray-applied two-part polyurea with an inner aramid weaved fabric on a CMU wall, and successfully stopped the CMU debris. Peak deflections were about 230 mm (9 inches). Stanley, Martinez, and Koenig (2005a) used a non-woven polypropylene geotextile fabric in between two layers of sprayed-on polyurethane applied to the interior and exterior surfaces of the CMU wall test specimen. In this case, the retrofitted wall did fail, although it would appear that the retrofit had provided some amount of early-time confinement and lateral restraint of the CMU blocks.

Retrofitting with Catcher Systems In contrast to the retrofit schemes described for walls above, which are related to strengthening them directly, catcher systems are placed behind the wall to catch any debris that might be generated by its breakup from a blast. Such a device may be composed of thin sheets of fabric (e.g., Kevlar® laminate), sheet metal, or polymer (e.g., as shown in Figure 17.11).

This form of catcher system may be augmented with cables for cases involving particularly high loads (e.g., impulses of thousands of psi-ms) or where anchorage conditions prohibit the use of the sheets by themselves. These systems are designated as cable catcher systems.

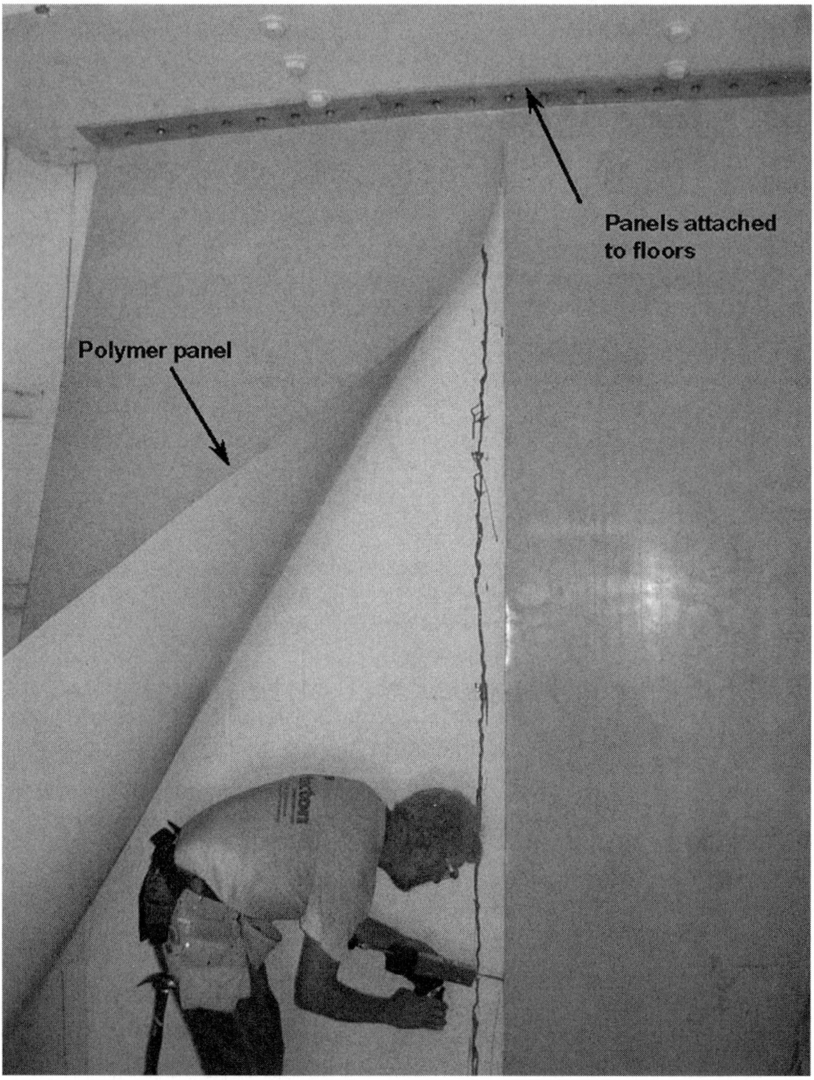

Figure 17.11 Example of a thin polymer catcher system being installed (Photo courtesy of Sherwin-Williams Company)

The intent of these concepts is just to capture the debris of the wall, not to strengthen the wall. These systems may be constructed in a variety of ways, and some designs have very large capacities. In addition, these systems are likely to impart less lateral load to the structural system of the building, which is significant where collapse of the framing system due to the lateral loading of the blast is of concern. The debris-catching solution is primarily intended for

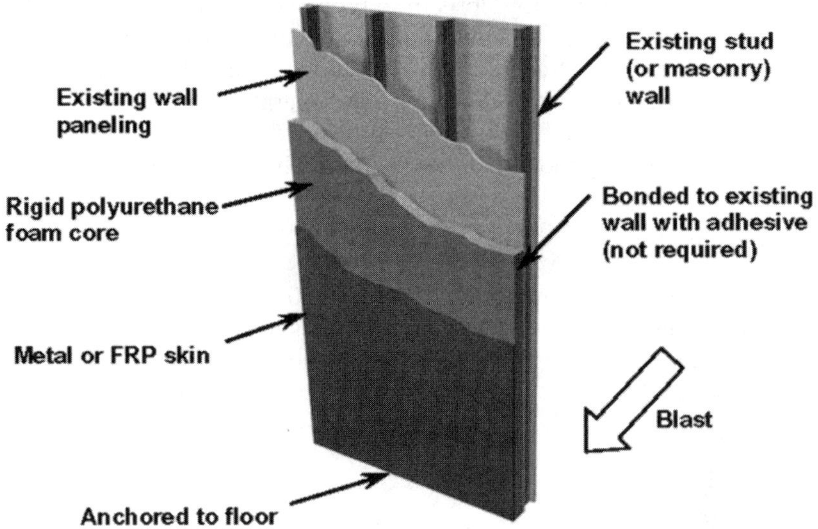

Figure 17.12 Example of the type of composite panels that may be used for retrofitting masonry and stud walls to enhance their blast resistance

non-load-bearing walls whose function is not critical to preventing failure of the structural system. However, these systems may also be applied to bearing walls, where the breadth of the wall failure incurred can be spanned by the remaining structural system.

In 2004, a series of reinforced and unreinforced polymer panels for retrofitting masonry walls was developed (Crawford, Lan, and Wu 2009). These panels are delivered to the site in the form of sheets, which are placed behind the wall and anchored to the floors to hold them in place, as shown in Figure 17.11. These panels have shown their worth against both vehicle and rocket blast threats in tests in Israel and the U.S.

Retrofitting with Composite Panels In 2000, large-capacity catcher systems were introduced and validated in blast tests (Crawford 2001) for retrofitting masonry or stud walls. These systems employed composite panels, like the one shown in Figure 17.12, that were composed of FRP or steel sheets bonded to a crushable core material, such as the rigid polyurethane foam shown in the figure. The panel is placed behind the wall and anchored to the floor, generally using a compliant anchorage. These systems have been proved successful in blast tests for both masonry and metal stud walls, and were shown to work for very high loads if the anchorage is designed appropriately (Crawford 2001, Crawford, Xin, and Bogosian 2002).

Retrofitting with Conventional Methods Conventional methods (e.g., shotcrete and dowels) may also be used to retrofit a masonry wall to enhance its blast resistance, or just to prevent its debris from propagating into the building. These systems should provide continuous support to the existing masonry wall to prevent localized breaching. Such systems would be expected to be less efficient than those already mentioned. Moreover, the generally large quantities of materials used to install these types of retrofits may add to the risks from blast by introducing considerably more material that could become a source of debris.

17.3.2 Stud Walls

Knox et al. (2000), and Crawford and Morrill (2003a, 2003b) also studied the retrofit of wood and steel stud walls, using composite materials, which provided an excellent means for adding blast protection to an existing stud frame building. In the Crawford and Morrill studies, both fabric and sheet metal composites, such as the one shown in Figure 17.12, were deployed.

17.4 FLOORS

Considerations of the blast protection afforded by the floors of a building are likely to involve three kinds of situations, which are related to the location of the detonation:

- Case 1: Considerations for detonations outside the building, where the detonation is some distance away. Here, enhancing the resistance of the floor may be unnecessary or may be accomplished using conventional strengthening techniques, mainly because the floor's response is neither very pronounced nor complex. The actual distance in question depends on the type of exterior wall, the floor's properties, and other factors.
- Case 2: Considerations for exterior detonations near the building, with no wall present (e.g., near a loading dock). Here, an upward shock could be imparted to a floor above the detonation and downward motion imparted to the floor below the detonation.
- Case 3: Interior detonation. Here, considerations for two situations are of issue: the blast above, and the blast below a floor. For an interior detonation, gas pressures may be of major concern. In this case, adding venting might be part of the retrofit strategy.

Retrofitting floors to provide protection from blast is likely to be more problematic than retrofitting any other structural component, mainly because of the large spans and loaded areas, their load resistance being primarily related to gravity (i.e., with limited resistance to uplift), and the possibility that the detonation may be quite close (e.g., a bomb resting on the floor). Moreover, in contrast to most other retrofit strategies, the weight of the floor retrofit can be an issue because of the floor's extent.

The following are three retrofit strategies for addressing the three threat situations mentioned above.

- Case 1. When the bomb is outside, the walls are likely to knock the blast down sufficiently so that the floor response may be ignored—especially since it's likely in such a case that the walls would be retrofitted first, and even less load would reach the floor. Moreover, the main concern is the net load acting on the floor, which may be negligible, even if the walls fail. However, in situations where the facades above and below a floor have very different blast resistance (e.g., ground floors that have no exterior walls), a substantial net upward (or downward) load on a floor can be generated, and the situation may require some form of mitigation.
- Case 2. If the bomb is relatively close to the outside of a building and opposite an opening in the wall (or possibly a glass facade), the floor, especially above the bomb, which would be loaded in its weak direction, may be a cause of concern. Here, however, the effects of gas pressure can probably be ignored.
- Case 3. For an interior detonation, both shock and gas pressure components of the blast may cause concern.

The risks associated with a floor hit by a blast depend greatly on the size and location of the bomb. For example, is the detonation from a bomb on the floor, or above/below the floor? This has mostly to do with proximity and whether the response is mainly localized or involves the whole of the floor, the magnitude of velocity (shock) imparted to the floor and possible debris generated from it, and whether the floor is loaded on its strong or weak axis.

Protection for floors against nearby bombs (i.e., within a few feet) is likely to be impractical. For example, for a bomb resting on a conventional floor, there are few (or no) reasonably priced ways to prevent the floor's breach. Where the blast is above the floor, then attaching a catcher system to the joists to prevent debris propagation could be a good strategy.

For situations involving a vented blast occurring beneath a floor that has a reasonable standoff, retrofitting the floor with an FRP may suffice. However, even if the floor remains relatively undamaged, there is likely to be significant shock imparted to it, which can cause severe injuries to those standing on it and launch furnishings, providing another source for injuries. Better protection can be provided by a blast shield, possibly hung below the actual floor, to either mitigate or prevent the airblast from striking the floor. In this circumstance, not only is the floor not damaged, but the velocity imparted by the shock, resulting in the launch of furnishings and people, is avoided altogether.

The selection of a retrofit strategy for floors is likely to primarily involve limiting the risks, rather than eliminating them. Three basic retrofit strategies, depending on the bomb's size and location, seem most applicable to addressing the vulnerability presented by floors:

- Strengthening the existing floor through the addition of ductility and capacity.
- Debris containment. This strategy is particularly useful for those floors in direct contact with or very near to the bomb.
- Shielding. Shielding a floor from the blast load avoids the issue altogether. This could also include constructing a second floor directly above/below the existing one that is directly exposed to the threat. Alternatively, the existing floor may be retrofitted so that it can provide the protection for a new floor.

Retrofit strategies for floors are likely to be much more difficult to implement than for walls, which would imply that the best strategy here is to prevent a bomb or an airblast from getting close enough to present a serious risk to the floors, and to use the walls of the building to protect the floors. Even for a loading dock, using roll-up doors or some other form of closure can probably deflect the blast enough so that a floor above it is sufficiently protected.

17.5 BEAMS/GIRDERS/CONNECTIONS

The retrofit strategies pertaining to enhancing the blast protection afforded by structural components that act as beams, girders, and the connections associated with them—hereafter collectively referred to as beams—are generally of less concern than the components already mentioned, because in many ways these components generally present less of a risk.

Retrofit strategies for beams to accommodate blast loads are likely to be relatively simple and probably similar to those that might be used in conventional retrofit strategies, especially for blast loads that are generated by charges that predominantly excite the fundamental modes of the beam.

In addition, whether the blast is pushing up or down on the beam can make a substantial difference in its response. Many beams are not as well attached in the upward direction, and may have similar weaknesses in terms of their ability to resist lateral loads. For closer charges, retrofit strategies are likely to be site-specific and not easily addressed with simplified tools and conventional retrofit strategies.

Blast effects data for beams (Crawford, Magallanes, and Morrill 2005, Magallanes and Morrill 2008), although sparse, indicate that the beams themselves are likely to be less of an issue than their support connections. For external blast, it is likely that the net loads pushing up/down on the beam will be negligible. However, loading on a beam can be quite localized (e.g., a bomb placed on the floor immediately above the beam), or broad-based; for example, if applied to a robust floor system that is strongly coupled to the beams framing it, the beam's loading would derive from a large tributary area.

A major concern for beams is their attachment (i.e., connections) to their support (e.g., columns). For many steel beams, especially in nonseismic designs, this may consist of only a shear tab, or for precast beams only a small ledger to which it is only lightly attached (e.g., the shear tab shown in Figure 17.7b).

Several retrofit concepts have been developed for seismic applications to improve the anchorage of beams/girders at their supports that would also work in blast environments, although they may need to be more robust. Another retrofit strategy that has been used (Crawford and Lan 2005) relies on cabling attached to the beams to carry the load, should their connection to the columns be damaged.

17.6 STRUCTURAL SYSTEM

The preceding sections addressed ways to retrofit structural components to enhance their survivability and protective capability in the event of a blast. It is also important, perhaps even more so, to evaluate the structural system's performance, especially once the retrofits are in place.

However, evaluating the risks to the system and describing retrofit strategies for addressing them is beyond the scope of this chapter. With regard to assessing risks to the system, however, it is important to recognize that retrofitting structural components can produce deleterious effects in terms of the system's performance. For instance, there would be a need to ensure that the impact of the additional loading imparted to the system by any enhancement in the protection be evaluated—for example, the added lateral load imparted to the system by the addition of blast-resistant walls could exacerbate the lateral response of a building system sufficiently to cause its failure from excessive story drift.

17.7 SUMMARY

Information was presented in this chapter pertaining to retrofitting a facility's structural framing system and components to enhance the protection afforded by them in the event that a blast occurs. Each of the major components of the structural system was addressed, including columns, walls, beams, and floors. Several strategies were described for mitigating risks from blast for each type of component.

17.7.1 Inexact Science

The ideas presented in this chapter concerning the retrofit of structural components to enhance the blast protection afforded by them can only, due to space limitations, reflect a general notion of strategies and methods that may be pursued in this endeavor. Moreover, only a few of the ways this might be accomplished could be cited, and all the concerns and complexities that might arise in an actual application could not be addressed. What is intended, however, is to apprise those desirous of developing retrofit strategies and designs of some of the main issues and methods that have been found useful and effective, and to describe extant test data and studies that are available.

Developing retrofit designs for any structure is often more art than engineering because each situation is likely to be unique, to have issues of unintended consequences, to raise difficulties of identifying the extant design, and so forth. In addition, blending with the existing structure, architecture, and functionality is likely to present a host of conflicting constraints, making the retrofit even more difficult.

It is likely that any retrofit design will not be able to offer the same certainty of function as a new design. Moreover, achieving too high a certainty is likely to result in a poorer level of protection or perhaps none at all—that is, there is a certain paralyzing effect that can easily be reached by demanding too much certainty. On the other hand, there is a particular need in this type of engineering to be certain that the design is not fatally flawed. Adopting a good enough standard may offer the best course for this type of work, the main issue here being the difficulty of defining such a standard ahead of time (e.g., for the purposes of generating a contract and design budget).

17.7.2 Complexities

Retrofitting a structure to enhance its resistance to blast may require an inordinate level of skill and experience. The limited amount of information presented in this chapter can only allude to some of the considerations that are needed to develop these retrofits and their integration with the existing structure, and should be viewed as just an introduction to the subject.

It is important to recognize that, in many cases, the intensity and rapidity of the blast load generate behaviors substantially different than generally encountered in structural design problems. Moreover, the forces and material damage incurred are often difficult to determine without using HFPB finite element models, and for most retrofit designs, they are likely to be much higher than ordinarily dealt with, requiring a familiarity with material and dynamic response behaviors in this high loading rate regime.

The level of complexity and difficulty encountered in developing retrofit designs is likely to be highly exacerbated when the blast load is from a large, close charge. These situations can be addressed, but they require considerations in addition to those that can be alluded to here. Moreover, in these situations, many simplified engineering models commonly used for blast effects analysis are likely to be deficient and misleading in their prediction of response.

REFERENCES

American Concrete Institute, Committee 440. 2002. *Design and Construction of Externally Bonded FRP Systems for Strengthening Concrete Structures* (ACI 440.2R-02). Farmington Hills, MI: American Concrete Institute.

Baylot J. T., B. Bullock, T. R. Slawson, and S. C. Woodson. 2005. Blast response of lightly attached concrete masonry unit walls. *Journal of Structural Engineering* 131 (8): 1186–1193.

Baylot J. T., S. T. Dennis, and S. C. Woodson. 2000. Response of 1/4 scale concrete masonry unit (CMU) walls to blast. In Emerging Technologies in Fluids, Structures, and Fluid/Structure Interactions, Vol. 1. Proceedings of the 2000 ASME Pressure Vessels and Piping Conference, Seattle, WA, July 23–27, 2000. New York: American Society of Mechanical Engineers.

Carney, P. and J. J. Myers. 2003. Shear and flexural strengthening of masonry infill walls with FRP for extreme out-of-plane loading. Proceedings, Architectural Engineering Institute Annual Meeting, Austin, TX, 2003.

________. 2005. Out-of-plane static and blast resistance of unreinforced masonry wall connections strengthened with FRP. In *7th International Symposium on Fiber-Reinforced Polymer Reinforcement for Concrete Structures*, ACI Symposium Publication 230, 229–248. Farmington Hills, MI: American Concrete Institute.

Conley C. H. and P. F. Mlakar. 1999. Blast resistance of CFRP reinforced masonry. Presented at the 70th Shock and Vibration Symposium, Albuquerque, NM, November 15–19, 1999.

Crawford J. E. 2001. Addressing workplace protection issues more effectively. Proceedings of the Society of American Military Engineers Symposium on Comprehensive Force Protection, Charlestown, SC, November 1–2, 2001.

Crawford J. E., D. D. Bogosian, and J. W. Wesevich. 1997. Evaluation of the effects of explosive loads on masonry walls and an assessment of retrofit techniques for increasing their strength. Presented at the 8th International Symposium on the Interaction of the Effects of Munitions with Structures, Defense Special Weapons Agency, McLean, VA, April 1997.

Crawford J. E. and S. Lan. 2005. Design and implementation of protection technologies for improving blast resistance of buildings. Presented at the Enhancing Building Security Seminar, Singapore, March 2005.

Crawford J. E., S. Lan, and Y. Wu. 2009. Design methods for enhancing the blast resistance of masonry walls and windows with bonded and unbonded polymer sheets. Proceedings of the 8th International Conference on Shock and Impact Loads on Structures, Adelaide, Australia, December 2–4, 2009.

Crawford J. E., J. J. Magallanes, and K. B. Morrill. 2005. Vulnerability of steel framing systems. Proceedings of the 12th International Symposium on the Effects of Munitions with Structures, New Orleans, LA, September 26–30, 2005.

Crawford J. E., L. J. Malvar, B. W. Dunn, and D. J. Gee. 1996. Retrofit of reinforced concrete columns using composite wraps to resist blast effects. Proceedings of the 27th Department of Defense Explosive Safety Seminar, Las Vegas, NV, August 19–22, 1966.

Crawford J. E., L. J. Malvar, J. M. Ferritto, K. B. Morrill, B. W. Dunn, and P. Bong. 2003. Description of Design Software for Retrofitting Reinforced Concrete Columns to Improve Their Resistance to Blast. (Report TR-01-16.2). Burbank, CA: Karagozian and Case. Limited Distribution.

Crawford J. E., L. J. Malvar, K. B. Morrill, and J. M. Ferritto. 2001. Composite retrofits to increase the blast resistance of reinforced concrete buildings. Proceedings of the 10th

International Symposium on Interaction of the Effects of Munitions with Structures, San Diego, CA, May 2001.

Crawford, J. E. and K. B. Morrill. 2002. New highly ductile designs for preventing injurious wall debris due to a terrorist bombing. Proceedings of the Protection of Structures Against Hazards, Singapore, November 2002.

________. 2003a. Retrofit techniques using polymers and fiber reinforced polymers for preventing injurious wall debris. Proceedings of the 6th International Symposium on Fibre-Reinforced Polymer Reinforcement for Concrete Structures, FRPRCS-6, Singapore, July 8–10, 2003, pp. 1199–1208.

________. 2003b. Retrofit techniques for strengthening stud walls subjected to nearby blast loads. *Proceedings,* 74th Shock and Vibration Symposium, San Diego, CA, October 27–31, 2003. Limited Distribution.

Crawford J. E., J. Wesevich, J. Valancius, and A. Reynolds. 1995. Evaluation of jacketed columns as a means to improve the resistance of conventional structures to blast effects. Proceedings, 66th Shock and Vibration Symposium, Biloxi, MS, October 30–November 2, 1995.

Crawford J. E., X. Xin, and D. D. Bogosian. 2002. Retrofit techniques for conventional walls to prevent their causing injurious debris due to a terrorist bombing. Proceedings of the 17th International Symposium on Military Aspects of Blast and Shock, Las Vegas, NV, June 10–14, 2002.

Davidson J. S., J. W. Fisher, M. I. Hammons, J. R. Porter, and R. J. Dinan. 2005. Failure mechanisms of polymer reinforced concrete masonry walls subjected to blast. *Journal of Structural Engineering* 131 (8): 1194–1205.

Davidson J. S., J. R. Porter, R. J. Dinan, M. I. Hammons, and J. D. Connell. 2004. Explosive testing of polymer retrofit masonry walls. *Journal of Performance of Constructed Facilities* 18 (2): 100–106.

Defense Special Weapons Agency (DSWA). 1998. *Design and Analysis of Hardened Structures to Conventional Weapons Effects* (TM 5-855-1, AFPAM 32-1147(1), NAVFAC P-1080, DAHSCWEMAN-97). Washington, DC: The Departments of the Army, Air Force, Navy, and The Defense Special Weapons Agency.

Dennis S. T., J. T. Baylot, and S. C. Woodson. 2002. Response of 1/4 scale concrete masonry unit (CMU) walls to blast. *Journal of Engineering Mechanics* 128 (2): 134–142.

Federal Emergency Management Agency, Building Performance Analysis Team. 1996. *The Oklahoma City Bombing: Improving Building Performance through Multi-Hazard Mitigation* (FEMA 277). Washington, DC: Federal Emergency Management Agency.

Hammons M. I., K. J Knox, and J. R. Porter. 2002. Polymer materials for blast retrofit of masonry construction. Proceedings, 30th Department of Defense Explosives Safety Seminar, Atlanta, GA, August 13–15, 2002.

Johnson C. F. and T. R. Slawson. 2004. Concrete masonry units walls retrofitted with fiber reinforced elastomeric systems for blast loads. Presented at the 75th Shock and Vibration Symposium, Virginia Beach, VA, October 27–31, 2004. Limited Distribution.

Johnson C. F., T. R. Slawson, T. K. Cummins, M. B. Hueste, and L. Beason. 2003. Concrete masonry units (CMU) static and dynamic wall experiments with elastomeric retrofits. Presented at the 74th Shock and Vibration Symposium, San Diego, CA, October 27–31, 2003. Limited Distribution.

Johnson C. F., T. R. Slawson, T. K. Cummins, and J. L. Davis. 2004. Concrete masonry unit walls retrofitted with elastomeric systems for blast loads. Presented at the 24th Army Science Conference, Orlando, FL, November 29–December 2, 2004.

Kersul A. M. and D. Sunshine. 2002. Structural collapse blast mitigation. Presented at the 17th International Symposium on Military Aspects of Blast and Shock, Las Vegas, NV, June 10–14, 2002.

Knox K. J., M. I. Hammonds, J. R. Porter, and T. T. Lewis. 2000. Polymer materials for structural retrofit. Proceedings, 29th Department of Defense Explosives Safety Seminar, New Orleans, LA, July 18–20, 2000.

Magallanes J. M. and K. B. Morrill. 2008. Fast running model for the residual capacity of steel columns damaged by blast & fragment loads. Proceedings of the 79th Shock and Vibration Symposium, Orlando, FL, October 26–30, 2008.

Malvar L. J. 1998a. Durability of composites in concrete. Presented at First International Conference on Durability of Composites for Construction, Sherbrooke, Quebec, Canada, August 5–7, 1988, pp. 361–372.

________. 1998b. Review of static and dynamic properties of steel reinforcing bars. *ACI Materials Journal* 95 (5): 609–616.

Malvar L. J., J. E. Crawford, and K. B. Morrill. 2007. Use of composites to resist blast. *Journal of Composites for Construction* 11 (6): 601–610.

Malvar L. J., K. B. Morrill, and J. E. Crawford. 1999. *CTS-1 Retrofit Designs: For Reducing the Vulnerability of an Office Building to Airblast* (Report TR-98-39.2). Burbank, CA: Karagozian & Case.

________. 2004. Numerical modeling of concrete confined by fiber reinforced composites. *Journal of Composites for Construction* 8 (4): 315–322.

Malvar L. J. and C. A. Ross. 1998. Review of static and dynamic properties of concrete in tension. *ACI Materials Journal* 95 (6): 735–739.

Malvar L. J. and J. W. Wesevich. 1996. *Peak Compression Membrane Response Enhancements for One-Way and Two-Way Slab Resistance Functions* (TM-96-1.2). Burbank, CA: Karagozian & Case.

Morrill, K. B., J. E. Crawford, and J. M. Magallanes. 2007. "Development of Simplified Tools to Predict the Blast Response of Steel Beam-Column Connections," Proceedings of the 2007 ASCE/SEI Structures Congress, Long Beach, CA, May 16–19, 2007.

Morrill K. B., L. J. Malvar, and J. E. Crawford. 1999. Retrofit design procedure for existing RC buildings to increase their resistance to terrorist bombs. Proceedings of the 9th International Symposium on Interaction of the Effects of Munitions with Structures (IEMS), Berlin, Germany. May 3–7, 1999.

Morrill K. B., L. J. Malvar, J. E. Crawford, and S. W. Attaway. 2000. RC column and slab retro-fits to survive blast loads. Proceedings of the ASCE Structures Congress, Philadelphia, PA, May 8–10, 2000.

Morrill K. B., L. J. Malvar, J. E. Crawford, and J. M. Ferritto. 2004. Blast resistant design and retrofit of reinforced concrete columns and walls. Presented at ASCE Structures Congress, Nashville, TN, May 23–25, 2004.

Morrill K. B., L. J. Malvar, J. E. Crawford, G. Hegemier, and F. Seible. 2001. Full-scale testing of reinforced concrete column retrofits to resist blast loads. Proceedings of the 10th International Symposium on Interaction of the Effects of Munitions with Structures, San Diego, CA, May 2001.

Morrill K. B., K. Nicolaisen, and T. Hutchinson. 2005. *Laboratory Testing of Kevlar® Reinforced Polyurea©* (Report TR-04-52.1). Burbank, CA: Karagozian & Case.

Muszynski L. C. 1998. Explosive field tests to evaluate composite reinforcement of concrete and masonry walls. In Fiber Composites in Infrastructure, Proceedings of the Second International Conference on Composites in Infrastructure, Tucson, Arizona, January 5–7, 1998, pp. 277–284.

Muszynski L. C. and M. R. Purcell. 2003. Use of composite reinforcement to strengthen concrete and air-entrained concrete masonry walls against air blast. *Journal of Composites for Construction* 7 (2): 98–108.

Myers J. J., A. Belarbi, and K. A. El-Domiaty. 2003. Blast resistance of un-reinforced masonry walls retrofitted with fiber reinforced polymers. Proceedings for The North American Masonry Society 2003 Annual Meeting, Clemson, South Carolina, pp. 312–323.

________. 2004. Blast resistance of FRP retrofitted un-reinforced masonry (URM) walls with and without arching action. *The Masonry Society Journal* 22 (1): 9–26.

Patoary M. K. H. and K. H. Tan. 2003. Blast resistance of prototype in-built masonry walls strengthened with FRP systems. Proceedings of the 6th International Symposium on Fibre-Reinforced Polymer Reinforcement for Concrete Structures, FRPRCS-6, Singapore, July 8–10, pp. 1189–1198.

Perea, A. and K. B. Morrill. 2002. Component high explosive testing of bare and retrofitted columns. Presented at the 17th International Symposium on Military Aspects of Blast and Shock, Las Vegas, NV, June 10–14, 2002.

.Piepenburg, D., R. Martinez, R. Del Frate, and K. Morrill. 2002. Blast Resistance Upgrade for Glazing, Window Frame, and Wall Systems—Non-Load Bearing In-Fill Walls with Blast Retrofits applied to the Wall, Window Frame and Glaze (Report No. SAIC-03/1001). San Diego, CA: Science Applications International Corporation.

________. 2003. *Blast Resistance Upgrade for Glazing, Window Frame, and Wall Systems – Load Bearing Walls with Blast Retrofits applied to the Wall, Window Frame and Glaze* (Report No. SAIC-03/1003). San Diego, CA: Science Applications International Corporation.

Simmons, L. 2005. Arena fragment experiments using mortars. Presented at the 76th Shock and Vibration Symposium, Destin, FL, October 30–December 4, 2005.

Stanley, M., R. Martinez, and J. Koenig. 2005a. *UL-Like Testing of Commercial Off-the-Shelf Products that Enhance the Blast and Ballistic Resistance of Structures, Quick Look Report 6* (New Mexico Tech, Energetic Materials Research and Testing Center, TR-05-08). Burbank, CA: Karagozian & Case. Limited Distribution.

________. 2005b. *UL-Like Testing of Commercial Off-the-Shelf Products that Enhance the Blast and Ballistic Resistance of Structures, Quick Look Report 7* (New Mexico Tech, Energetic Materials Research and Testing Center, TR-05-14). Burbank, CA: Karagozian & Case. Limited Distribution.

Stanley, M., J. Metzger, R. Martinez, and J. Koenig. 2005. *UL-Like Testing of Commercial Off-the-Shelf Products that Enhance the Blast and Ballistic Resistance of*

Structures, Quick Look Report 5 (New Mexico Tech, Energetic Materials Research and Testing Center, TR-05-03). Burbank, CA: Karagozian & Case. Limited Distribution.

Stanley, M., J. Metzger, K. B. Morrill, and R. Martinez. 2005a. *UL-Like Testing of Commercial Off-the-Shelf Products that Enhance the Blast and Ballistic Resistance of Structures, Quick Look Report 2* (New Mexico Tech, Energetic Materials Research and Testing Center, TR-04-48). Burbank, CA: Karagozian & Case. Limited Distribution.

________. 2005b. *UL-Like Testing of Commercial Off-the-Shelf Products that Enhance the Blast and Ballistic Resistance of Structures, Quick Look Report 3* (New Mexico Tech, Energetic Materials Research and Testing Center, TR-04-50). Burbank, CA: Karagozian & Case. Limited Distribution.

________. 2005c. *UL-Like Testing of Commercial Off-the-Shelf Products that Enhance the Blast and Ballistic Resistance of Structures, Quick Look Report 4* (New Mexico Tech, Energetic Materials Research and Testing Center, TR-04-55). Burbank, CA: Karagozian & Case. Limited Distribution.

Tan K. H. and M. K. H. Patoary. 2004. Strengthening of masonry walls against out-of-plane loads using fiber-reinforced polymer reinforcement. *Journal of Composites for Construction* 8 (1): 79–87.

INDEX